T0281586

Understanding the impacts of climate change on economic behavior is an important aspect of deciding when to take policy actions to prevent or mitigate its consequences.

This book applies advanced economics methodologies to assess the impact of climate change on potentially vulnerable aspects of the US economy: agriculture, timber, coastal resources, energy expenditure, fishing, outdoor recreation. It is intended to provide improved understanding of key issues raised in the recent Intergovernmental Panel on Climate Change (IPCC) reports. This new study includes measurement of possible benefits as well as the damages. It concludes that some climate change may produce economic gains in the agriculture and forestry sectors, whereas energy, coastal structures, and water sectors may be harmed. In any event, the impacts appear to be small relative to the size of the economy. This clearly has implications for future greenhouse gas policies.

The book will serve as an important reference for the scientific, economic, and policy community interested in climate change issues. The book will also be of interest to natural resource/environmental economists as an example of economic valuation techniques. The volume will clearly be of main importance to researchers and policy-makers in the United States, but will also be influential in the rest of the world owing to the importance of the US economy, and as a model for assessment of impacts on other economies worldwide.

The Impact of Climate Change on the United States Economy

The Impact of Climate Change on the United States Economy

Edited by

Robert Mendelsohn

School of Forestry and Environmental Studies, Yale University
New Haven, Connecticut, USA

James E. Neumann

Industrial Economics Incorporated
Cambridge, Massachusetts, USA

CAMBRIDGE
UNIVERSITY PRESS

PUBLISHED BY THE PRESS SYNDICATE OF THE UNIVERSITY OF CAMBRIDGE
The Pitt Building, Trumpington Street, Cambridge, United Kingdom

CAMBRIDGE UNIVERSITY PRESS
The Edinburgh Building, Cambridge CB2 2RU, UK
40 West 20th Street, New York NY 10011–4211, USA
477 Williamstown Road, Port Melbourne, VIC 3207, Australia
Ruiz de Alarcón 13, 28014 Madrid, Spain
Dock House, The Waterfront, Cape Town 8001, South Africa

http://www.cambridge.org

First published 1999
First paperback edition 2004

Typeset in Ehrhardt MT 11/15 pt QuarkXpress [SE]

A catalogue record for this book is available from the British Library

Library of Congress Cataloguing in Publication data

The Impact of Climate Change on the United States Economy / edited by
  Robert Mendelsohn, James E. Neumann.
     p.   cm.
   ISBN 0 521 62198 4 hardback
   1. Climatic changes – Economic aspects – United States. 2. United
States – Economic policy – 1993– I. Mendelsohn, Robert O., 1952– .
II. Neumann, James E., 1962– .
QC981.8.C51449 1999
330.973–dc21– 97-35704 CIP

ISBN 0 521 62198 4 hardback
ISBN 0 521 60769 8 paperback

Contents

vii

Contributors

RICHARD ADAMS Department of Agriculture
and Resource Economics, Oregon State
University, Ballard Extension Hall 330B,
Corvallis, OR 97331-3601

KELLEY J. BRYANT University of Arkansas,
Southeast Research and Extension Center,
Box 3508, Monticello, AR 71656

JOHN M. CALLAWAY UNEP Centre, Risoe
National Laboratory, PO Box 49, DK-4000
Roskilde, Denmark

RICHARD CONNER Department of Agriculture
and Economics, Texas A & M University,
Agriculture Building, College Station, TX 77843

JOHN CRESPI University of California at Davis,
Davis, CA 95616

BRUCE L. DIXON Department of Agricultural
Economics and Rural Sociology, University of
Arkansas, Agriculture Building, Fayetteville, AR
72701

ROBERT E. EVENSON Economic Growth
Center, Yale University, 27 Hillhouse Avenue,
Box 1987 Yale Station, New Haven, CT 06520

JOHN GATES Department of Resource
Economics, University of Rhode Island, Lippitt
Hall, Kingston, RI 02881

BRIAN HURD Hagler Bailly Consulting, Inc.,
1881 9th Street, Boulder, CO 80302

PAUL KIRSHEN Tellus Institute, 11 Arlington
Street, Boston, MA 02116-3411

ANGELIQUE KNAPP Industrial Economics
Incorporated, 2067 Massachusetts Avenue,
Cambridge, MA 02140

JOHN B. LOOMIS Department of Agriculture
and Resource Economics, Colorado State
University, C-306 Clark, Fort Collins,
CO 80523-0002

MARLA MARKOWSKI Industrial Economics
Incorporated, 2067 Massachusetts Avenue,
Cambridge, MA 02140

PATRICK MARSHALL Industrial Economics
Incorporated, 2067 Massachusetts Avenue,
Cambridge, MA 02140

BRUCE A. McCARL Department of Agricultural
Economics, Texas A & M University, College
Station, TX 77843

ROBERT MENDELSOHN Yale School of
Forestry and Environmental Studies, 360
Prospect Street, New Haven, CT 06511

WENDY N. MORRISON Economics Department,
Middlebury College, Monroe Hall, Middlebury,
VT 05753

JAMES E. NEUMANN Industrial Economics
Incorporated, 2067 Massachusetts Avenue,
Cambridge, MA 02140

WILLIAM NORDHAUS Department of
Economics, Yale University, 28 Hillhouse
Avenue, New Haven, CT 06511

DENNIS OJIMA Natural Resource Ecology
Laboratory, Colorado State University, Fort
Collins, CO 80523

CYNTHIA ROSENZWEIG Goddard Institute for
Space Studies, 2880 Broadway, New York, NY
10025

KATHLEEN SEGERSON Department of
Economics, The University of Connecticut,
341 Mansfield Road, Storrs, CT 06269-1063

DAIGEE SHAW Institute of Economics,
Academia Sinica, Nanking, Taipei, Taiwan,
11529, Republic of China

BRENT SOHNGEN Department of Agricultural
Economics, Ohio State University, 2120 Fyffe
Road, Columbus, OH 43210-1067

JOEL SMITH Hagler Bailly Consulting, Inc.,
1881 9th Street, Boulder, CO 80302

GARY W. YOHE The John Andrus Center for
Public Affairs, Wesleyan University,
Middletown, CT 06459

Acknowledgements

Not surprisingly, a project designed to address as complex a topic as the future impacts of climate change on the United States requires the concerted efforts of many individuals. Since the inception of the project in mid-1993, this research has demanded a great deal from project directors, principal investigators, research assistants, peer reviewers, editors and typists, (and, from time to time, patient and forbearing family members). We consider ourselves lucky to have worked with such dedicated and professional individuals over the last several years; it goes without saying that the project owes its success to their efforts.

We are of course grateful to the Electric Power Research Institute for the principal funding of this research, but owe particular thanks to overall Project Director Thomas Wilson as well as Richard Richels and Victor Niemeyer for their direction, guidance, insightful comments, and patience through the course of the project. We also wish to acknowledge Robert Unsworth of Industrial Economics and Joel Smith of Hagler-Bailly Consulting for their instrumental roles in recruiting participants and guiding the inception and early vision of the project.

We wish to thank the major contributors to this volume, recognized as authors and co-authors of the chapters that follow, for their substantial investment in the project and their dedication to producing high-quality research. The eagerness of these principal investigators to forge cross-disciplinary links and to integrate their work into the larger project was an essential ingredient and source of creative energy that has greatly improved the analysis of this complex and diverse topic. The tireless efforts of their research assistants, acknowledged individually by the principal authors, were a major asset to the project. We also would like to recognize the efforts of Lou Pitelka and Timothy Kittel for providing and interpreting the ecosystem modeling results that are an important input to many of our analyses.

The project benefited greatly from the continued involvement of peer reviewers. These individuals provided helpful advice to improve the integration and usefulness of the final product and provided constructive comments that greatly improved the analysis of specific economic sectors. In addition to those reviewers acknowledged by the principal investigators, we would like to thank A. Myrick Freeman, John Reilly, Roger Sedjo, Kenneth Frederick, John Houghton, Jae Edmonds, and Jake Jacoby for their contributions as overall project reviewers, as well as the anonymous reviewers recruited by Cambridge University Press.

Finally, the volume has been greatly improved through the professional editorial and word processing efforts of Marieke Ott, Lara Sheer, Joanne Kelleher, Stephen Perroni, Margaret Cella, Daniel Fuentes, and several individuals at Cambridge University Press.

Our goal as editors has been to provide an accurate and integrated work that can serve as a useful reference. Most of all, we have sought to provide thought-provoking insights on this topic of interest to us all. We hope that readers find the presentation to be interesting and enlightening, and we hope that all the individuals involved in this effort find that their contributions are accurately reflected; any remaining errors are our responsibility.

Robert Mendelsohn

James E. Neumann

1   Introduction

ROBERT MENDELSOHN, JOEL B. SMITH,
AND JAMES E. NEUMANN

In the absence of abatement measures, emissions of greenhouse gases are likely to grow over the next century largely from the burning of fossil fuels. As a result, atmospheric concentrations of carbon dioxide and other greenhouse gasses will continue to increase. The most recent IPCC (1996a) report links such increases to climate change. This poses a difficult choice for policy-makers. How much should society sacrifice to slow and possibly reverse the steady increase in greenhouse gas emissions?

Although not without controversy, there is a growing consensus among economists that near-term reductions in greenhouse gases could result in substantial costs. For example, many models suggest that the annual costs of stabilizing emissions could exceed 1–2 percent of GDP in OECD countries (IPCC, 1996a). Immediate reductions in emissions could add to costs if economies have little time to adjust to the change in policy. At the same time, global changes in climate could have undesirable impacts on both managed lands and unmanaged ecosystems. Examples of managed lands include agriculture, timber and water resources. Effects on unmanaged ecosystems could include effects on human health and biodiversity. As a result, choices made in the next few decades to either reduce emissions or continue the current pace of emissions growth have large and widespread ramifications.

The rational approach to greenhouse gas policy is to weigh the benefits of different control policies against their costs. But what are the benefits of reducing greenhouse gas emissions? What damages will be avoided if emissions are reduced and how do they compare with the costs of control? Unfortunately, the economic benefits of policies to reduce greenhouse gas emissions remain unclear. Some progress has been made, however, on the simpler question of estimating damages from a doubling of greenhouse gases. Existing national-level estimates of the economic impacts of climate change are largely expert judgments (Nordhaus, 1991; Cline, 1992; Fankhauser, 1995; Tol, 1995) based on a small set of comprehensive sectoral studies (Smith and Tirpak, 1989; Rosenberg, 1993). These early studies identified the sectors of the economy and the aspects of quality of life that are sensitive to climate change (IPCC, 1996b). They also provided an initial benchmark quantifying the impact of doubling greenhouse gases. These estimates of economic damage concluded that doubling would result in global damages equal to from 1.5 to 2 percent of GDP (IPCC,

1

1996c). Damages in the US were estimated to be between 1 and 2 percent of US GDP (IPCC, 1996c).

This book reexamines the link between climate change and damages by developing a new set of methods to measure the impacts on daily life from climate change. After engaging in several meetings to identify a set of needed studies, leading authors in each sector were recruited to conduct state-of-the-art studies. Based on results from climate research (IPCC, 1990; Houghton *et al.*, 1992), these authors were given an initial set of climate changes and future economic conditions. They were then asked to develop new methodologies and applications of existing approaches to improve estimates of impacts in their sectors. These strategies were implemented in a series of coordinated studies to quantify the damages from climate change. Beginning with a broad range of climate scenarios, the consequences to each sensitive sector of the economy were quantified. Effects on agriculture, coastal structures, energy, timber, water, and commercial fishing were all measured.

This book provides a detailed and comprehensive assessment of the effect of warming on the economy. In addition, several chapters quantify the effect of climate change on water quality and recreation. Although these latter two effects are illustrative of the kinds of changes in quality of life due to climate change, many important quality of life studies have not yet been completed and so are not included here. Specifically, health, aesthetics, and nonmarket environmental changes are not evaluated. The book consequently does not provide a comprehensive assessment of all impacts on the United States.

The studies presented in this book develop several improvements in the methodology of measuring global warming impacts. The research advances the state-of-the-art in impact assessment by:

> more fully including adaptation to climate change;
> developing "natural experiments", which compare economic activity across a cross-section of climate zones;
> employing dynamic modeling techniques to capture transitional responses by capital intensive sectors;
> generating more comprehensive welfare measures of affected sectors;
> using a consistent set of assumptions about future economic and population growth;
> estimating the response function of sectors to a broad range of climate outcomes.

In the next section, we critically assess the existing literature identifying its overall strengths and weaknesses. In Section 1.2, we highlight the major methodological innovations in this study. The chapter concludes with a brief review of each subsequent chapter.

1.1   Literature review

There is an extensive literature that links economic activities to greenhouse gas emissions to their atmospheric concentration to climate change, but there are far fewer studies that link climate change to economic impacts (IPCC, 1996a,b,c). This young but growing literature on economic impacts provides the foundation upon which this book builds. The early literature made three important contributions. First, it identified the market and nonmarket sectors which might be sensitive to climate change. Climate change is expected to affect society by affecting parts of the economy: agriculture, coastal resources, energy, timber, and fisheries, and aspects of the quality of life: aesthetics, human health, and ecosystems (including recreation and species loss). By defining the problem and identifying these key effects, the literature allowed this study to focus scarce resources on the most important impacts. Second, the early literature developed some initial techniques to measure potential severity. These methods measured what would happen to the current economy and society if the climate suddenly changed. Third, these potential severity studies revealed that it was possible that climate change impacts in several key sectors, including coastal resources, human health, and ecosystems, could be serious.

The most comprehensive empirical effort to measure national climate change impacts in the literature was conducted by the USEPA (US Environmental Protection Agency) at the request of Congress (Smith and Tirpak, 1989). This study identified economic sectors and nonmarket services that are sensitive to climate change. In each sector, empirical studies of scientific effects and the resulting economic impacts were conducted. For example, the agriculture study began with an agronomic simulation model which predicted reductions in yield from climate change and then entered these predictions in a large scale agricultural model which predicted supply and price effects for the country (Adams *et al.*, 1990). The sea level rise study projected the total quantity of wetland and dryland which would be inundated and how much it would cost to protect all developed drylands (Titus *et al.*, 1991). The energy sector study examined how warming would affect the demand for electricity and what utilities would have to spend to meet this additional demand (Linder and Inglis, 1989). By studying a host of damages which were expected to be caused by climate change, the USEPA study pulled together a wide range of information about impacts for the first time.

Another important question facing impact analysis is whether studies of impacts

3

can be carried out sector by sector or whether a systematic model of the entire economy is needed. If there are economy-wide impacts on prices, wages, and interest rates, a general equilibrium model may be needed to measure impacts accurately. Although Rosenberg (1993) does not employ a general equilibrium model, the study does construct an integrated analysis of impacts across sectors within a four state region (the "MINK" region, including the states of Missouri, Iowa, Nebraska, and Kansas). The MINK study concluded that climate impacts were too small to generate important interaction effects across sectors and thus require a general equilibrium approach. One must be cautious not to over-generalize from the MINK study, because sectors in some regions, though not MINK, may be closely linked (for example, agriculture and water in arid areas) and because MINK evaluated only a relatively mild climate scenario (less than a 2 °C increase). However, the MINK study does indicate that it is reasonable to conduct individual sector-specific impact studies as long as one controls for obvious interactions with other sectors.

Another issue addressed by the literature is how to measure climate sensitivity. Because climates have not noticeably changed over the last two centuries, it is difficult to measure the sensitivity of market and nonmarket sectors to climate change. Even with climate predictions becoming more moderate, the forecast rate and magnitude of climate change is unprecedented in human history (IPCC, 1996a). In order to understand what might happen to society if climate changes, it is necessary to look for other experiences or circumstances which are similar to climate change. The ideal impact "experiment" replicates what society will face if the climate changes and then measures what happens. Ultimately, this leads to a valuation, a measurement of the harm done. In the process of generating this estimate, the assessment postulates a plausible mechanism, linking cause and effect. Not only should an assessment provide a final estimate of damages for specific scenarios, but it should also provide sufficient detail about the process so that people can judge what confidence they should place in the damage estimate.

One important methodology to measure climate sensitivity is controlled experiments where individuals (crops, trees, buildings, etc.) are placed in artificial settings and their response to altered climate conditions and higher carbon dioxide concentrations are measured (Idso *et al.*, 1987; Kimball, 1982; Strain and Cure, 1985). These controlled experiments have been invaluable in identifying the mechanisms through which greenhouse gases and climate change will affect different sectors. Further, the empirical results have been incorporated in a number of simulation models to predict the impact of selected climate change scenarios (Rosenzweig and Parry, 1994; Rosenthal *et al.*, 1995).

Another important strategy is to compare the behavior of individuals (especially people) who currently live in different climates (see Mendelsohn *et al.*, 1994) or who face extreme climate events such as floods or droughts (Glantz, 1988). Because these experiments are not controlled, they can be marred by unwanted uncontrolled variation which can bias the results. Further, aspects of the environment which remain the same across space, such as carbon dioxide levels, are difficult to capture in cross-sectional experiments. However, in their favor, these natural experiments are conducted in field conditions where the real climate experiment will occur.

Adaptation

A controversial issue throughout the impact literature is the amount of adaptation to include in impact studies. The early studies, by focusing upon what sudden climate change would do to the current economy, placed a low weight on including adaptation. There are two types of adaptation to consider, private and public. Private adaptation is an action by an individual (firm) for the benefit of that individual (firm). Public adaptation is an action by a group or government whereby the group acts to protect the group's interest. Economists argue that victims will purchase private adaptation if the benefits to them (the damages removed or gains made) exceed the costs.[1] Efficient private adaptation is likely to occur, even if there is no official (government) response to global warming. Impact assessments need to capture private adaptation in order to represent this likely social response. It is less clear whether public adaptation will be efficient. In some cases, groups may decide to do costly public adaptation even when the benefits are small. In other cases, groups may fail to purchase public adaptation even when the benefits exceed the costs because the members of the group are not sufficiently cohesive to act in the group's best interest. Thus, although it is likely that efficient private adaptation will occur, it is not clear whether efficient public adaptation will be forthcoming.

Early impact studies included some adaptation but made limited attempts to include an efficient response. Specifically, many studies may have omitted important private mitigation efforts. For example, Adams *et al.* (1990) allowed farmers to adjust the existing mix of crops grown in their region and markets to adjust prices but did not allow new crops to be introduced from other regions nor did it consider many of the adaptations that can be made by farmers to offset adverse climate change effects. Crosson and Katz (1991) and Kaiser and Drennen (1993) demonstrate that efficient farm level adaptation can mitigate a sizable fraction of the potential damages caused

---

[1] Private actions may not account for damages to third parties. For example, an individual may make the rational decision to build a seawall to protect his property from sea level rise, but the seawall could destroy migrating wetlands which the individual might ignore. We classify cases, such as this, which involve externalities, as part of public adaptation.

5

by global warming. Some studies consider more adaptation than may actually be implemented. Rosenzweig and Parry (1994) examine two levels of adaptation in their world food supply study. In their higher level of adaptation, they assumed all farmers irrigated when necessary, regardless of cost or availability. It is likely that farmers, especially in third world countries, would not actually irrigate if it was not profitable.

With sea level rise, Titus *et al.* (1991) assumed all developed land would be protected against sea level rise. In many cases, however, developed land may not be worth protecting from sea level rise because the costs of protection exceed the value of the protected property. Further, the costs of protection can be lowered dramatically if one delays protection decisions until they are necessary. In a study of Long Beach Island, New Jersey, Yohe (1991) demonstrates that efficient protection decisions can lower the cost of sea level rise dramatically compared to complete and immediate protection. Although it is only a case study, the Yohe example indicates that efficient adaptation is important. However, with sea level rise, protection is often a group decision since a set of landowners must often agree before effective barriers can be erected. Further, protecting one area can have adverse effects on neighboring locations and these decisions need to be coordinated. With sea level rise a wide range of constraints and motivations affect public expenditures on coastal lands – these factors may influence the pace and nature of adaptation, and so it is reasonable to debate whether efficient options will be undertaken.

Dynamic versus static

Another debate which runs through the literature concerns whether or not impact models should be dynamic. Climate change is continuous but relatively slow. It is estimated that the climate change associated with the doubling of greenhouse gases will take 70 years to be realized (IPCC, 1996a). Whether or not impact models must be dynamic depends upon how rapidly the affected sector adjusts relative to this underlying rate of climate change. Some sectors are slow to adjust because they involve large inventories or stocks of resources which take many years to alter. Both forestry and coastal resources are slow to adjust relative to climate change because the timber inventory and housing stock take several decades to adjust. Dynamic forces are likely to be important in these sectors. Whether dynamic analysis is important in other sectors, such as agriculture and energy, is less clear.

Early impact studies relied heavily upon comparative static equilibrium analyses. In some cases, researchers explore a series of equilibrium analyses along a path of climate change, but the models contain limited dynamic properties. For sectors that adapt quickly, equilibrium models are reasonable. For example, equilibrium analyses may be perfectly adequate for modeling agriculture, because farmers appear to adjust to changing conditions within a few years. Most agricultural climate studies are equilibrium analyses (Kaiser and Drennen, 1993, is a notable exception).

Sectors that cannot adapt rapidly, however, may have dynamic dimensions which are important to capture in impact assessment. In assessing impacts on forests and developed coastal resources, it is important to model how capital stocks change over time in response to a path of climate change. Most of the forestry and sea level rise studies, however, simply compare current conditions to what would happen if the climate suddenly changed to a new equilibrium (Smith and Tirpak, 1989; Titus *et al.*, 1991; Callaway *et al.*, 1994). Because they are comparative static analyses, they cannot capture potentially important dynamic responses.

Representative studies

Impact studies are expensive to conduct and it is not always possible to analyze every aspect of a sector or every site of impacts. It is important when analyzing only a sample of impacts in a sector, that the sample be representative. For example, if one wanted an estimate of the impact of climate on outdoor leisure, one should not generalize from a study of snowskiing to the sector as a whole. Winter sports are a small fraction of outdoor leisure and the impact of climate on winter sports is not likely to be representative of impacts on summer sports. Similarly, one would have to be cautious in generalizing from an electricity study to the energy sector as a whole since electricity is utilized more in cooling than heating compared to other major fuels. The early literature on impacts tended to gravitate to effects which people anticipated would be deleterious. This focus on the harmful aspects of change can give a misleading impression of sector-wide impacts.

National estimates

Policy-makers want to know the magnitude of the benefits (damages avoided) from control programs they are considering today. Few studies, however, have measured effects across a sufficient geographic area to generate empirical measurements of national sector impacts. For example, in the Smith and Tirpak (1989) study, national empirical values were developed only for coastal resources, electricity, and agriculture. There were no national valuations or damage estimates for other sectors. In order to provide national damage estimates given limited empirical results, it was necessary to make expert judgments. The first such judgment was developed by Nordhaus (1991), who reviewed the Smith and Tirpak USEPA study and other available information and made informed guesses concerning the magnitude of sectoral impacts. Other experts (Cline, 1992; Titus, 1992; Fankhauser, 1995; Tol,

1995) reviewed this same material and made their own judgments. The results of these judgments are presented in Chapter 12.

Although these experts (Nordhaus, Cline, Titus, Fankhauser, and Tol) relied on the same background information, they came to very different sector level conclusions. For example, although all five analysts rely heavily on the same empirical study of electricity (Linder and Inglis, 1989) their estimates of energy sector damages range from $0.5 to $9.9 billion annually. In agriculture, they rely heavily on Adams *et al.* (1990) and yet predict damages from $1.1 to $17.9 billion. Given the same scientific evidence of forest decline, these authors estimate timber damages ranging from $0.7 to $43.6 billion annually. Even in market sectors which have received the most empirical research, "expert judgments" of the magnitude of sectoral impacts have a surprisingly large range. This reliance on expert judgment rather than empirical measurement in developing national-level estimates is one of the greatest weaknesses of this early literature (e.g. Chapter 6 in IPCC, 1996c).

Nonmarket effects

Nonmarket impacts are difficult to measure and hard to value. Nonetheless, changes in weather, natural ecosystems, health, recreation, and water quality all are potentially important. These impacts could have large consequences for the quality of life of many people. Unfortunately, there is significant uncertainty in the natural sciences concerning how climate change will affect natural ecosystems and human health. Further, even if the science were understood, it is difficult to assign an economic value to these nonmarket effects.

A number of mechanisms by which global warming may affect health have been identified. First, reductions in aggregate or local food supply can result in malnourishment. Global agricultural studies indicate that world food supplies will not be threatened by the level of climate change foreseen over the next century (Kane *et al.*, 1992; Rosenzweig and Parry, 1994; Darwin *et al.*, 1995). However, in places with extensive poverty and subsistence agriculture, failures of local food supply could result in increased rates of local malnutrition (Rosenzweig and Parry, 1994). Second, changes in ecosystems could alter disease vectors allowing some diseases to spread beyond their current boundaries. Martens *et al.* (1995) demonstrate that climate change could enlarge the potential geographic scope of malaria. Infectious vector-borne diseases such as dengue fever, tsetse fly morsitans, and arboviral encephalitis could all be affected by global warming (IPCC, 1996b). Third, warmer temperatures might induce heat stress resulting in heart attacks and pulmonary failure. Kalkstein (1989) found that populations (especially the elderly) in northern US cities have higher daily

8

mortality rates during heat waves than southern cities. He concluded that climate change could increase heat stress mortality. Relying heavily on Kalkstein's heat stress results, Cline, Fankhauser, Titus, and Tol all predict that climate will generate sizable human health damages in the United States.

These judgments about sizable health effects from climate change may be premature. Uncertainties about the role of climate variability and human adaptation to heat stress make it difficult to predict the magnitude of the effect. For example, if daily temperature variation declines, use of air conditioning increases, or housing stock improves, vulnerability to heat stress could decline in the future. In addition, the value to assign to these premature deaths is problematic. It is uncertain whether those who die from heat stress mortality would have lived only a few days more (and therefore would place a relatively low value on this loss) or would have lived for decades (and would place a high value on this loss).

Ecological studies of natural systems suggest that these systems will be different if the climate changes. The gap models (Smith and Tirpak, 1989) and the biogeographical models (VEMAP, 1995) all suggest that tree species will retreat from their southern boundaries and expand beyond their current northern boundaries. It is likely that there will be some noticeable shifts between grasslands, deserts, and forests. Some studies conclude that grassland and open forests will expand (Smith and Shugart, 1993), others find that productive closed forests will expand (Prentice *et al.*, 1992), and others predict that overall ecosystem productivity will increase (Melillo *et al.*, 1993). In addition, it is likely there will be subtle shifts of species composition within existing systems. What is poorly understood is how quickly these systems will change and what will happen during the transition period. One possibility is that a large fraction of the forest will die back because of increased fires and pests, potentially leaving large tracts of dead trees as forests gradually adapt by migrating to new and more suitable locations. An alternative possibility is that standing trees will survive during the transition which will largely affect new stands. The composition of the forest will shift towards early succession species but the forests will remain intact during the transition. Each of these scenarios, in turn, would have significant effects on animal populations as their habitat increases or shrinks.

Even if all the physical changes which will occur in natural systems over time were understood, it is still difficult to determine what value to place on these effects. Each of the impacts discussed above is likely to be valued quite differently by society. However, there are no universal measures of ecosystem value. Complex changes in ecosystems across locations and time are difficult for even one individual to assess. Within society, there is a wide range of values held concerning ecosystems. Determining what aggre-

9

gate value society should place on a complex set of ecological changes is a formidable task. Given that the issue has received scant attention to date, it is no surprise that there are no clear answers.

1.2   Key methodological improvements in this study

The economic approaches employed in this series of parallel sector studies reflect several improvements in measuring climate sensitivity. First, all of these studies attempt to capture the potential for adaptation to mitigate impacts. Throughout history, there is strong evidence that societies learn to adapt to harsh environments by adjusting their behavior. Each of the sector studies assesses the extent to which economic agents could adapt to climate change given current technology. For example, owners of coastal structures are assumed to make economically rational decisions about whether to protect coastal structures from rising sea level or gradually abandon them. Farmers and foresters are assumed to choose crops, planting, and harvest methods suitable for the new climates that they are experiencing. By moderating their behavior to fit the changing environment, people can and most likely will mitigate some of the possible harms and increase the potential benefits from change.

A second important innovation in this book is that several studies rely on natural climate experiments. Although there are few examples of climates changing over time that we can readily use to measure sector responses, nature is full of examples where agents adapt to different climates over the landscape. By comparing behavior in one location with one climate to another location with a different climate, we can learn a great deal about how people might adapt to climate change in the long run. As each locality adapts to the environment they experience, they customize their behavior to their climate. For example, by observing the energy expenditures, leisure activities, and farming values of town A (which experiences 25 °C temperatures) and comparing them to a similar town B (which experiences 30 °C temperatures), one can learn how a 5 °C temperature increase may affect town A. These cross-sectional comparisons reveal long run changes in which firms and people adapt to their new environment.

Third, some sectors, such as coastal structures and timber, are characterized by large capital stocks which are difficult to adjust over time. These sectors are quite vulnerable to the rate of change of climate because it is difficult to change large housing stocks or vast timber stocks quickly. In order to understand what would happen in these sectors, it is critical to build dynamic models that explicitly examine the rate of climate change and how quickly the sectors can respond. The timber and coastal property (sea level rise) studies are the first impact models in these sectors to explicitly address the rate of climate change and find that these sectors are sensitive to the rate of climate change and confirm the importance of using a dynamic approach.

Fourth, these new studies are more comprehensive than earlier research. For example, the agriculture study extends previous analyses of grains to include livestock effects and selected fruit and vegetable crops. Not only does this capture a larger fraction of the agricultural sector, but it explicitly includes farming activities that predominate in warmer environments. The energy analysis extends earlier research in electricity to include all fuels used for heating and cooling. Because electricity is used primarily for cooling, extending the analysis to all fuels provides a more balanced treatment of both heating and cooling. The recreation study extends earlier work on skiing to include summer outdoor recreation activities. Although warming should shorten the winter season, it should also tend to lengthen the summer season when most outdoor recreation occurs. A study of commercial fishing effects is included for the first time. The water study provides national water impacts that include consumptive uses, hydroelectricity and water quality effects.

Fifth, the new studies were carefully designed to be consistent across sectors. Before the individual sector studies began, the authors conducted several planning meetings to develop consistent economic and climate scenarios and to anticipate interactions across sectors. For example, as temperature rises, agriculture turns to more irrigation and so requires a bigger share of water consumption. The projected increase in irrigation in the agricultural sector model was included in the water sector model. As forests become more productive, the forestry model projects that forest land will increase slightly in some marginal agricultural areas. In turn, the agricultural model predicts large baseline productivity increases from technological change, suggesting that the agricultural sector can afford to lose some marginal lands. Similar economic assumptions were also used across all scenarios. For example, each study examined both current economic conditions and projected conditions for 2060. The projections assumed that US GNP (Gross National Product) would grow to $20.8 trillion by 2060 and population would grow to 294 million. All estimates are presented in 1990 US dollars unless otherwise stated.

Sixth, the climate scenarios were based on a broad range of projections. The most recent report of the IPCC (1996a) projects temperature increases of from 1 to 4 °C over the next century. Beyond 2100, temperatures could rise even higher. The climate scenarios in this book were chosen in order to reveal the nature of how damages relate to different magnitudes of climate change. Temperature increases for the United States of 1.5, 2.5, and 5.0 °C were included. For each temperature increase, precipitation was assumed to increase by zero, 7 percent, and 15 percent, for a total of nine climate scenarios. Carbon dioxide was assumed to increase to 530 ppmv (parts per million, volume) (710 ppmv in the timber study) which is consistent with a doubling of all greenhouse gases from preindustrial times. The study does not attempt to assess which of these climate scenarios is more likely, it is merely trying to indicate how sensitive the economy is to different outcomes. For most of the studies, temperature and precipitation changes were assumed to be uniform across the continental United States and across seasons.

Previous impact research, however, has relied on predictions from individual General Circulation Models (GCMs), which are sophisticated climate simulation models that predict changes in temperature and precipitation (among other variables) at the regional level. (For more information, see IPCC, 1996a.) For example, the timber study (Chapter 5) relies on climate predictions from three individual GCMs. GCM model predictions were also explored for the agriculture, energy, and recreation studies so that researchers could compare the results with past studies. The individual GCM analyses indicate that overall impacts are sensitive to differing geographic and seasonal distributions of climate change.

Both dynamic studies (timber and sea level rise) also specify a path of climate change from current to future conditions. In the timber model, temperature is assumed to increase linearly with time (IPCC, 1996a). In the sea level rise model, the seas are assumed to rise at a quadratic rate with time (Titus, *et al.*, 1991). Three scenarios are tested: rising sea levels of 0.33, 0.66, and 1.0 meter by 2100.

1.3   Organization of the book

This research is intended to serve as a resource for policy-makers and researchers interested in the economic dimensions of climate change. Toward that end, the book includes detailed accounts of the methods and results of each individual sector study. Chapters 2 to 11 form the heart of the book, containing detailed reports of each of the individual sector studies. A summary of the previous literature and a synthesis of the new results are reported in the concluding chapter.

No single climate impact instills more fear in people's minds than widespread starvation caused by failing farms. Because of the central importance of agricultural systems to society's continued survival and because of the clear connection between agriculture and climate, one of the most critical impacts of global warming is on agriculture. Given its prominent potential magnitude and its central role in policy, we have taken extra pains to study agriculture more thoroughly than other sectors. There are three separate studies of agriculture presented, each using a different approach.

Chapter 2 presents an expanded agronomic–economic study by Richard Adams,

Bruce McCarl, Kathleen Segerson, Cynthia Rosenzweig, Kelly Bryant, Bruce Dixon, Richard Conner, Robert Evenson, and Dennis Ojima. The analysis begins with predictions of how climate affects potential yields, introduces a farm choice model, and concludes with predictions of aggregate supply and price effects. The agronomic–economic approach has been widely applied in both the United States and abroad. Its close reliance on experimental evidence and sophisticated simulation models makes it an attractive tool. In this chapter, the widely used ASM (Agricultural Sector Model) is improved by exploring farm level adaptation, including warm-climate crops, and adding livestock effects.

Chapter 3, by Robert Mendelsohn, William Nordhaus, and Daigee Shaw, extends earlier research on the Ricardian model by including interannual and diurnal climate variation. The Ricardian method measures climate sensitivity by comparing farm land values across climate zones. Chapter 4, by Kathleen Segerson and Bruce Dixon, is also a cross-sectional empirical study. This study examines farms across the Great Plains and estimates actual yield changes as a function of climate for specific crops.

In keeping with the existing literature, we rely primarily upon the agronomic–economic model to measure climate impacts. The two other studies, however, provide important support for the agronomic model by testing key assumptions. One of the most serious criticisms of the agronomic–economic approach is that past studies have failed to capture efficient farmer adaptation. The burden for the agronomic models is that all adaptations must be explicitly modeled to be included. The modeler must anticipate all the ways that farmers may adapt. Although this is technically possible, it is demanding and early studies often gave adaptation short shrift. The Segerson–Dixon study examines cross-sectional farm evidence to test whether the predicted climate impacts on crop yields by the agronomic model are accurate. The study suggests that the productivity predictions of the agronomic models are reasonable but slightly pessimistic. The Ricardian study also examines actual farms and seeks to measure how net revenues vary with climate. The Ricardian model also suggests that the agronomic–economic model is reasonable but slightly too pessimistic. Both empirical studies suggest that the agricultural system is likely to do better than the agronomic model providing confidence that the agronomic–economic model is not overly optimistic. The three studies together strongly support the result that mild (but not severe) warming will be beneficial to US agriculture.

Chapter 5, authored by Brent Sohngen and Robert Mendelsohn, is a study of climate impacts on timber markets. This research employs a dynamic economic modeling approach to capture gradual intertemporal adjustments in forests. The study carefully links climate scenarios with a broad set of ecological models to economics. The ecological models suggest future scenarios may be amenable to forests. If forests do in fact expand and grow more productive, the economic model predicts sizable benefits. Chapter 6 addresses water resource impacts. This study, written by Brian Hurd, Mac Callaway, Joel Smith, and Paul Kirsten, evaluates several case studies and then extrapolates to produce one of the first national estimates of the effect of climate change on water. The study considers a range of geographic areas, including for the first time the Northeast and Southeast United States. The study finds that consumptive uses of water such as agriculture and industry will not be severely affected. The bulk of the damages will be to the nonconsumptive sectors: water quality and hydropower.

Chapter 7 covers sea level rise impacts on coastal property. This study, authored by Gary Yohe, James Neumann, and Patrick Marshall, is a careful dynamic analysis of 30 US coastal areas using a cost–benefit framework to assess protect or abandon decisions on a site-by-site basis. Sea level rise will cause damages but dynamic adaptation can keep these costs far lower than earlier estimates. Chapter 8, on energy resources, is an extensive empirical analysis of energy demand by both commercial and residential properties across the United States. A major advance in this study, written by Wendy Morrison and Robert Mendelsohn, is the coverage of all fuels used in heating and cooling rather than electricity alone. Despite the inclusion of all fuels, however, the results of earlier studies suggesting that warming would cause damages to the United States are confirmed in this study.

Chapters 9 to 11 cover commercial fishing and recreation, providing the first national estimates for these sectors. Chapter 9, written by Marla Markowski, Angelique Knapp, James Neumann, and John Gates, examines commercial fishing and finds that these effects are likely to be small given the small size of this sector. Though small, the commercial fishing impacts are also highly uncertain because little is known concerning how warming will affect the oceans and how this will in turn affect fisheries. The recreation chapters conclude that a small benefit associated with warming is likely. Chapter 10, written by Robert Mendelsohn and Marla Markowski, is a cross-sectional study comparing recreation in different states. The study includes summer recreation activities for the first time and finds that warming will increase fishing and boating significantly. Chapter 11, written by John Loomis and John Crespi, analyzes recreation using a benefits transfer approach. The authors also find that summer activities are likely to benefit from warming and that this effect will exceed damages from lost skiing opportunities. The net results of both the cross-sectional and benefits transfer approaches are consistent.

Chapter 12 integrates the results of each of the sector studies. The synthesis highlights the important methodological advances of the studies in this book, reviews the individual sectoral results, discusses their implications for greenhouse abatement policies, and notes the remaining holes in our knowledge about impacts and where additional research dollars would best be spent.

The research presented in this book provides new insights into how climate change will affect the US economy and the quality of life. The research provides repeated support of the importance of adaptation. Adaptation mitigates the impacts of environmental change in every sector studied. The research also demonstrates that modest warming will entail benefits for the United States in some sectors. The US agriculture, forestry, and outdoor recreation sectors are all projected to benefit from a slightly warmer, wetter, CO_2-enriched world. These benefits outweigh the damages measured in the coastal, water, and energy sectors suggesting small amounts of warming could be good for the US economy. The research, however, does not measure all relevant nonmarket benefits such as health effects, species loss, and human amenity impacts, so nothing definitive can be said about the net effect of climate change on the quality of life in the United States. The research also does not extend beyond US borders. Extending the techniques used in these analyses, especially to developing countries, is an important future research direction. Nonetheless, this book represents an important step forward in our quest to understand and measure the implications of climate change for us and future generations.

References

Adams, R.M., Rosenzweig, C., Peart, R.M., Ritchie, J.T., McCarl, B.A., Glyer, J.D., Curry, R.B., Jones, J.W., Boote, K.J. and Allen, L.H. Jr. 1990. Global Change and U.S. Agriculture. *Nature* 345: 219–24.

Callaway, M., Smith, J. and Keefe, S. 1994. *The Economic Effects of Climate Change for U.S. Forests*. Report to US Environmental Protection Agency. Boulder, Colorado: Hagler Bailly Consulting, Inc.

Cline, W. 1992. *The Economics of Global Warming*. Washington, DC: Institute of International Economics.

Crosson, P. and Katz, L. 1991. *Report IIA: Agricultural Production and Resource Use in The MINK Region With and Without Climate Change*. DOE/RL/01830T-H7: Washington, DC: US Dept. of Energy.

Darwin, R., Tsigas, M., Lewandrowski, J. and Raneses, A. 1995. *World Agriculture and Climate Change: Economic Adaptations*. Agricultural Economic Report Number 703. Washington, DC: US Department of Agriculture.

Fankhauser, S. 1995. *Valuing Climate Change – The Economics of The Greenhouse*. London: EarthScan.

Glantz, M.H. 1988. *Societal Responses to Regional Climatic Change – Forecasting by Analogy*. Boulder, Colorado: Westview Press.

Houghton, J.T., Callander, B.A. and Varney, S.K. 1992. *Climate Change 1992 – The Supplementary Report to the IPCC Scientific Assessment.* WMO/UNEP Intergovernmental Panel on Climate Change. Cambridge: Cambridge University Press.

Idso, S., Kimball, B., Anderson, M. and Mauney, J. 1987. Effects of Atmospheric CO_2 Enrichment on Plant Growth: The Interactive Role of Air Temperature. *Agriculture and Ecosystem Environments* **20**: 1–10.

IPCC. 1990. *Climate Change: The IPCC Scientific Assessment.* Houghton, J.T., Jenkins, G.J. and Ephraums, J.J. (eds.). Cambridge: Cambridge University Press.

IPCC. 1996a. *Climate Change 1995: The Science of Climate Change,* Houghton, J.T., Filho, L.G., Callander, B.A., Harris, N., Kattenberg, A. and Maskell, K. (eds.). Cambridge: Cambridge University Press.

IPCC. 1996b. *Climate Change 1995: Impacts, Adaptations, and Mitigation of Climate Change: Science-Technical Analyses.* Watson, R., Zinyowera, M., Moss, R. and Dokken, D. (eds.). Cambridge: Cambridge University Press.

IPCC. 1996c. *Climate Change 1995: Economic and Social Dimensions of Climate Change.* Bruce, J., Lee, H. and Haites, E. (eds.). Cambridge: Cambridge University Press.

Kaiser, H. and Drennen, T. (eds.) 1993. *Agricultural Dimensions of Global Climate Change.* Delray Beach, FL: St. Lucie Press.

Kalkstein, L.S. 1989. The Impact of CO_2 and Trace Gas-Induced Climate Changes Upon Human Mortality. In: *The Potential Effects of Global Climate Change on the United States.* Smith, J.B. and Tirpak, D. (eds.). Appendix G: Health EPA-230-05-89-057. Washington, DC: US Environmental Protection Agency.

Kimball, B. 1982. Carbon Dioxide and Agricultural Yield. *Agronomy Journal* **75**: 779–88.

Kane, S., Reilly, J. and Tobey, J. 1992. An Empirical Study of the Economic Effects of Climate Change in World Agriculture. *Climatic Change* **21**: 17–35.

Linder, K.P. and Inglis, M.R. 1989. The Potential Effects of Climate Change on Regional and National Demands for Electricity. In: *The Potential Effects of Global Climate Change on the United States,* Appendix H: Infrastructure. Smith, J.B. and Tirpak, D. (eds.). EPA-230-05-89-058. Washington, DC: US Environmental Protection Agency.

Martens, W.J., Rotmans, J. and Niessen, L.W. 1995. Climate Change and Malaria Risk: An Integrated Modelling Approach. *Environmental Health Perspectives.* **103**: 458–64.

Mearns, L.O., Rosenzweig, C. and Goldberg, R. 1992. Effect of Changes in Internannual Climatic Variability of CERES – Wheat Yields: Sensitivity and 2 X CO_2 General Circulation Model Studies. *Agricultural and Forest Meteorology* **62**: 159–89.

Melillo, J.M., McGuire, A.D., Kicklighter, D.W., Moore, B., Vorosmarty, C.J. and Schloss, A.L. 1993. Global Climate Change and Terrestrial Net Primary Productivity. *Nature* **363**: 234–40.

Mendelsohn, R., Nordhaus, W. and Shaw, D. 1994. The Impact of Global Warming on Agriculture: A Ricardian Analysis. *American Economic Review* **84**: 753–71.

Mitchell, J.F.B., Johns, T.C., Gregory, J.M. and Tett, S.F.B. 1995. Climate Response to Increasing Levels of Greenhouse Gases and Sulphate Aerosols. *Nature* **376**: 501–4.

Nordhaus, W. 1991. To Slow or Not to Slow: The Economics of The Greenhouse Effect. *Economic Journal* **101**: 920–37.

Prentice, C., Cramer, W., Harrison, S., Leemans, R., Monserud, R. and Solomon, A. 1992. A Global Biome Model Based on Plant Physiology and Dominance, Soil Properties, and Climate. *Journal of Biogeography* **19**: 117–34.

Rosenberg, N.J. 1993. Towards an Integrated Impact Assessment of Climate Change: The MINK Study. *Climatic Change* **24**: 1–173.

Rosenthal, D., Gruenspecht, H. and Moran, E. 1995. Effects of Global Warming on Energy Use for Space Heating and Cooling in the United States. *Energy Journal* **16**: 77–96.

Rosenzweig, C. and Parry, M.L. 1994. Potential Impact of Climate Change on World Food Supply. *Nature* **367**: 133–8.

Rutherford, T. 1992. *The Welfare Effects of Fossil Carbon Reductions: Results from a Recursively Dynamic Trade Model*. Working Paper, No. 112, OECD/GD(92)89. Paris: Organization for Economic Cooperation and Development.

Smith, J. and Tirpak, D. 1989. *The Potential Effects of Global Climate Change on the United States: Report to Congress*. EPA-230-05-89-050. Washington DC: US Environmental Protection Agency.

Smith, T.M. and Shugart, H.H. 1993. The Transient Response of Terrestrial Carbon Storage to a Perturbed Climate. *Nature* **361**: 523–6.

Strain, B.R. and Cure, J. (eds.) 1985. *Direct Effects of Increasing Carbon Dioxide on Vegetation*. DOE/ER-0238. Washington, DC: US Department of Energy.

Titus, J.G. 1992. The Cost of Climate Change to the United States. In: *Global Climate Change: Implications, Challenges and Mitigation Measures*. Majumdar, S.K., Kalkstein, L.S., Yarnal, B., Miller, E.W. and Rosenfeld, L.M. (eds.). Easton, PA: Pennsylvania Academy of Science.

Titus, J., Park, R., Leatherman, S., Weggel, J., Greene, M., Mausel, P., Brown, S., Gaunt, C., Trehan, M. and Yohe, G. 1991. Greenhouse Effect and Sea Level Rise: The Cost of Holding Back the Sea. *Coastal Management* **19**: 171–204.

Tol, R. 1995. The Damage Costs of Climate Change Toward More Comprehensive Calculations. *Environmental and Resource Economics* **5**: 353–74.

VEMAP. 1995. Vegetation/Ecosystem Modeling and Analysis Project (VEMAP): Comparing Biogeography and Biogeochemistry Models in a Continental Scale Study of Terrestrial Ecosystem Responses to Climate Change and CO_2 Doubling. *Global Biogeochemical Cycles* **9**: 407–37.

Yohe, G. 1991. Uncertainty, Climate Change, and the Economic Value of Information: An Economic Methodology for Evaluating the Timing and Relative Efficacy of Alternative Response to Climate Change. *Policy Sciences* **24**: 245–69.

2   Economic effects of climate change on US agriculture

RICHARD M. ADAMS, BRUCE A. MCCARL,
KATHLEEN SEGERSON, CYNTHIA ROSENZWEIG,
KELLY J. BRYANT, BRUCE L. DIXON, RICHARD CONNER,
ROBERT E. EVENSON, AND DENNIS OJIMA

Agriculture was one of the first economic sectors studied in climate change impact research because of its importance to human survival and its well known sensitivity to climate (see d'Arge, 1975; Kokoski and Smith, 1987; Dudek, 1988; Adams, *et al.*, 1989; Adams *et al.*, 1990).[1] Although these studies provide a methodological basis for studying the agricultural impacts of climate change, there are some important shortcomings in this literature.

First, early studies focused on conventional agricultural crops such as grain (e.g. corn and wheat) and soybeans. Results of these studies suggest that some regions of the United States, such as the Southeast, may suffer substantial economic losses if production of grains shifts to more northerly latitudes. However, since these southern regions are major producers of heat tolerant crops such as cotton, sorghum, fruits, and vegetables, failure to include such heat tolerant crops in previous analyses may overstate potential economic losses. In addition, the effects on livestock have been assessed through effects on the price of feed grains; direct effects of climate change on livestock weight gain and other performance measures are not addressed.

Second, previous analyses have incorporated only limited possibilities for farm-level adaptations or adjustments to climate change. There are several ways that farmers may be able to adjust. For example, if other inputs such as fertilizer or irrigation water are substitutes for "climate" in production, then farmers may be able to adjust input mixes to maintain or at least offset reductions in output levels in the face of adverse climate change. In addition, farmers may be able to adjust production processes through changes in the timing of planting and/or harvesting of crops. Farmers may also be able to adjust by changing their crop mix, possibly introducing crops not previously grown in their area. Adjustment may also be possible through research and plant breeding to offset some adverse climate effects.

[1] Helms *et al.* (1996) provide an overview of recent research on agricultural impacts. Rosenzweig and Parry (1994) and Reilly *et al.* (1994) examine the effects on world food production and trade.

18

A few recent studies illustrate the importance of adaptation (Kaiser *et al.*, 1993; Mendelsohn *et al.*, 1994). However, Kaiser *et al.* model only an individual representative farm and do not consider aggregate or market-level impacts of adaptation. In addition, it is based on "simulated" adjustment rather than empirical evidence on actual responses to differing climates. Mendelsohn *et al.* examine changes in land values as well as farmers' revenues using county-level data that incorporate adaptations to climate, as reflected in current production practices. While the study demonstrates the nature of adaptations to climate variables, the results do not address potential changes in prices. Other studies (Council on Agricultural Science and Technology, 1992; Crosson, 1993) discuss the prospects for crop migration and research targeted to adaptation options (e.g. plant breeding for high-temperature tolerance, drought tolerance), both for current crops and in-migrating crops. However, these studies do not provide estimates of migration possibilities or evidence regarding potential payoff from research targeted specifically at climate change. Studies that ignore these adjustment possibilities are likely to overstate the costs of climate change.

Third, previous studies do not consider the economic effects of changes in forage production on natural or improved range lands. The climate change assessments by Adams *et al.* (1989, 1990, 1995) do allow for some changes in the productivity of pasture and haylands, which in turn change livestock/feed balances. However, no changes in forage production or carrying capacity associated with the large amounts of public and private rangelands of the western United States are included in previous Adams *et al.* studies. Forage production is important in the livestock/feed balance relationship which affects regional production patterns for livestock. Since livestock amount to about one-half of the total farm-gate value of agricultural production in many states, changes in livestock/feed relationships may have a substantial economic impact. The production of forage, like that of dryland crops, is influenced heavily by variability in the timing and magnitude of precipitation. Long-term changes in forage productivity or carrying capacity would have implications for the structure of livestock enterprises. Reductions in forage available for cattle and sheep imply increased demand for feed grains and hay, which in turn may affect the cost of production of other livestock commodities, such as pork or chicken.

Livestock enterprises can be affected not only by changes in precipitation but also by changes in temperature. Extreme temperatures affect livestock performance (e.g. reduction in weight gain per unit of food intake). Thus, even if precipitation is not reduced, temperature increases beyond some level will reduce performance and increase the time required for cattle and other livestock to reach given weight levels. Such changes in livestock weight gain affect the profitability of livestock enterprises.

This chapter extends previous analyses of the economic effects of climate change on agriculture to address the limitations found in existing studies. In particular, this project extends previous work by (1) incorporating other crops such as fruits and vegetables into the regional crop alternatives for the Southeast and other southerly locations; (2) considering the impacts of additional farmer adaptations to climate change; (3) allowing for crop migration into regions where those crops are not currently being grown; (4) incorporating changes in forage production and livestock performance; and (5) assessing the potential for technological change, as manifested in present and future yields, to offset climate change.

The basic model used to analyze the impacts of climate change is the ASM (Agricultural Sector Model) which is a spatial equilibrium model of the US agricultural sector that has been used in many analyses of the interaction between agriculture and the environment, including the work by Adams *et al.* (1989, 1990, 1995) on climate change. It allows for disaggregate analysis of regional impacts of change, with endogenous price adjustments. Section 2.1 presents an overview of ASM. The model is the basis of most of the existing quantitative estimates of the economic impacts of climate change on US agriculture.

Several changes were made to ASM for this research. First, citrus and tomatoes (a proxy for vegetable production in general) have been added to the model. Incorporating these crops allows an analysis of the effect of climate change on selected high-valued fruits and vegetables that might benefit from temperature increases. Second, the model has been modified to allow migration of crops (northward) into other production areas. The treatment of farmer adaptations, changes in livestock performance, adjustments in water availability, and changes in exports of US agricultural commodities are handled by adjustments in existing coefficients and parameters of the model. Third, the livestock sector data have been updated. Finally, a dynamic component was added, allowing simulations into the future (for the 2060 analysis).

With these changes, ASM was then used to simulate the impacts on the agricultural sector of alternative climate change scenarios. Before the ASM can be run, exogenous inputs must be prepared using a variety of sources. Section 2.2 describes the climate change and economic scenarios that were used. Section 2.3 presents the predicted impacts on (1) the yields of crops and forage (including the role of technological change in future yields); (2) animal grazing requirements and performance; (3) crop migration potentials; and (4) changes in water resource availability using a variety of sources. The economic impacts on prices, production, and welfare are reported in Section 2.4. The chapter concludes with a summary of findings and a few observations.

20

2.1   Overview of the agricultural sector model

The ASM is a spatial equilibrium model formulated as a mathematical programming problem (Takayama and Judge, 1971). The model represents production and consumption of primary agricultural products including both crop and livestock products. Processing of agricultural products into secondary commodities is also included. The production and consumption sectors are assumed to be made up of a large number of individuals, each of whom operates under competitive market conditions. This leads to a model which maximizes the area under the demand curves less the area under the supply curves. The area between baseline supply and demand curves equals the baseline economic welfare. Similarly, the area between supply and demand curves after a posited climate change equals the new economic welfare. The difference between these two areas equals the change in economic welfare, equivalent to the annual net income lost or gained by agricultural producers and consumers as a consequence of global climate change. Both domestic and foreign consumption (exports) are included.

The model integrates a set of micro- or farm-level crop enterprises for multiple production regions which capture agronomic and economic conditions with a national (sector) model. Specifically, producer-level behavior is captured in a series of technical coefficients that portray the physical and economic environment of agricultural producers in each of the 63 homogeneous production regions in the model, encompassing the 48 contiguous states. These regions are then aggregated to 10 macro regions, as defined by the US Department of Agriculture (USDA) (Figure 2.1). Like earlier studies, irrigated and non-irrigated crop production and water supply relationships are included in the analysis. Availability of land, labor, and irrigation water is determined by supply curves defined at the regional level depicted in Figure 2.1. Farm-level supply responses generated from the 63 individual regions are linked to national demand through the objective function of the sector model, which features demand relationships for various market outlets for the included commodities (see Chang and McCarl, 1993, for details of ASM).

Features have been added to the ASM to allow dynamic updating. These involve the ability to project yields, domestic demand, imports, and exports for major commodities. Quantities of cropland, pasture, AUMs (animal unit months), labor, and water as well as the prices of inputs are also projected. The basic mechanisms for this updating fall in two classes: items that are updated based on time (trends) and those updated based on yield changes. The time updated items include yield levels, demands, import levels and supplies, and quantities of available inputs. In all cases

21

Figure 2.1 Farm production regions in the United States.
Source: USDA Economic Research Service.

these are updated by a formula $(1 + r_i)^t$ where r_i is the annual rate of change for item I.[2]

The other major updating feature involves input adjustments related to yield levels. Such updating is done for crop input uses, crop profits, livestock feed use, and livestock profits using an elasticity expressing the response of input usage to a percentage change in yield. Input usage is changed by the percentage change in yield multiplied by that elasticity. Elasticities are based on (1) results derived by Evenson and (2) estimation from a 15-year period (from the mid 1970s to the early 1990s) from selected crops. Livestock feed use is assumed to be directly proportional to yield increases (thus, if there is a 10 percent increase in milk output there is a 10 percent increase in assumed feed consumption).

The overall procedure then is to first input the desired year, project all the demand, yield import and export figures, then update input use via the elasticity of input use change with respect to yield change multiplied by the projected yield change.

---

[2] The r_i terms have been estimated using 40 years of agricultural statistics to determine the annual percentage rate at which those items have increased over time. The model takes the base yield starting in 1990 and t, the desired year, and then multiplies the base number by $(1 + r)^{t-1990}$ to obtain the updated estimates.

2.2   Climate change scenarios

In this study a broad set of climate change scenarios are examined both to avoid the limitations inherent in the use of any single scenario and to identify the sensitivity of climate impacts across a wide range of potential temperature and precipitation changes. The climate change effects are estimated for a 2060 economy in order to understand effects in the economic context in which climate change is expected to occur. The 2060 agricultural assessment requires projections of commodity demand, resource availability, and agricultural yields, all of which increase the uncertainty surrounding these estimates. Results are also reported for a 1990 economy, partially to be consistent with past literature and also to reveal the impact of assumptions made to generate the 2060 economy.

Two types of climate change scenarios are analyzed: uniform and individual General Circulation Model (GCM) projections. The uniform scenarios assume uniform national changes in temperature and precipitation changes across all regions of the United States and each season. Sixty-four uniform scenarios are tested using a combination of four temperature, four precipitation, and four atmospheric CO_2 concentration changes. For the purposes of discussion, however, we focus on the nine scenarios that are analyzed throughout the remainder of this book. These nine scenarios include 1.5 °C, 2.5 °C, and 5.0 °C changes in temperature along with 0 percent, 7 percent, and 15 percent changes in precipitation. Two GCM scenarios used in most earlier studies were also selected (GISS, or Goddard Institute for Space Studies and GFDL-R30, the R-30 run from the Geophysical Fluid Dynamics Laboratory at Princeton University). The individual GCM scenarios introduce some regional and seasonal variation in the change in temperature and precipitation.

Although the project examines a set of possible atmospheric CO_2 concentrations (355, 440, 530, and 600 ppmv), there is not sufficient space to describe all the results. This chapter focuses on the results for a 530 ppmv atmospheric CO_2 concentration (the remaining forecasts are described in Appendix A2). Two forecasts are described in detail. A central case is developed with a warming of 2.5 °C and a 7 percent precipitation increase. A more severe climate change case is also presented with a 5.0 °C increase in temperature with no change in precipitation. These two "case studies" are used in a series of sensitivity analyses concerning the role of farmer adaptations and export (world food production) assumptions.

The two GCM-based climate change scenarios offer a useful point of comparison with some previous economic analyses of climate change. The effects of the GISS climate forecasts on US agriculture were recently assessed by Adams *et al.* (1995). Comparing the GCM results in this paper with Adams *et al.*, one can test the import-

23

ance of the changes in adaptation and mitigation opportunities made to ASM in this research. The second GCM-based climate forecasts for which yield data are available (GFDL-R30) is not the same climate forecast from the Geophysical Fluid Dynamics Laboratory employed by Adams *et al.* (they used the GFDL-QFLUX run). GFDL-R30 forecasts a somewhat harsher climate change for the United States than GFDL-QFLUX. Average annual warming in the United States in QFLUX is 4.0 °C, whereas in R30 average annual warming is 4.4 °C. In some important crop production areas (e.g. the Corn Belt), warming is approximately 5.0 °C. Indeed, these climate changes more closely resemble earlier UKMO (United Kingdom Meteorological Office) forecasts. Although not directly comparable with earlier studies based on GFDL, the GFDL-R30 results help provide a sense of the distribution of effects one could get with different climate forecasts.

2.3   Changes in crops, forage, and livestock performance

A basic input into ASM is the yield level (output per acre) for each crop (including forage) in the model. The effects of climate change are modeled primarily as changes in these yield levels.[3] Estimates of yield changes were developed for this research by investigators at Goddard Institute for Space Studies, the University of Arkansas, Colorado State University, Texas A&M University, the University of Connecticut, and Yale University. A variety of methods were used, including crop simulation models, a plant–soil ecosystem model and regression analysis. In addition, a nutritional balance model was used for estimates of changes in livestock performance. A description of the methodologies used in each of these supporting studies is given below.

Crop yield changes

Major grains

Wheat, corn, and soybeans are the three most important US crops in both the domestic and export markets. Estimates of yield changes for these crops were provided by Cynthia Rosenzweig of Goddard Institute for Space Studies and Columbia University. These estimates were generated using the SOYGRO (Jones *et al.*, 1988),

---

[3] Climate change can also affect the demand for water in the model. Changes in water demand are estimated for both dryland and irrigated crops using the plant simulation models. Water supplies are altered (from 1988–92 levels) in these analyses, based on estimates provided by the water study (Chapter 6).

CERES-Maize (Ritchie *et al.*, 1989), and CERES-Wheat (Godwin *et al.*, 1989) dynamic crop simulation models. These models have been widely validated and are applicable to estimating crop yields at many locations. The crop models use daily maximum and minimum temperatures, precipitation, and solar radiation as climate inputs. In addition, they require various management parameters. These include farmer-controlled factors such as cultivar (variety of a specific crop) and planting date, as well as edaphic (soil-related) factors. The simulations incorporate current management practices at the study sites. Physiological effects of increased CO_2 on crop growth and water use have been incorporated into the models, based on experimental literature (Acock and Allen, 1985). The degree to which these experimental CO_2 results hold up under field conditions is uncertain. The crop models assume optimal pest management and no changes in technology over time.

Seventeen US sites were selected to represent the major agroclimatic regions in the United States. Thirty years (1951–80) of daily climate data were used to produce estimates of current yield and irrigation demand at each site. Simulations were then conducted for each crop at each site for all 64 possible combinations of changes in temperature, precipitation, and atmospheric CO_2 concentration. In order to estimate changes in water use, simulations with full irrigation were run at all sites. The wheat and maize models were run with a continuous year-to-year water balance for both base and changed conditions, while the soybean model was run non-continuously (i.e. soil moisture was re-filled to the drained upper limit each year). No adaptations, such as changes in planting date or cultivar, were assumed in the estimates.

At the base CO_2 level of 355 ppmv, all three crops experienced yield decreases with increasing temperatures, although the magnitudes of the reductions varied with the crop. Higher temperatures during the growing season speed annual crops through their development, allowing less grain to be produced. Additionally, demand for water is increased. Wheat had the largest simulated reduction in yield of the three crops, reaching 25 percent for a 2.5 °C temperature increase and no precipitation change. However, the reduction was only 8 percent when this temperature increase was coupled with a 15 percent increase in precipitation. Most of the relationships appear to be linear, with a few exceptions. With increased CO_2 levels, improvement is seen in maize and wheat yields, but not enough to offset the negative impacts of the higher temperatures. Because soybeans have a higher positive yield response to increased CO_2, the negative impacts of increased temperature are offset to a greater degree for soybeans than for wheat or maize.

Yield changes for the GCM-based climate forecasts varied dramatically across regions. For example, corn (maize) yields in the Corn Belt decreased by 34 percent under GFDL-R30, while in the Delta states there was no change in corn yields (from

25

the base). Similarly, soybean yields in the Corn Belt decreased by 29 percent under GFDL-R30, while in the Lake States they increased by 47 percent. All of these changes include the CO_2 effects on yields; as with the uniform climate change scenarios, increases in CO_2 mitigate some of the negative effects of climate change in isolation.

Cotton and sorghum

Previous work (e.g. Adams *et al.*, 1989, 1990, 1995) assumed that yield changes for other crops such as cotton, sorghum, and hay would equal the average yield changes predicted for corn, wheat, and soybeans. For this study, specific yield change estimates for cotton and sorghum were generated using EPIC (erosion productivity impact calculator). EPIC is a simulation model developed to determine the relationship between soil erosion and soil productivity (Williams *et al.*, 1984). EPIC simulates these processes, as well as crop yields.

EPIC has the ability to simulate different crops. Here, EPIC was used to simulate cotton and sorghum (as well as improved pasture) yield response to climate change and changes in atmospheric CO_2. One EPIC dataset was constructed for each of six locations: North Platte, Nebraska; Dodge City, Kansas; Lynchburg, Virginia; Memphis, Tennessee; Lubbock, Texas; and Fresno, California. Of the six locations, the first three are too far north for economic production of cotton presently. The last three locations are in three of the current major cotton producing areas in the United States. These six locations allow estimation of representative changes in cotton, sorghum, and hay production as climate and atmospheric CO_2 change.

An EPIC dataset consists of weather data, soil data, and crop management data for a specific location. The weather data used in these base EPIC datasets were taken from weather stations near these cities. The CO_2 level was set at 355 ppmv. These datasets represent the baseline. Fertilizer use for cotton and sorghum production was allowed to fluctuate to meet changing crop needs under different scenarios. Irrigated as well as dryland yields were simulated for Texas. Only irrigated yields were simulated for California. Sorghum in Kansas was assumed to be irrigated and the remaining crop/location combinations were simulated under dryland conditions.

Cotton and sorghum yields were simulated for the different temperature/precipitation change scenarios with 530 ppmv CO_2 and for the GCM-based climate changes at each of the six locations. The predicted yield response of cotton and sorghum to climate change and changes in atmospheric CO_2 were uniform in terms of their direction of change. Cotton and sorghum yields for all locations increased as CO_2 and precipitation increased, and decreased as temperature increased. Yields of these crops decreased as temperature increased because the crop reached maturity in fewer days, i.e. a shorter growing season resulting in fewer and/or smaller seeds or fruit.

Irrigated crop production was simulated for California and Texas, and for sorghum in Kansas. Irrigation water use decreased as rainfall increased. Irrigation water use also decreased as CO_2 increased because increased CO_2 increases crop water use efficiency. Most of the climate scenarios resulted in less water use than the baseline. For example, water use on sorghum in Kansas declined by 40 percent from the baseline for the 15 percent precipitation/5.0 °C temperature scenario. This was the largest decrease in irrigation water use for any of the irrigated scenarios.

Tomatoes, citrus, and potatoes

The yield change estimates for fruits and vegetables were provided by Cynthia Rosenzweig of Goddard Institute for Space Studies and Columbia University. Estimates were provided for citrus (Valencia oranges), potatoes, and tomatoes. In each case, crop simulation models were used to simulate plant growth under a variety of conditions. In some cases, the simulations were combined with information obtained from a literature review to provide estimates of yield changes. In all cases, they assume current management practices and no adaptation in planting dates or cultivars in response to climate change. Fertilization, irrigation, and pest control are generally considered to be at optimal levels. Because these crops are generally irrigated, the effects of changes in precipitation were not simulated.

Citrus: Estimated changes in citrus yields were quantified using a dynamic simulation model developed for sweet oranges (Ben-Mechlia and Carroll, 1989a,b). The model simulates the stages of plant growth and the effect of temperature increases during these stages. Simulations were run for eight sites where citrus is currently grown commercially and 14 potential sites.

The results indicate that temperature increases may cause some decrease in citrus production in the southern-most producing sites primarily due to the shortening or complete lack of a dormant period that is required for acceptable yields in citrus. However, the major trend will be some increase in production at the more northerly current sites. In addition, a slight expansion of the area of citrus production in the southern states could occur, although many regions which may develop climates suitable for citrus production would not have the sandy, well-drained soils that orange trees require.

The effect of increased levels of atmospheric CO_2 on citrus yields were not modeled explicitly due to a lack of physiological data. Based on a literature review, yields were estimated to increase by 50 percent at a doubling of the current CO_2 levels. However, because the changes are poorly substantiated in the present literature, they should be interpreted as extremely preliminary estimates.

Potatoes: Simulations of yield changes for potatoes were performed using SIM-POTATO, a physiologically based model (Hodges *et al.*, 1992). A total of 12 sites, nine of which are currently moderate to high level producers of autumn potatoes, were chosen for simulations. Temperature increases caused a decline in potato yield at all sites, with reductions as high as 50 percent with the 5 °C temperature rise. These losses are generally due to the sensitivity of the potato to temperature in its tuberization response, with currently grown cultivars requiring cool temperatures to begin tuber formation and growth. However, potatoes appear to contain considerable genetic variability, suggesting possibilities for their adaptation to warmer climates. In addition, adjustments in planting dates to accommodate the later arrival of autumn may offset some of the losses. As with citrus, the effects of increased CO_2 on potato yields was not modeled explicitly, but rather taken from the existing literature. Experimental results suggest that the yield will increase by approximately 15 percent if atmospheric CO_2 doubles.

Tomatoes: Finally, yield changes for tomatoes were based on an adaptation of a generic crop simulation model, CROPGRO (Hoogenboom *et al.*, 1992). A number of sites were chosen from each of the main tomato producing regions of the United States. As temperature increased, simulated yield increased in most cases to an optimum of 2.5 °C above actual means. At increased levels of CO_2 some of the yield increases were reduced. However, the patterns of an optimum yield with moderate temperature increases remained unchanged. The pattern of increased yields at 1.5 °C or 2.5 °C increases in temperature regardless of the latitude of the site suggests that actual planting dates are well calibrated for each region. Formal calibration of the model by its developers, however, is not yet complete. As a result, conclusions regarding the potential effects of changes in temperature on yield are tentative. Nonetheless, the results suggest that increases in both temperature and atmospheric CO_2 levels are likely to result in increased tomato yields at most sites.

Yield changes with adaptation

The estimated yield changes discussed above assume that farmers use management practices that are currently prevalent at each site. They do not incorporate the possibility of adjustments in management practices or technological change to offset the negative impacts of climate change (or to enhance potential positive impacts). In this sense, they can be viewed as very short-term predictions. However, over time, adjustments are likely to occur. Failure to allow for them will overstate the negative (or understate the positive) effects of climate change.

Two approaches were used to estimate how yields might change if adjustments occur. Both use regression analysis to examine how yields have actually changed in response to climate differentials, although the datasets and theoretical underpinnings of the two approaches are quite different. With the help of Robert Evenson, the first approach uses state-level pooled data on yields, temperature differentials, soil differentials, and research expenditures (as well as other variables) to examine the extent to which crops can migrate geographically in response to climate differentials and the extent to which research (e.g. plant breeding research) can be expected to mitigate the negative impacts of climate change. Drawing on the work of Kathleen Segerson and Bruce Dixon (see Chapter 4), the second approach is based on neoclassical duality theory. It uses county-level cross-sectional data on yields, output prices, seasonal temperatures, seasonal precipitation levels, and soil characteristics (as well as other variables) to estimate how yields vary with temperature and precipitation.[4] These two approaches are explained in more detail below.

The role of crop migration and agricultural research

When a change in temperature, CO_2, or other environmental factor occurs, crop yields will change in the short-run. Over time, two factors will modify these short-run yield changes. The first is crop migration and/or expansion of the range over which crops are grown. The current pattern of crop production in different temperature/soil regions reflects the comparative advantages of different crops. A rise in temperature will induce crop migration. For example, crops currently produced in warmer locations may migrate "northward" as temperatures rise in cooler regions to levels now experienced in warmer locations. This migration will be limited by edaphic and rainfall conditions which serve as "barriers" to the crop migration induced by temperature changes. If crop yields decline substantially when migration over the edaphic barrier occurs, migration will not take place.

The second factor that will modify short-run yield changes is the responsiveness of the research system (both public and private) to the temperature change. Plant breeders can put more weight on temperature tolerance of plants in their crossing and selection strategies. They can also facilitate crop migration through programs to achieve tolerance to different edaphic and rainfall conditions.

There is no direct evidence regarding the likely scope for yield loss modification through these two factors because we have had little experience with temperature increases in the United States in recent decades. It is possible, however, to draw some inferences from indirect evidence associated with regional (state) differences in

---

[4] This second approach, in its reliance on cross-sectional yield and climate data for the midwest, generates results concerning effects on producers' profits which can be compared with those estimated recently by Mendelsohn et al., 1994 which are described in Chapter 3.

temperature, soils and rainfall. This indirect evidence is associated with the yield change comparisons between regions (in this case, states) that differ in temperature and soil conditions. If there are few temperature and soil barriers to the transmission of a technological improvement that has been realized in one state to other states differing in temperature and soil conditions, then historical yield changes will be highly correlated between states. If these barriers are high, yield changes will not be transmitted from one state to another.

Robert Evenson specified and estimated a yield transmission equation using crop production data for 20 US states for the 1956–86 period. Estimates for four crops (wheat, corn, soybeans, and cotton) were obtained. The estimates for these crops showed that yield transmission over temperature barriers was consistent with the short-run biophysical crop yield estimates when only temperature barriers were included in the specification. However, when soils barriers were also included, the temperature effects on transmission were small. This indicates that crops will migrate easily (i.e. without substantial yield losses) within the same geoclimate zone. For corn and cotton, research programs enhanced transmission. This result indicates that research programs are likely to respond to temperature rises at least for these two crops even if crop migration does not occur. The estimates for the soils barrier, on the other hand, showed significant yield transmission losses across these barriers except for soybeans, where they were small. Research programs did not enhance transmission over the soils barriers.

Overall, the results suggest that the key barriers to yield transmission with existing varieties are the geoclimate barriers. Thus, it seems likely that most crops (except possibly soybeans) will not migrate large distances; i.e. across entire regions. However, movement from subregion to subregion, where varieties are already genetically similar, may occur. In particular, migration might be expected along the northern-frost borders, since there is room for this migration without crossing multiple geo-climate boundaries. The results also suggest, however, that research systems may be quite important in mitigating temperature effects within regions. This will effectively limit incentives to migrate crops. As a result, in the analysis of migration with ASM, crop migration is limited to relatively small movements (200 miles) from present crop production areas.

Duality-based estimates

The second approach to estimating adjustment possibilities is based on neoclassical duality theory. Neoclassical producer theory provides a prediction of how producers (farmers) make production decisions in response to exogenous factors, such as input and output prices, and environmental and technological constraints. Duality theory provides a methodology for predicting those decisions from observations on producers' costs or profits.

Using neoclassical theory, the long-run effect of a climate change can be estimated in one of two ways. First, a supply or yield function can be estimated directly with a dataset containing observations on yields, input and output prices, site characteristics, and climate variables. Alternatively, the economic impact can be predicted from an estimated profit or cost function. For this report, only estimates based on the direct estimation of the yield equations will be discussed since these are the estimates that provided input into the adaptation runs of ASM.

Yield equations were estimated for corn, wheat, and soybeans using county-level data from the 1987 Census of Agriculture for 12 midwest states. The functional form was quadratic in the climate variables (seasonal temperature and precipitation) to allow for non-monotonicity but linear in all other terms where monotonicity would be expected (e.g. output prices). The estimated equations were then used to predict yield changes under the alternative climate scenarios. The estimates all assume the 1987 level of CO_2. Thus, a total of 15 scenarios were run for each crop. For each crop–scenario combination, we estimated the yield change at five sites in the midwest.[5] These sites were the counties corresponding to the midwest sites used in deriving the short-run estimates of yield changes for corn, wheat, and soybeans based on biophysical crop models. Comparison of the two sets of estimates thus provides an indication of the extent to which farmers can be expected to adjust their production processes to mitigate negative impacts of climate change.

The results suggest that mitigation is possible, particularly for corn and wheat. For both of the crops, temperature increases averaged over the sites growing that crop induced long-run yield reductions that were smaller than the short-run estimates. For some sites, long-run wheat yields were predicted to increase rather than decrease as a result of temperature increases, since the relationship between yields and temperature was estimated to be non-monotonic and those sites were on the upward-sloping portion of the curve. In addition, the differences in the short-run and long-run estimates were larger for a 2.5 °C warming than for a 5.0 °C warming, indicating a greater potential for adaptation to mitigate negative impacts for smaller temperature increases.

The results for precipitation increases were somewhat different. The general result was that the long-run estimates were more pessimistic than the short-run biophysical estimates, again with the exception of soybeans. In general, the short-run estimates

---

[5] For some sites, current production levels of some crops was very low or zero. For example, soybeans and corn are not grown in Fargo. While estimated yield changes were calculated for these crop–site combinations, these estimates were not used in the analysis.

indicate that increased precipitation will increase yields. The long-run yield changes estimated from the yield equations were generally less positive and in some cases negative. The negative impacts reflect the fact that in the estimated yield equations the increases in April precipitation were generally yield reducing. In addition, the differences between the two sets of estimates may be attributable to differences in the treatment of irrigation. Because of data limitations, the long-run estimates do not distinguish between yields on irrigated and non-irrigated acreage.

While the above methodology provided site-specific estimates of yield changes, the estimates for any individual site–crop–climate combination are not sufficiently precise to provide substitutes for the short-run estimates. We therefore believe that the best use of the results is as evidence that adaptation to climate change is likely to offset some of the otherwise negative impacts.

To provide a rough estimate of the magnitude of the offset, we combine information from three sources. The first is a comparison of the short-run and long-run yield change estimates discussed above (i.e. a comparison of the biophysical estimates and the estimates from the estimated yield equations). The second is information on adaptation potential based on alternative runs of the biophysical simulation models under varying assumptions about adaptation (i.e. Level 1 and Level 2 adaptation). These alternative adaptation levels are described in Rosenzweig and Parry (1994) and are based on their previous work using different GCM-based scenarios. The third source of information is an analysis regarding the potential for adjustment through technological change, conducted as part of this overall research effort.

A central finding gleaned from these three information sources is that while adaptation is very site and crop specific, a reasonable first approximation is that adaptation could potentially offset roughly half of the negative impacts of a moderate climate change. However, the evidence suggests that adjustment possibilities are smaller for larger temperature changes. No evidence is available on the potential for adaptation to further enhance positive impacts. Based on these conclusions, the adaptation runs of ASM incorporate the following changes: (1) for the 2.5 °C scenario, negative yield change estimates derived from the biophysical models are reduced by one-half, and (2) for the 5.0 °C scenario, negative yield change estimates are reduced by 25 percent. No adjustments are made when predicted yield change estimates from the biophysical models were positive. These changes are applied to all crops in all regions of ASM.

Changes in forage production and livestock performance

Estimates of yield changes on pasture land were obtained from two sources. Estimates for the Southeast US were generated using the EPIC crop simulation model (described above). Estimates for natural (unimproved) grassland sites west of the Mississippi River were generated using a plant–soil ecosystem model, the CENTURY model (Parton *et al.*, 1992). These two sets of estimates were then combined to provide estimates of forage changes for all regions within ASM.

EPIC was used to simulate bermuda grass yield response to climate change and changes in atmospheric CO_2 for the 64 weather scenarios at each of five locations. Changes in crop yield were calculated from the baseline for all of the temperature, precipitation, and CO_2 combinations described in the scenarios above. Crop growth was simulated for 30 years for each climate scenario at each location. Averages of these 30 observations on yields were used to generate a percentage change in crop yield by crop, scenario, and location. All improved pasture land was assumed to be rainfed, so no water demands were estimated.

Yield response of improved pasture to climate change and changes in atmospheric CO_2 varied by location. Florida and Louisiana yields responded positively to increased precipitation, increased temperature, and increased CO_2 in every case. Alabama, North Carolina, and Tennessee yields responded positively to increased CO_2, but reactions to changes in precipitation and temperature were mixed depending on the particular combination of the three climatic variables. Few of the scenarios had lower yields than the base, and the ones that did were small reductions (less than 5 percent). While Alabama, Louisiana, and Florida had some large yield changes, North Carolina and Tennessee had yield changes of only 7 percent or less.

The forage changes in the western United States were estimated using the CENTURY model (Parton *et al.*, 1992, 1993). CENTURY is a general model of the plant–soil ecosystem that has been used to represent carbon and nutrient dynamics for different types of ecosystems. The model simulates the dynamics of grassland systems, and implements land management options that influence the level of grazing, fire frequency, and nitrogen deposition. See Ojima *et al.*, 1993 or Parton *et al.*, 1992, 1993 for details of the CENTURY model.

Modifications were made to the plant production parameters for both C_3- and C_4-type grasslands under doubled atmospheric CO_2 by changing production relative to potential evapotransportation (PET) and to nitrogen use efficiency (NUE). To analyze the sensitivity of grassland ecosystems to modified climate and atmospheric CO_2 levels, simulations were performed for 12 grassland sites west of the Mississippi River. For each site, a total of 64 simulations were run, corresponding to the 64 possible combinations of the temperature, precipitation, and CO_2 values. For each site, a current 30-year weather file of monthly precipitation and monthly mean maximum and minimum temperatures was created using existing weather station data from the site itself or from a nearby meteorological station. Climate perturbations were applied uniformly to each monthly value of the contemporary climate input file. The climatology of the sites ranged from a low in annual rainfall of 14.5 cm (Bakersfield, California) to a high of 83.5 cm (Manhattan, Kansas). The mean annual temperatures ranged from a low of 4.8 °C (Fargo, North Dakota) to a high of 18.6 °C (Bakersfield, California).

At all sites, increases in precipitation were predicted to increase net primary production. The sites with the least response to changes in rainfall were the two driest sites (Bakersfield, California and El Paso, Texas). This is due to the relatively low perturbation in actual rainfall imposed at these sites. The sites with the greatest response to changes in precipitation were sites with annual rainfall ranging from 50 to 60 cm (Abilene, Texas; Fargo, North Dakota; and North Platte, Nebraska), with a slightly smaller impact at the highest rainfall site.

The temperature effect did not display a consistent pattern of changes in net primary production (vegetative yields, measured in grams per square meter). The mean responses ranged from -3.22 per °C for Bakersfield, California, to $+2.08$ per °C for Manhattan, Kansas. Eight of the sites had a negative response to temperature changes, indicating that as temperature increases net primary production declines. Production at the other four sites (Boise, Idaho; Fort Collins, Colorado; Manhattan, Kansas; and North Platte, Nebraska) responded positively to increased temperature. The overall effect of CO_2 increases across the sites was positive. The combined effects were additive across the factors.

The 17 location-specific estimates of changes in forage production generated for the Southeast and for the 12 western sites were used to estimate regional changes in forage production for the production regions in ASM. Forage production for each simulated site was converted to lbs/acre expected from each of the 64 climate condition combinations. For the 12 locations simulated using the CENTURY model, this was achieved by converting the g/m^2 of net primary production. For the five locations simulated using EPIC, the reported tons/acre were converted to lbs/acre. Percentage changes were calculated by comparison with the baseline.

In developing estimates of forage production changes for the regions in ASM, consideration was given to the number of cattle in each region, and their relative proximity to each of the 17 sites for which forage production was simulated. Weights were then assigned for each of the sites relative to the proportion which its production would contribute to the estimated average production for each region.

Baseline enterprise budgets in the ASM represent current input and cost information for each activity in each region for the following livestock enterprises: beef cow/calf, sheep, weaning stocker steers, yearling stocker steers, weaning stocker heifers, yearling stocker heifers and dairy cows. These were modified to reflect changes in forage production. Changes in forage production/availability would be

34

expected to affect the livestock production budgets by changing (1) the acres of grazable forage required per animal per year; (2) the amounts of non-grazed feeds required per year; (3) the amount of salable product produced per animal per year; or (4) some combination of all of these. The amounts that each of these will change varies by region and by producer within a region. Furthermore, to estimate the relative degree that, for example, a supplemental feed such as grain might substitute for reduced availability of grazed forage requires knowledge of the quality (energy and protein content) of both the forage and the grain. While the quality of the grain is likely to be uniform across regions, the quality of the forage is expected to be highly variable.

The information required to estimate substitution effects and the changes that might be expected in per animal production as a reduction of changes in forage production was not available. As a result, the amount of grazable forage available/required in the livestock enterprise budgets in ASM was modified to simulate climate induced changes in forage production.

To capture the effects of climate change on livestock enterprises, the following changes were made in ASM. First, for all enterprises, the availability of AUM (animal unit months) of forage from public lands was changed for each region in proportion to the percentage change in forage production under each climate change scenario. This represents a change in the resource constraints in the model rather than a change in the enterprise budgets.

Second, for the beef cow/calf, sheep, dairy cows and stockers in Texas, the number of acres required per head was changed to reflect climate-induced changes in forage production per acre in each region. The changes in acres of pasture required per head were estimated by first multiplying the production per acre in the base climate combination by the acres per head in the original budget for each enterprise. This product, representing the total pounds of pasture forage available per animal, was then divided by the production per acre from each subsequent climate scenario to determine the acres of pasture that would be required under that scenario.

Third, for stockers in Texas, the cost per head for wheat pasturage was changed in proportion to estimates of climate-induced changes in wheat production per acre for Abilene, Texas. The stocker cattle enterprise budgets for regions in Texas included the cost per head for grazing wheat pasture during the winter months. For winter wheat, the estimated yield changes for wheat production for Abilene were used as a proxy for the percentage change in the cost of wheat pasturage per head for all the stocker cattle enterprises in Texas.

The above changes reflect climate-induced changes in forage production. In addition, modifications due to the direct effects of climate on livestock (cattle) production and costs were estimated. These include the effects of elevated summer temperatures on intake of forage and supplemental feeds (appetite depressing) and, secondarily, the decreased energy requirements for body maintenance due to warmer winters. Information on these effects was obtained by Jerry Stuth, range animal nutrition and grazing specialist in the Department of Rangeland Ecology and Management at Texas A&M University. Based on a review of relevant literature and the use of NUTBAL, a nutritional balance analysis tool developed at Texas A&M, we estimated the changes in production efficiency (lbs of primary product produced per head) that would result as a combination of the lower weight gains and decreased demand for forage due to increased summer heat and the higher weight gains due to warmer winter weather. In all cases, however, the negative effects of hotter weather in the summer would be expected to outweigh the positive effects of warmer winters.

Based on this analysis, adjustments were made in estimates of primary production to reflect the effect of temperature change on livestock performance. These adjustments are in addition to the adjustments due to impacts on forage production which are region- and enterprise-specific. The largest adjustment was a 10 percent reduction in primary production for cow/calf and dairy enterprises in Appalachian, Southeast, Delta States, Southern Plains, and Texas under a 5.0 °C temperature increase. The smallest change was a 1 percent reduction in primary production from stocker enterprises in this same region under 1.5 °C warming.

2.4   Economic results

The preceding discussion presents background information on procedures and assumptions used to generate key inputs for the ASM-based climate change analysis.[6] The scenarios evaluated with ASM include (1) two baseline configurations of ASM (1990 economic and agronomic conditions; 2060 economic and agronomic conditions); (2) the nine uniform temperature–precipitation combinations that include a central case (2.5 °C temperature increase, 7 percent precipitation increase) and a more severe case (5 °C increase, 0 percent precipitation change), analyzed using 1990 and then 2060 economic projections for each scenario; and (3) two GCM climate forecast scenarios (GISS and GFDL-R30), again using 1990 and 2060 economic projections for each scenario. Also presented are a series of sensitivity analy-

---

[6] Changes in yields and water use for the two case analyses are contained in the appendices to this chapter. Yield changes for the entire set of analyses performed here (165 ASM runs) are available on disk from the authors. Additional runs including federal farm provisions are not reported here because of questions about whether they would be in place in the long-run. The farm programs tend to increase damages by encouraging inefficient responses by farmers to changes in conditions.

ses exploring producer adjustment (mitigation) options based on projected adjustment possibilities and changes in exports based on changes in world food production obtained from Rosenzweig and Parry (1994). In this section, we present the economic consequences of each of these scenarios.

Baseline

The baseline solution to ASM, assuming no climate change, is important because the economic consequences of climate change are measured as changes from this baseline. ASM calculates the maximum social welfare (the sum of consumer and producer surplus) for each climate scenario. By comparing the results of the new climate scenarios with the corresponding baseline, one can estimate the welfare effect of the climate change. The endogenous prices and quantities predicted by the 1990 baseline correspond closely to observed 1990 prices and quantities providing confirmation of the model's validity (Chang and McCarl, 1993). It is not possible to validate the welfare values for the 2060 ASM.

The importance of the baseline can be seen clearly by comparing the results using the 1990 economy versus the 2060 economy. Previous analyses using ASM measured the consequences of climate change relative to 1990 base values. Economic and other forces will transform agriculture over the next 70 years making the agricultural sector in 2060 dramatically different from that in the 1990s. While the 2060 baseline is highly uncertain, the analysis reveals it to be quite different from 1990. Further, the choice of baseline affects the magnitude of the impacts.

In subsequent tables we focus on changes in welfare, using the sum of consumer (both foreign and domestic) and producer surplus as a measure of welfare. We also report indices of national price and quantity changes and changes in national resource use, including changes in land (irrigated and dryland) and water use at the national level. Changes in regional crop production are also presented to provide important insights into shifts in the crop market shares. Taken together, these results provide an indication of the economic consequences of the yield and other changes imposed on the ASM. This set of economic effects was selected to facilitate a comparison with previous climate change research, which has focused primarily on the welfare changes and changes in regional production (comparative advantage) across alternative climate change scenarios.

Welfare effects

The economic implications of each of the nine uniform temperature and precipitation combinations are presented in this chapter using both the 1990 and 2060 base models. Thus, a total of 18 ASM solutions are presented. The changes in welfare

Table 2.1. *Net welfare from climate change; 2060 (billions of 1990 dollars)*

Climate scenario	Consumer surplus	Producer surplus	Foreign surplus	Total
1.5 °C–0% P	45.4	−8.2	10.0	47.2
1.5 °C–7% P	53.5	−8.4	10.0	55.1
1.5 °C–15% P	62.9	−9.3	9.0	62.5
2.5 °C–0% P	37.3	−4.7	5.3	37.9
2.5 °C–7% P	47.7	−6.2	6.0	47.4
2.5 °C–15% P	55.6	−6.5	5.7	54.8
5.0 °C–0% P	−11.4	20.9	−9.3	0.2
5.0 °C–7% P	12.2	10.1	−9.4	12.9
5.0 °C–15% P	29.4	2.3	−7.5	24.2

Note:
Assumes 530 ppmv for CO_2.

Table 2.2. *Net welfare from climate change; 1990 (in billions of 1990 dollars)*

Climate scenario	Consumer surplus	Producer surplus	Foreign surplus	Total
1.5 °C–0%P	8.7	3.4	3.7	15.8
1.5 °C–7% P	11.2	4.8	4.1	20.0
1.5 °C–15% P	12.7	6.2	4.4	23.3
2.5 °C–0% P	6.8	0.3	3.4	10.5
2.5 °C–7% P	9.7	1.6	3.6	14.9
2.5 °C–15% P	10.8	2.6	4.0	17.3
5.0 °C–0% P	−10.9	5.8	−2.2	−7.2
5.0 °C–7% P	−4.3	1.6	−0.8	−1.7
5.0 °C–15% P	0.7	1.0	0.8	2.5

Note:
Assumes 530 ppmv for CO_2.

estimated using the 2060 economy projections are reported in Table 2.1 and changes using the 1990 economy context are presented in Table 2.2.

The results in Table 2.1 differ substantially from those presented in the IPCC summary of damage estimates for this sector (IPCC 1996, see Table 6.4, page 203). The expert judgments presented by the IPCC indicate a range of annual damage estimates of $1.1 billion to $17.5 billion for warming in the range of 2.5 to 4 °C. This contrasts sharply with our results which, with the exception of the 5 °C warming case with no change in precipitation, indicate uniform climate change results in net national benefits for the agricultural sector. In our study, overall production levels are projected to increase, leading to reductions in prices and generating benefits for consumers at home and abroad. Under the central case, 2.5 °C, 7 percent precipitation increase and 2060 economic projection, welfare increases by $47 billion, or less than 3 percent of the total base value for the agricultural sector. Under the more severe 5 °C, no precipitation change scenario, the net impact is essentially no change in 2060 aggregate welfare. Under this latter scenario, the benefits attributable to carbon fertilization just offset the damages from the higher temperatures. That is, the direct effect of CO_2 on agriculture and plant growth just offsets the effect of temperature change.

Differences in these results compared to estimates summarized by the IPCC appear to be attributable to three factors: (1) the climate scenario, including the projected concentration of carbon dioxide; (2) inclusion of the carbon fertilization effect; and (3) refinements in ASM to capture additional opportunities for adapting to climate change. The large impact of carbon fertilization is clear in this study as well as the earlier work of Adams *et al.* (1989, 1990, 1995). For example, the earlier work indicates a range of estimates from approximately $11 billion in damages to approximately $10 billion in benefits, depending largely on the projection of future atmospheric carbon dioxide levels and the resulting carbon fertilization effect. Nordhaus (1991) presents this range of estimates in his overview of damages, and analysts appear to have interpreted this treatment as implicitly giving both numbers equal weight. Cline (1992) dismisses the long-term impact of carbon fertilization and focuses more intently on the high damages estimates. Scientific research on carbon fertilization, however, strongly indicates a beneficial impact on crops (IPCC, 1996b). We consequently focus on those results that reflect carbon dioxide concentrations (530 ppmv) and the resulting carbon fertilization effect consistent with a doubling of greenhouse gases.

The estimates of the sensitivity of agricultural production to changes in temperature reveal an interesting pattern. In general, the benefit estimates in our study appear to be maximized with a mild warming of about 1.5 °C. Beyond a 1.5 °C change, the benefits fall at an increasing rate as temperatures continue to rise. Additional precipitation appears to be strictly beneficial. The magnitude of benefits associated with more precipitation appears to be independent of temperature.

A comparison of the 1990 and 2060 results suggests that climate change has a similar pattern and relative magnitude of impacts under both economic scenarios. In both cases, benefits are maximized with a 1.5 °C warming and fall thereafter. In both cases, benefits increase with additional precipitation. The difference between the results for the two economic scenarios comes from changes in demand and population projections that shift the relative shares of producer, consumer, and foreign surplus. In addition, the benefits in 2060 are larger because the baseline sector is larger. It is also important to note, as stated in the previous section, that the 2060 results are based on long-term economic projections that are highly uncertain, which in turn make the 2060 results more uncertain.

A broader set of welfare changes are presented in the tables in Appendix A2, to provide information on the economic consequences of various temperature, precipitation, and CO_2 combinations. Comparing alternative CO_2 levels, it is apparent that welfare in this sector increases steadily with more carbon dioxide. Carbon fertilization substantially increases plant productivity and this results in expanded farm outputs.

Effects on prices, quantities, and land

The changes in welfare reported in Tables 2.1 and 2.2 are driven by endogenous changes in crop production and crop prices within the ASM. Specifically, an optimal ASM solution (i.e. one which maximizes economic surplus) for a given ASM configuration or scenario reflects a set of quantities and prices for the economic activities in ASM. In this report, we capture the endogenous changes in crop prices and quantities with Fisher indices of aggregate crop and livestock production and their companion price indices. Fisher indices weight composites of goods given initial prices. The indices are presented in Table 2.3 for the central and severe cases using both 1990 and 2060 economic scenarios.

The indices are consistent with the welfare effects reported in Tables 2.1 and 2.2. With the central case (for both 1990 and 2060), aggregate crop and livestock production increases, which gives rise to price declines. The nature of the decline reflects the generally inelastic demand for agricultural commodities. Conversely, the severe climate case yields slight reductions in crop and livestock production for 1990, which translates into increases in the prices of these commodities. The price changes observed here drive the changes in the welfare measures.

Table 2.3 also contains information on land (both irrigated and dryland) and water use under the various scenarios. Crop land use declines in three of the four cases, with the decline greatest under the central cases. Thus, favorable climate (environmental quality) is a substitute for land in the agricultural production function. Under the 1990 severe case, irrigated crop land expands substantially, as irrigation is used to mitigate for the hotter and drier climate assumed in these cases. As expected, water use also increases, though by much less than the change in acreage. This reflects the improvements in water efficiency inherent in shifts in crop location in the ASM. It is not clear whether expansion of irrigated acreage would be sustainable with no increases in total precipitation.

Climate change is also expected to change agricultural production patterns across the landscape. Previous agronomic studies suggest that global warming will have both

Table 2.3. Changes in prices, quantities, and land

| Scenario number | Fisher indices: Base = 100 | | | | % Change from base | | | |
| | Prices | | Quantity | | Resources | | | |
	Crops	Livestock	Crops	Livestock	Irrigated land	Dry-land	All crop land	Water use
Central 1990	80.93	95.50	127.09	132.29	−0.66	−6.01	−5.06	−23.49
Severe 1990	115.33	107.87	99.28	95.33	9.92	−5.57	−2.83	1.60
Central 2060	69.92	97.72	115.59	100.17	−15.19	−20.50	−19.40	−27.69
Severe 2060	112.50	100.11	101.95	99.47	−30.05	8.63	0.61	−24.26

Table 2.4. *Regional index estimates for crop production*

US region	1990 Central case	1990 Severe case	2060 Central case	2060 Severe case
Northeast	112.87	112.73	44.59	83.49
Lake States	163.79	94.68	165.91	122.66
Corn Belt	124.98	73.53	106.28	82.99
Northern Plains	152.77	143.15	113.54	148.75
Appalachia	103.74	77.02	96.48	59.02
Southeast	110.67	74.63	138.65	98.26
Delta States	78.71	71.32	91.30	70.68
Southern Plains	83.40	66.37	75.17	59.00
Mountain States	127.33	129.24	121.97	115.75
Pacific Coast	138.52	144.90	134.64	129.76

positive and negative effects on production depending on location. Crop production may be enhanced in cooler areas, as warming reduces climate barriers to production. Conversely, some areas may become too warm for production of current crops. Unless heat tolerant crops are available, such regions will see a decline in agricultural output. Along with these aggregate welfare effects, it is interesting to see how regional crop production will adjust to climate change. At the regional level, changes in crop production indicate possible changes in comparative advantage. Table 2.4 reports changes in crop production for the central and severe cases with both 1990 and 2060 economic conditions. The table provides index numbers of total crop production for the 10 major production regions in ASM. These index numbers are measured against the appropriate base-level production (base production equals 100).

The results reported in Table 2.4 are similar to earlier findings in Adams *et al.* (1989, 1990, 1995). There is a general pattern observed in the four case studies with an expansion in more northerly agricultural regions and a corresponding decline in the southern latitude regions. For example, in the central case, all regions experience expansion in total crop production except the Southern Plains and Delta States regions. In the more severe case, gains are confined to the Northeast, Northern Plains, Mountain States, and Pacific Coast, with losses observed at southern latitude regions. The same pattern holds for both economic scenarios except that the Northeast is much harder hit in 2060 than in 1990.

One difference between the present results on regional productivity and those reported in earlier work by Adams concerns the Southeast. In these earlier studies, the Southeast experienced large reductions in crop production under all GCM scenarios.

42

In some cases, Southeast production was reduced by 50 percent. In the current analysis, the Southeast actually increases crop production under the central case scenario (for both 1990 and 2060). Under the severe case scenario, the Southeast experiences slight to moderate reductions, but these are much smaller than in the earlier assessment.

This difference is due to two factors. The first is the addition of heat tolerant crops (citrus, tomatoes) to the Southeast crop mix. In the aggregate, increases in these crops mitigate or offset the negative effects of temperature increases on crops such as soybeans and corn. A second factor is that the previous studies use GCM climate forecasts, which are, on average, warmer than the central case conditions. However, the more adverse case here is actually warmer (5 °C) than the GISS (4.2 °C) and GFDL (4.4 °C) average global temperature forecasts. Thus, the uniform increases used here are likely to be less important in explaining the findings for the Southeast than is the addition of heat tolerant crops to ASM. An implication of this finding is the importance of including a reasonable range of alternative crop options in agricultural assessments. Failure to include these crops will overstate the economic losses due to climate change.

Sensitivity analyses: role of farmer adaptations and export assumptions

Two areas of uncertainty in assessing the economic consequences of climate change on US agriculture are (1) the ability of farmers to adapt to long-term changes in climate and (2) the effect of changes in global food production and demand due to climate change. In terms of adaptation, historical evidence suggests that US farmers readily adopt new technology; they also adapt rapidly to institutional and other changes. There is also evidence that farmers adapt effectively to climatic variations (see Chapters 3 and 4). However, the level of adaptation across large temperature changes has not been documented in the current record and remains a source of uncertainty. Changes in world food production under climate change will influence US agriculture, given the large share that some US commodities have of the world markets. Assumptions regarding export demand elasticities in ASM have an impact on total welfare. Given the potential importance of these two factors in determining welfare estimates generated by ASM, a series of sensitivity analyses are performed here.

Role of farmer adaptations

To explore the possible role of adaptation in mitigating the yield effects of climate change, yield changes generated by the plant simulation modeling approach were compared with those reported by the duality yield equations which reflect

Table 2.5. *Welfare with farmer adaptation (1990 $ billions)*

Scenario	Consumer surplus	Producer surplus	Foreign surplus	Total surplus
1990 Central case	9.5	2.0	3.5	14.9
1990 Severe case	−7.8	4.2	−1.3	−4.8
2060 Central case	48.9	−6.7	5.8	48.1
2060 Severe case	0.4	13.3	−6.2	7.3

Note:
On farm adaptation assumed to climate change.

producer responses to present temperature and precipitation gradients (see Chapter 4). Based on the combination of results from the duality study, the effects of adaptations from the crop simulation models and the evidence regarding partial factor productivity, the magnitude of potential adaptations (mitigation) were estimated to be 50 percent for the central case and 25 percent for the severe case. To test the potential effect of these adaptations, yields in ASM were adjusted accordingly. The results of this sensitivity analysis are reported in Table 2.5.

The changes in welfare measured for these alternative specifications of the scenarios are generally consistent with the magnitude of the difference in yield adjustments (of 50 and 25 percent). Specifically, the loss in welfare under the severe case for 1990 has been reduced from over $7 billion to less than $5 billion under the new yield effects. Similarly, the severe case for 2060 now shows a net gain of over $7 billion, compared with a $0.2 billion gain in the absence of these adaptation adjustments. For the 1990 and 2060 central cases, the role of adaptation is less important (both show very slight gains with adaptation). The reason for the small effect in this case is that only negative yield changes are modified for adaptation in these runs. There are far fewer negative yield changes in the central case than in the severe case.

The significance of these findings is that assumptions regarding farmer adaptation play a major role in the welfare effects estimated from ASM. Assessment of whether the adjustments used here are "reasonable" or plausible under future climate change is beyond the scope of this effort. However, future assessments need to explore the potential for farmer adaptations, along with the larger issue of the role of technological change.

Role of exports assumptions on welfare estimates
The United States is a major exporter of feed grains and other agricultural commodities. For some domestically produced crops, over 50 percent of production is

Table 2.6. *Welfare impacts with export changes (1990 $ billions)*

Climate change/ GCM export scenario	Consumer surplus	Producer surplus	Foreign surplus	Net welfare
1990 Central w/GISS	9.0	2.3	5.8	17.2
1990 Central w/GFDL	8.9	2.4	4.4	15.6
1990 Central w/UKMO	9.1	2.7	32.1	43.9
1990 Severe w/GISS	−5.5	4.2	1.0	−0.2
1990 Severe w/GFDL	−5.0	3.7	−0.0	−1.3
1990 Severe w/UKMO	−5.4	4.5	27.0	26.1
2060 Central w/GISS	46.7	−5.4	11.4	52.7
2060 Central w/GFDL	46.7	−5.6	1.0	42.1
2060 Central w/UKMO	47.7	−4.2	79.7	123.3
2060 Severe w/GISS	11.7	10.7	−3.6	18.8
2060 Severe w/GFDL	12.3	8.2	−10.8	9.7
2060 Severe w/UKMO	8.9	12.2	65.8	86.8

Note:
Global GCM forecast used to adjust US export demand.

exported. Given the importance of world supply and demand conditions in determining US exports, an ideal assessment of the effects of climate change on US agriculture should reflect concomitant changes in world food production. Such an effort would require analyses of global supply responses under climate change, an effort far beyond the scope of this project. The analyses reported above use import/export demand relationships (elasticities) based on historical levels.

We test the effect of global changes in agricultural production on the welfare effects of the central and severe cases using recent estimates of changes in agricultural trade patterns under climate change (Rosenzweig and Parry, 1994). Three GCMs are utilized to predict changes in the demand for US exports. Except for GISS, the estimates are tied to different GCM forecasts and thus are not strictly comparable to any scenarios used here. (The average temperature increases in GISS, GFDL, and UKMO are 4.0 °C, 4.2 °C and 5.0 °C, respectively.) Nonetheless, the directions of the changes are comparable. The changes in US exports for each GCM, by commodity group, as reported in Rosenzweig and Parry, are used to adjust export demands in the ASM. After calibration to insure internal consistency in prices and quantities, the ASM is solved for the four central cases, for a total of 12 ASM solutions. The welfare changes for each of these sensitivity analyses are reported in Table 2.6.

Changes in US exports, as forecast by Rosenzweig and Parry (1994), in general have a minimal effect or increase welfare gains. For example, 2060 welfare under the central

Table 2.7. *Welfare estimates for GCM scenarios (1990 $ billion)*

Climate change	Consumer surplus	Producer surplus	Foreign surplus	Total surplus
1990 GISS	10.4	−1.9	3.5	12.0
1990 GFDL-R30	−20.4	9.4	−5.2	−16.2
2060 GISS	20.6	45.4	50.6	116.6
2060 GFDL-R30	−65.7	52.2	−3.4	−16.9

case is reduced slightly from $47 billion to $42 billion (GFDL), or increases to $53 billion (GISS) or $123.9 billion (UKMO). The severe case changes from $0.2 billion to $10 billion (GFDL), $19 billion (GISS), and $87 billion (UKMO). Not surprisingly, the more severely that foreign countries are affected by climate change relative to the United States, the higher the US welfare. The bulk of these gains are to producers and foreign consumers; the impact on US consumers is slight (due to the excess demand characteristic of agricultural exports). Similar conclusions can be drawn using the 1990 economy assumption.

GCM-based climate effects

The changes in economic welfare for the GCM scenarios are reported in Table 2.7. These results reflect crop yield changes across regions arising from the regional temperature and precipitation changes forecast by the GISS and GFDL-R30 general circulation models. Unlike the uniform climate change scenarios, large differences in temperature and precipitation exist across the regions, particularly with GFDL-R30 forecasts. These GCM-based results can thus provide insights into the potential importance of these regional differences in terms of national–level economic consequences.

The change in 1990 welfare associated with the GISS forecast is approximately $12 billion, or 20 percent higher than the GISS welfare change reported in Adams *et al.* (1995). This 20 percent increase in welfare is due to the modification in ASM performed in the current assessment. Specifically, the inclusion of more crops and additional adaptation options, such as crop migration, allows the agricultural sector to more fully exploit the generally beneficial regional climatic condition forecast by GISS. Under 2060 conditions, the welfare change increases dramatically, to about $116 billion. The results using GFDL-R30 are more pessimistic, suggesting damages of $17 billion in 2060. The large temperature increases in some important crop production areas (e.g. the Corn Belt) translate into large reductions in yields.

2.5   Summary and conclusions

This chapter extends previous analyses of the economic effects of climate change on agriculture. Five important improvements are made: (1) new crops are incorporated such as fruits and vegetables to represent regional crop alternatives in more southerly locations; (2) farmer adaptation to climate change is explored; (3) crop migration across regions is included even where crops are not currently being grown; (4) changes in forage production and livestock performance are introduced; and (5) the potential for technological change is assessed.

This updated model is then used to assess a set of uniform climate change scenarios and two GCM scenarios. Assuming carbon dioxide levels of 530 ppmv, these climate scenarios are examined for both a 1990 and a 2060 economy. A series of sensitivity analyses are also developed, exploring the role of farmer adaptations and export (world food production) assumptions.

Estimation of the economic consequences of these scenarios required predictions of the impacts of climate change on (1) yield levels for crops and forage; (2) animal grazing requirements and performance; (3) crop migration potentials; (4) technology-based changes in yields; and (5) changes in water resource availability. These predicted changes were then used in an economic model of US agriculture. The economic model provides estimates of changes in social welfare, crop prices and quantities, resource use, and other measures of economic performance arising from the climate scenarios.

The analyses of the various climate change scenarios reveal a range of potential economic effects on the welfare of consumers and producers. The welfare results show some similar patterns in both 1990 and 2060. Agricultural welfare strictly increases in the United States with a 1.5 °C warming. Further warming reduces this benefit at an increasing rate. Additional precipitation is strictly beneficial. The welfare gain from a 1.5 °C warming with 7 percent precipitation is $55 billion, using a projected 2060 economy, and $20 billion using a 1990 economy context. Further warming by 2.5 °C reduces these benefits to $47 billion. A 5.0 °C warming with 7 percent precipitation leads to benefits of only $13 billion. The results for the 1990 economy are similar but smaller. The magnitude of welfare gains for the central case is only $15 billion. Severe warming results in levels of welfare below the baseline (damages relative to no warming at all) of $2 to $7 billion depending upon the precipitation level.

The cross-sectional evidence presented in Chapter 4 indicates that farmers may adapt to climate change thus mitigating some of its adverse effects. This finding, coupled with the importance of technology and related assumptions underlying the 2060 analyses, supports the inclusion of such features in future economic assessments.

Sensitivity analyses on export assumptions reinforces the importance of world trade (exports) on the welfare of the US agricultural sector. Global climate change is likely to increase the demand for US commodities. As the analysis here shows, increased export demand increases US welfare.

The GCM-based analyses indicate that if climate changes according to the GISS forecasts, net welfare increases by $12 billion for the 1990 base. This value is approximately 20 percent greater than that found in previous analyses using GISS and the ASM (Adams *et al.*, 1995). The difference (increase) is due to changes in ASM, such as the addition of crops and other mitigation opportunities, which allow the agricultural sector to exploit more fully the new climate conditions. The GFDL-R30 analysis reveals losses of over $14 billion (measured against the 1990 base). These losses arise from the harsher climate conditions under this GCM. The GCM temperature and precipitation results reflect regional and seasonal variation, but if aggregated to a national average climate change, the welfare changes for the GCM scenarios bracket the comparable uniform climate change estimates. The GCM results also indicate the broader range of impacts possible with different regional and seasonal forecasts.

The results generated by these analyses provide an improved set of estimates of the economic effects of climate change on agriculture. This study addresses some problems found in earlier studies, but also suffers from some similar shortcomings arising from the long-term nature of the problem. However, by including a larger number of crops and livestock activities and more adaptation possibilities than previous climate change assessments, this study is the most comprehensive study to date of the effects of climate change on US agriculture.

References

Acock, B. and Allen, H. 1985. Crop Responses to Elevated Carbon Dioxide Concentrations. In: *Direct Effects of Increasing Carbon Dioxide on Vegetation*, B. Strain and J. Cure (eds.). DOE/ER-0238. Washington, DC: US Department of Energy.

Adams, R.M. 1989. Global Climate Change and Agriculture: An Economic Perspective. *American Journal of Agricultural Economics* 71: 1272–9.

Adams, R.M., Rosenzweig, C., Peart, R.M., Ritchie, J.T., McCarl, B.A., Glyer, J.D., Curry, R.B., Jones, J.W., Boote, K.J. and Allen, J.H. Jr. 1990. Global Climate Change and US Agriculture. *Nature* 345(6272): 219–24.

Adams, R.M., Fleming, R., McCarl, B.A. and Rosenzweig, C. 1995. A Reassessment of the Economic Effects of Climate Change on US Agriculture. *Climatic Change* 30: 147–67.

Ben-Mechlia, N. and Carroll, J.J. 1989a. Agroclimatic Modeling for the Simulation of Phenology, Yield and Quality of Crop Production: 1. Citrus Response Formulation. *International Journal of Biometeorology* 33: 36–51.

Ben-Mechlia, N. and Carroll, J.J. 1989b. Agroclimatic Modeling for the Simulation of Phenology, Yield and Quality of Crop Production: 2. Citrus Model Implementation and Verification. *International Journal of Biometeorology* **33**: 52–65.

CAST. 1992. *Preparing U.S. Agriculture for Global Climate Change*. Task Force Report 119, Ames, IA: Council on Agricultural Science and Technology.

Chang, C.C. and McCarl, B.A. 1993. *Documentation of ASM: The U.S. Agricultural Sector Model*. Texas A&M University, Texas: Department of Agricultural Economics.

Cline, W. 1992. *The Economics of Global Warming*. Washington DC: Institute of International Economics.

Crosson, P. 1993. Impacts of Climate Change on the Agricultural Economy of the Missouri, Iowa, Nebraska and Kansas (MINK) Region. In: *Agricultural Dimensions of Global Climate Change*, Kaiser, H. and Drennen, T. (eds.). Delray Beach, FL: St. Lucie Press.

d'Arge, R.C. 1975. *The Climate Impact Assessment Program (CIAP): Monograph 6*. Washington DC: Department of Transportation.

Dudek, D.J. 1988. Climate Change Impacts Upon Agriculture and Resources: A Case Study of California. In: *The Potential Effects of Global Climate Change in the United States*, Smith, J.B. and Tirpak, D.A. (eds.). Washington, DC: US Environmental Protection Agency.

Godwin, D., Ritchie, J., Singh, U. and Hunt, L. 1989. *A User's Guide to CERES-Wheat–v2.10*. Muscle Shoals, AL: International Fertilizer Development Center.

Helms, S., Mendelsohn, R. and Neumann, J. 1996. The Impact of Climate Change on Agriculture. *Climatic Change* **33**: 1–6.

Hodges, T., Johnson, S.L. and Johnson, B.S. 1992. A Modular Structure for Crop Growth Simulation Models: Implemented in the SIMPOTATO Model. *Agronomy Journal* **84**: 911–15.

Hoogenboom, G., Jones, J.W. and Boote, K.J. 1992. Modeling Growth, Development and Yield of Grain Legumes Using SOYGRO, PNUTGRO and BEANGRO: a Review. *American Society of Agricultural Engineers* **35**: 2043–56.

IPCC. 1996b. *Climate Change 1995: Impacts, Adaptations, and Mitigation of Climate Change: Science–Technical Analyses*. Watson, R., Zinyowera, M., Moss, R. and Dokken, D. (eds.). Cambridge: Cambridge University Press.

Jones, J.W., Boote, K.J., Jagtap, S.S., Hoogenboom, G. and Wilkerson, G.G. 1988. *SOYGRO v5.41: Soybean Crop Growth Simulation Model User's Guide*. Florida Agricultural Experiment Station Journal No. 8304, University of Florida: IFAS.

Kaiser, H., Riha, S., Wilkes, D. and Sampath, R. 1993. Adaptation to Global Climate Change at the Farm Level. In: *Agricultural Dimensions of Global Climate Change*, Kaiser, H. and Drennen, T. (eds.). Delray Beach, FL: St. Lucie Press.

Kokoski, M.F. and Smith, V.K. 1987. A General Equilibrium Analysis of Partial Equilibrium Welfare Measures: The Case of Climate Change. *American Economic Review* **77**: 331–41.

Mendelsohn, R., Nordhaus, W.D. and Shaw, D. 1994. The Impact of Global Warming on Agriculture: A Ricardian Approach. *American Economic Review* **84**: 753–71.

Nordhaus, W. 1991. To Slow or Not to Slow: The Economics of the Greenhouse Effect. *Economic Journal* **101**: 920–37.

Ojima, D.S., Parton, W.J., Schimmel, D.S., Scurlock, J.M. and Kittel, T.G. 1993. Modeling the Effects of Climatic and Carbon Dioxide Changes on Grassland Storage of Soil Carbon. *Water, Air, And Soil Pollution* **70**: 643–57.

Parton, W.J., McKeown, B., Kirchner, V. and Ojima, D. 1992. *CENTURY Users Manual*. Fort Collins, CO: Natural Resource Ecology Laboratory, Colorado State University.

Parton, W. J., Scarlock, J.M., Ojima, D.S., Gilmore, T.G., Schlores, R.J., Schimmel, D.S., Kirchner, T., Menaut, J.C., Seastedt, T., Moya, T.G., Kamnalrut, A. and Kinyamario, J.L. 1993. Observations and Modeling of Biomass and Soils Organic Matter Dynamics for the Grassland Biome Worldwide. *Global Biogeochemical Cycles* **7**: 785–809.

Reilly, J., Hohmann, N. and Kane, S. 1994. Climate Change and Agricultural Trade: Who Benefits and Who Loses? *Global Environmental Change* **4**: 24–36.

Ritchie, J., Singh, U., Godwin, D. and Hunt, L. 1989. *A User's Guide to CERES-Maize v2.10*. Muscle Shoals, AL: International Fertilizer Development Center.

Rosenzweig, C. and Parry, M. 1994. Potential Impacts of Climate Change on World Agriculture. *Nature* **367**: 133–8.

Takayama, T. and Judge, G. 1971. *Spatial and Temporal Price and Allocation Models*. Amsterdam: North Holland Publishing Company.

Williams, J.R., Jones, C.A and Dyke, P.T. 1984. A Modeling Approach to Determining the Relationship Between Erosion and Soil Productivity. *Transactions of the ASAE*: 129–44.

Appendix A2

Table A2.1. *Change in welfare from 1990 base, in 1990 $ (billions)*

Year	Precipitation change	Temperature change °C	CO$_2$ ppmv	Change in consumer surplus	Change in producer surplus	Change in foreign surplus	Change in total surplus
1990	−10	None	355	−10.87	4.87	−2.31	−8.30
1990	−10	None	440	2.16	−1.00	1.36	2.52
1990	−10	None	530	8.31	1.96	3.59	13.85
1990	−10	None	600	10.53	5.54	4.30	20.37
1990	−10	1.5	355	−18.17	9.33	−4.49	−13.32
1990	−10	1.5	440	−2.70	1.86	0.22	−0.62
1990	−10	1.5	530	6.18	0.07	3.37	9.62
1990	−10	1.5	600	7.16	4.12	4.12	15.40
1990	−10	2.5	355	−28.97	13.97	−8.13	−23.13
1990	−10	2.5	440	−11.09	6.17	−2.51	−7.42
1990	−10	2.5	530	2.17	0.12	2.04	4.33
1990	−10	2.5	600	6.12	0.81	3.50	10.44
1990	−10	5.0	355	−171.10	46.19	−47.23	−172.14
1990	−10	5.0	440	−57.69	25.66	−17.23	−49.26

Table A2.1. (*cont.*)

Year	Precipi- tation change	Temperature change °C	CO$_2$ ppmv	Change in consumer surplus	Change in producer surplus	Change in foreign surplus	Change in total surplus
1990	−10	5.0	530	−23.23	12.11	−5.83	−16.95
1990	−10	5.0	600	−11.34	6.27	−2.23	−7.30
1990(Base)	None	None	355	0.00	0.00	0.00	0.00
1990	None	None	440	7.77	−0.91	2.73	9.59
1990	None	None	530	11.35	3.95	4.35	19.66
1990	None	None	600	12.67	7.23	5.17	25.07
1990	None	1.5	355	−6.76	4.55	−1.58	−3.79
1990	None	1.5	440	4.36	−0.78	1.94	5.52
1990	None	1.5	530	8.70	3.36	3.72	15.79
1990	None	1.5	600	10.78	6.47	4.61	21.86
1990	None	2.5	355	−14.17	7.92	−3.98	−10.23
1990	None	2.5	440	−0.75	1.46	0.19	0.91
1990	None	2.5	530	6.78	0.26	3.43	10.47
1990	None	2.5	600	8.34	3.31	3.83	15.48
1990	None	5.0	355	−70.93	29.59	−21.53	−62.86
1990	None	5.0	440	−30.70	15.34	−9.30	−24.67
1990[b]	None	5.0	530	−10.89	5.82	−2.18	−7.24
1990	None	5.0	600	−2.61	2.36	0.52	0.28
1990	7	None	355	6.25	−2.45	1.59	5.39
1990	7	None	440	11.11	0.01	3.24	14.36
1990	7	None	530	13.59	5.22	4.73	23.54
1990	7	None	600	15.41	7.44	5.43	28.28
1990	7	1.5	355	0.95	0.97	0.10	2.02
1990	7	1.5	440	8.49	−0.77	2.59	10.32
1990	7	1.5	530	11.15	4.77	4.12	20.03
1990	7	1.5	600	13.57	7.10	4.82	25.49
1990	7	2.5	355	−6.10	4.42	−2.17	−3.85
1990	7	2.5	440	5.01	−0.31	1.37	6.08
1990[a]	7	2.5	530	9.67	1.64	3.55	14.86
1990	7	2.5	600	11.14	4.13	4.17	19.44
1990	7	5.0	355	−43.86	20.79	−14.32	−37.39
1990	7	5.0	440	−18.95	10.37	−6.48	−15.06
1990	7	5.0	530	−4.34	3.38	−0.77	−1.73
1990	7	5.0	600	3.19	0.50	1.66	5.35
1990	15	None	355	8.46	−1.32	2.54	9.69
1990	15	None	440	12.73	1.50	3.68	17.90
1990	15	None	530	14.72	6.08	5.12	25.92
1990	15	None	600	16.26	8.31	5.57	30.14

Table A2.1. (*cont.*)

Year	Precipi-tation change	Temperature change °C	CO_2 ppmv	Change in consumer surplus	Change in producer surplus	Change in foreign surplus	Change in total surplus
1990	15	1.5	355	5.25	−0.75	1.50	6.00
1990	15	1.5	440	10.56	0.53	3.17	14.25
1990	15	1.5	530	12.66	6.21	4.44	23.31
1990	15	1.5	600	14.70	8.08	5.22	28.01
1990	15	2.5	355	−0.13	1.65	−0.18	1.35
1990	15	2.5	440	7.88	−0.62	2.54	9.79
1990	15	2.5	530	10.75	2.59	3.99	17.33
1990	15	2.5	600	12.53	5.03	4.67	22.22
1990	15	5.0	355	−29.79	14.89	−9.91	−24.81
1990	15	5.0	440	−12.14	7.35	−4.01	−8.81
1990	15	5.0	530	0.68	1.03	0.76	2.46
1990	15	5.0	600	6.69	−1.71	2.73	7.71

Notes:

[a] central case

[b] severe case

Table A2.2. *Change in welfare from 2060 base, in 1990 $ (billions)*

Year	Precipitation change	Temperature change °C	CO_2 ppmv	Consumer surplus	Producer surplus	Foreign surplus	Total surplus
2060	−10	None	355	−23.65	12.34	−5.20	−16.52
2060	−10	None	440	5.96	−2.13	7.08	10.91
2060	−10	None	530	34.96	−9.18	12.21	37.99
2060	−10	None	600	44.65	−9.51	14.80	49.94
2060	−10	1.5	355	−53.74	36.43	−11.04	−28.35
2060	−10	1.5	440	−2.99	6.51	1.35	4.87
2060	−10	1.5	530	28.17	−4.83	10.47	33.81
2060	−10	1.5	600	31.29	−6.37	17.91	42.84
2060	−10	2.5	355	−97.52	68.11	−18.71	−48.11
2060	−10	2.5	440	−27.88	26.82	−7.69	−8.75
2060	−10	2.5	530	12.47	3.96	5.86	22.28
2060	−10	2.5	600	31.90	−3.07	8.72	37.55
2060	−10	5.0	440	−248.18	137.67	−43.62	−154.13
2060	−10	5.0	530	−77.61	56.80	−13.77	−34.58
2060	−10	5.0	600	−32.31	33.00	−4.39	−3.70
2060	None	None	355	0.00	0.00	0.00	0.00

Table A2.2. (*cont.*)

Year	Precipitation change	Temperature change °C	CO_2 ppmv	Consumer surplus	Producer surplus	Foreign surplus	Total surplus
2060	None	None	440	30.58	−7.09	6.28	29.77
2060	None	None	530	47.25	−9.07	11.38	49.56
2060	None	None	600	55.37	−11.60	16.03	59.80
2060	None	1.5	355	−3.67	6.85	−7.21	−4.02
2060	None	1.5	440	22.70	−3.03	3.18	22.85
2060	None	1.5	530	45.39	−8.18	10.00	47.21
2060	None	1.5	600	53.19	−8.39	12.77	57.57
2060	None	2.5	355	−23.86	20.73	−14.99	−18.12
2060	None	2.5	440	12.91	3.21	−4.38	11.75
2060	None	2.5	530	37.34	−4.74	5.32	37.91
2060	None	2.5	600	47.92	−7.01	8.87	49.78
2060	None	5.0	355	−204.41	117.86	−38.71	−125.26
2060	None	5.0	440	−81.69	61.94	−27.67	−47.42
2060[b]	None	5.0	530	−11.40	20.89	−9.33	0.15
2060	None	5.0	600	13.62	8.34	−3.16	18.79
2060	7	None	355	24.22	−5.57	−0.16	18.49
2060	7	None	440	41.08	−9.08	6.35	38.35
2060	7	None	530	56.24	−11.16	12.05	57.12
2060	7	None	600	63.19	−12.78	15.82	66.22
2060	7	1.5	355	17.77	−0.20	−6.00	11.57
2060	7	1.5	440	38.07	−5.32	2.17	34.93
2060	7	1.5	530	53.51	−8.44	10.03	55.10
2060	7	1.5	600	60.91	−10.45	13.91	64.37
2060	7	2.5	355	0.45	11.86	−13.82	−1.51
2060	7	2.5	440	30.30	−1.16	−3.93	25.22
2060[a]	7	2.5	530	47.66	−6.24	6.02	47.44
2060	7	2.5	600	56.25	−7.06	8.40	57.59
2060	7	5.0	355	−92.08	57.05	−37.38	−72.41
2060	7	5.0	440	−23.40	28.72	−24.94	−19.62
2060	7	5.0	530	12.17	10.14	−9.41	12.91
2060	7	5.0	600	31.10	1.88	−3.71	29.27
2060	15	None	355	31.17	−6.15	2.49	27.51
2060	15	None	440	50.97	−10.24	6.28	47.01
2060	15	None	530	63.82	−11.58	11.25	63.49
2060	15	None	600	70.53	−14.32	15.43	71.64
2060	15	1.5	355	26.19	−1.84	−1.68	22.67
2060	15	1.5	440	49.37	−7.26	2.52	44.63
2060	15	1.5	530	62.89	−9.33	8.98	62.53
2060	15	1.5	600	69.31	−11.49	12.74	70.56

53

Table A2.2. (*cont.*)

Year	Precipitation change	Temperature change °C	CO_2 ppmv	Consumer surplus	Producer surplus	Foreign surplus	Total surplus
2060	15	2.5	355	19.22	1.88	−7.67	13.43
2060	15	2.5	440	42.16	−4.01	−2.02	36.14
2060	15	2.5	530	55.59	−6.53	5.73	54.79
2060	15	2.5	600	63.05	−7.99	9.14	64.21
2060	15	5.0	355	−54.28	44.13	−31.72	−41.87
2060	15	5.0	440	−3.11	18.84	−19.91	−4.18
2060	15	5.0	530	29.35	2.32	−7.48	24.19
2060	15	5.0	600	39.64	0.05	−1.91	37.79

Notes:
[a] central case
[b] severe case

Table A2.3. *Net welfare as a function of precipitation, temperature, and CO_2 concentration: linear-dummy variable specification, 2060 ASM*

	Parameter estimates/summary statistics			
Variable	Parameter estimate	Standard error	"*t*" statistics	Prob > |t|
INTERCEP	1750.594	10.118	173.022	0.0001
P1 (−10%)	3.003	0.905	3.318	0.0016
P3 (7%)	1.854	1.293	1.434	0.1572
P4 (15%)	1.548	0.603	2.565	0.0131
T2 (+1.5)	−2.720	6.033	−0.451	0.6539
T3 (+2.5)	−5.521	3.620	−1.525	0.1330
T4 (+5.0)	−14.541	1.810	−8.034	0.0001
C2 (440)	0.086	0.021	4.175	0.0001
C3 (530)	0.125	0.017	7.295	0.0001
C4 (600)	0.130	0.015	8.623	0.0001

Notes:
$R^2 = 0.7964$; Adjusted $R^2 = 0.7625$

3   The impact of climate variation on US agriculture

ROBERT MENDELSOHN, WILLIAM NORDHAUS,
AND DAIGEE SHAW

This chapter explores the effect of climate on the value of US agricultural land using a Ricardian model. The research extends previous analyses by including both inter-seasonal and diurnal climate variation in addition to average temperature and precipitation variables. With these climate variation variables included, small increases in average temperature are predicted to be beneficial. Increases in interannual climate variation are predicted to be generally harmful to US agriculture but decreases in diurnal variation will be beneficial.

For centuries analysts have been interested in the impact of weather on crops in order to predict what crops to grow, when to plant and harvest, and what agricultural prices will be each year. With the growing likelihood that accumulating greenhouse gases will change the climate (IPCC, 1996), there has been growing interest in also measuring the impact of climate change on agriculture. Two distinct ways to measure the impacts of climate on agriculture have emerged in the literature: an agronomic approach and a Ricardian rent approach. The agronomic approach (Chapter 2; Adams *et al.*, 1989, 1990, 1995; Crosson and Katz, 1991; Rosenzweig and Parry, 1994) predicts changes in yield from crop simulation models such as CERES and SOYGRO and then enters these changes in mathematical models of agriculture production and consumption. The Ricardian approach (Johnson and Haigh, 1970; Mendelsohn *et al.*, 1994, 1996) uses an empirical cross-sectional approach and estimates the relationship between land prices and climatic, economic, and soil variables.

The agronomic approach, with its extensive reliance on specific crop models, has the advantage of being based directly on carefully controlled scientific experiments so that it can predict phenomena (such as carbon fertilization) that have not yet occurred in nature. The method is also capable of detailed displays of the links between climate, crop yields, and market equilibrium. The approach is popular among scientific analysts of climate impacts because it captures the tremendous detail of individual crop models. The approach, unfortunately, is somewhat mechanistic. The myriad adaptations that farmers might make to climate are difficult to model explicitly and so are often omitted, overestimating the damages from climate warming.

55

The Ricardian approach, by relying upon how farmers and ecosystems have actually adjusted to varying local conditions, incorporates adaptation readily. However, the Ricardian approach does not provide much information about the process of climate change or about conditions which are not evident in today's environment, such as carbon fertilization. The Ricardian approach has only recently been applied to climate change and so there is less experience of using this approach compared with the production function technique. Further, because it does not contain the minute detail captured in the crop response models, crop scientists have been slow to understand its merits. Each method has its own strengths and weaknesses and the two approaches complement each other.

This chapter begins by addressing several theoretical issues with the Ricardian model and specifically explores the bias introduced by assuming that prices remain constant (Section 3.1). The thrust of the chapter, however, lies in the extension of the empirical results to include the influence of climate variation (described in Section 3.2). Specifically, the study explores the impact of including interannual and diurnal variation in precipitation and temperature on US agricultural land values. The empirical study is described in detail in Section 3.3. These models are then used to assess the economic damages to US agriculture from several climate change scenarios in Section 3.4. The chapter concludes with some general observations.

3.1   Theory

This section summarizes the theoretical underpinnings of the Ricardian approach to climate modeling and explores a few extensions of this theory. We postulate a set of consumers with well-behaved utility functions (preferences for goods) and linear budget constraints. Assuming that consumers maximize their utility functions across available purchases and aggregating leads to a system of inverse demand functions for all goods and services:

$$
\begin{aligned}
P_1 &= D^{-1}(Q_1, Q_2, \ldots, Q_n, Y) \\
&\ \ \vdots \qquad \vdots \\
P_n &= D_n^{-1}(Q_1, Q_2, \ldots, Q_n, Y),
\end{aligned}
\tag{3.1}
$$

where P_i and Q_i are respectively the price and quantity of goods i, $i = 1, \ldots, n$, and Y is the aggregate income. Inverse demand functions describe the prices at which consumers are willing to purchase specific bundles of goods. The Slutsky equation is assumed to apply, so that (3.1) is integrable.

We also assume that a set of well-behaved production functions exist which link purchased inputs and environmental inputs into the production of outputs by a firm on a certain site:

$$Q_i = Q_i(K_i, E), \qquad i = 1, \dots, n. \tag{3.2}$$

In this equation, we use bold face to denote vectors or matrices. Q_i is the output of goods i, $K_i = [K_{i1}, \dots, K_{ij}, \dots, K_{iJ}]$ where K_{ij} is the purchased input j ($j = 1, \dots, J$) in the production of good i, and $E = [E_1, \dots, E_l, \dots, E_L]$ where E_l is an exogenous environmental input l ($l = 1, \dots, L$) into the production of goods, e.g. climate, soil quality, air quality, and water quality, which would be the same for different goods' production on a certain production site. Given a set of factor prices, R_j, for K_j, the exogenously determined level of environmental inputs, and the production function, cost minimization leads to a cost function:

$$C_i = C_i(Q_i, R, E). \tag{3.3}$$

Here, C_i is the cost of production of goods i, $R = [R_1, \dots, R_J]$, and $C_i(*)$ is the cost function. In this analysis, it is helpful to separate land from the vector of inputs, K. We assume that land, L_i, is heterogeneous with characteristics E and has an annual cost or rent of P_E. Companies are assumed to maximize profits given market prices:

$$\text{Max } P_i Q_i - C_i(Q_i, R, E) - P_E L_i \tag{3.4}$$

where P_i is the price of goods i. This maximization leads firms to equate prices and marginal costs as well as determine cost minimizing levels of production. We assume that there is perfect competition for land, which implies that entry and exit will drive pure profits to zero:

$$P_i Q_i - C_i(Q_i, R, E) - P_E L_i = 0. \tag{3.5}$$

If use i is the best use for the land given the environment E and factor prices R, the observed market rent on the land will be equal to the annual net profits from production of goods i_1. Solving for the value of land rent per acre yields:

$$P_E = [P_i Q_i - C_i(Q_i, R, E)] / L_i. \tag{3.6}$$

The land rent should be equal to the net revenue from the land. Land value, V_E, is equal to the present value of the stream of future net revenue, which can be described by:

$$V_E = \int_0^\infty P_E e^{-rt} dt = \int_0^\infty [P_i Q_i - C_i(Q_i, R, E)] e^{-rt} / L_i dt. \tag{3.7}$$

The discount rate is represented by r and time by t. By examining the relationship between land value and the environmental variable of interest, one can measure its

57

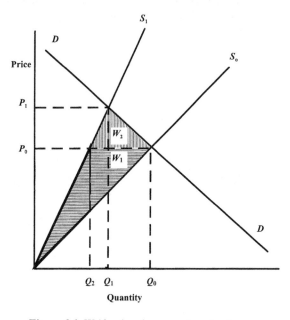

Figure 3.1 Welfare loss from supply reduction.

impact on the present value of net revenue. The essence of the Ricardian model is (3.7).

If an environmental factor reduces the stream of future land rents, land values will be reduced as well (note the similarity of this analysis and hedonic property studies, see Freeman (1979)). Reliance upon land values rather than land rents, however, introduces a potential source of additional problems. Land values will represent the present value of the rents using the parcel at its highest purpose. Although land may now be in agricultural use, it could be that its best future use may be industrial or urban. In order to control for nonagricultural influences, proxies for the development value of farmland must be included in the analysis.

Let us now examine the welfare value of an environmental change from an initial point E_A to a new point E_B. The change in annual welfare, W, from this environmental change is the change in net consumer surplus:

$$W(E_A - E_B) = \int_0^{Q_B} \sum D^{-1}(Q_i) dQ_i - \sum C_i(Q_i, R, E_B) -$$

$$[\int_0^{Q_A} \sum D^{-1}(Q_i) dQ_i - \sum C_i(Q_i, R, E_A)] \tag{3.8}$$

where $\int\sum$ is the line integral evaluated between the initial vector of quantities and the zero vector, $\mathbf{Q}_A = [Q_1(K_1, E_A), \ldots, Q_i(K_i, E_A), \ldots, Q_n(K_n, E_A)]$, $\mathbf{Q}_B = [Q_1(K_1, E_B), \ldots, Q_i$

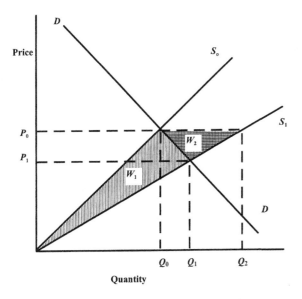

Figure 3.2 Welfare gain from supply expansion.

$(K_i, E_B),\dots,Q_n(K_n,E_B)]$, $C_i(Q_i,R,E_A)=C_i(Q_i(K_i,E_A),R,E_A)$, and $C_i(Q_i,R,E_B)=C_i(Q_i$ $(K_i, E_B),R,E_B)$. The above equation includes changes in both consumer and producer surplus. It is necessary to take this line integral as long as the environmental change affects more than one output. If only one output is affected, then (3.8) simplifies to the integral of the equation for a single item of goods. Note that as long as the Slutsky equation is satisfied, the solution to (3.8) is path-independent and unique.

If we assume that the changes in the environment will leave market prices unchanged,[1] then (3.8) can be expressed:

$$W(E_A - E_B) = PQ_B - \sum C_i(Q_i,R,E_B) - [PQ_A - \sum C_i(Q_i,R,E_A)] \qquad (3.9)$$

where $P=[P_1,\dots,P_i,\dots,P_n]$. In this case, consumer surplus is not affected. Substituting (3.6) into (3.9) yields:

$$W(E_A - E_B) = \sum (P_{EB} \times L_{EB} - P_{EA} \times L_{EA}) \qquad (3.10)$$

where P_{EA} is the value per acre of land area L_{EA} in environmental state A and P_{EB} is the value per acre of land area L_{EB} in environmental state B. The environmental state affects both the value per acre and the total number of acres in farmland. It follows that the present value of this welfare change is:

---

[1] If there is a nonmarginal change in market prices, one must also add changes in consumer surplus to find the total damages. The difficulty of including price changes should not be underestimated as it requires estimation of the international supply and demand for food.

Table 3.1. *Bias from holding prices constant*[a]

Demand elasticity	Supply elasticity		
	0.5	1.0	2.0
0.5	1.17	1.07	1.03
1.0	1.11	1.05	1.03
2.0	1.07	1.04	1.02

Notes:
[a] The table presents the ratio of the true welfare measure of damages from a 10% reduction in aggregate supply to the Ricardian welfare measure. The Ricardian method overestimates the benefits of a 10% increase in aggregate supply by a similar amount.

$$\int_0^\infty \sum W(E_A - E_B)e^{-rt}dt = \sum_i (V_{EA} - V_{EB}). \qquad (3.11)$$

Equation (3.11) is the definition of the *Ricardian estimate of the value of environmental changes*. Under the assumptions used here, *the value of the change in the environment is captured exactly by the change in aggregate land values.*

The strongest assumption above is that output prices remain constant. Suppose that this assumption is relaxed. Climate change is expected to lead to increases in the supply of some crops and decreases in the supply of others. For example, crops which prefer cooler environments, such as apples and winter wheat, may not do as well with climate warming. In contrast, heat-loving plants, such as tomatoes and citrus fruit, should be able to grow in wider settings. As supply expands (contracts) for the warm- (cool-) loving plants, prices will fall (rise).

In a warming scenario, the crops which benefit will fall in price and crops which grow less well will rise in price. For example, the supply function for cool-loving crop A could shift from S_0 to S_1 in Figure 3.1. We measure the loss in net revenue holding prices constant as W_1. In fact, there is an additional consumer surplus loss of W_2. The model understates the damages from the change in supply. Similarly, if supply expands from S_0 to S_1 as in Figure 3.2, holding prices constant, we estimate a benefit of W_1. This overstates the benefits because prices fall to P_1 from P_0. The size of this over-estimate is equal to W_2.

Given that the welfare estimates of the Ricardian model are biased, it is important to estimate the size of this bias. Suppose that demand and supply price elasticities take on values within a plausible range for agriculture. What will be the size of

$W_1 + W_2$, the true measure of welfare, relative to W_1, the Ricardian measure? Mendelsohn and Nordhaus (1996) examine these ratios for a simple model with linear supply and demand functions. The results are given in Table 3.1. Assuming that global warming causes a 10 percent change in the aggregate supply of goods, the table estimates the error associated with the Ricardian measure of welfare. With typical unitary price-elasticities, the error is about 5 percent of the Ricardian measure. With price-inelastic demand and supply functions of 0.5, the error can be as large as 17 percent and with price-elastic demand and supply functions of 2.0, the error falls to 2 percent. With smaller changes in aggregate supply, the effect shrinks. Given that most models of aggregate supply predict very small changes in aggregate quantities of food as a result of warming,[2] the Ricardian measures of welfare should be accurate.

3.2 Data

In this section, we extend the Ricardian technique developed by Mendelsohn *et al.* (1994) to examine the impact of climate variation on US agriculture. We rely on data from the 1982 US Census of Agriculture to obtain much of the data on farm characteristics in each county. Although the analysis conducted in this study relies upon 1982 data, a similar analysis was conducted on 1978 data with similar results. The results appear to be robust over time. Nonetheless, it would be helpful if future analysts update the Ricardian estimate using more recent census data.

For the most part, the data reflects actual county averages, so that there are no major geographic issues involved in obtaining the census information on these variables. The *County and City Data Book*, and the computer tapes of that data, are the source for much of the agricultural data used here, including farmland and building values, and information on market inputs for farms in every county in the United States. In addition, we include social, demographic, and economic data on each of the counties drawn from the *County and City Data Book*.

Data about soils were extracted from the National Resource Inventory and other USDA surveys with the kind assistance of Daniel Hellerstein and Noel Gollehon of the US Department of Agriculture. For each county, we have average measures of salinity, clay content, sand content, soil permeability, available water capacity, flood probability, soil erosion, slope length, whether or not the land is a wetland, and numerous other variables that are not used in this analysis.

[2] See, for example, predictions in Chapter 2 which are only for the United States. Predictions for the world are even less severe because of trade between countries (Kane *et al.*, 1992).

Climatic data is available by weather stations rather than by county. The climate data was obtained from the National Climatic Data Center, which gathers data from 5511 meteorological stations throughout the United States. The data include information on precipitation and temperature for each month from 1951 to 1980. This analysis includes data on normal daily mean temperatures and normal monthly precipitation for January, April, July, and October, representing each season of the year. Interannual variation in precipitation and temperature in each of the four months is measured as the difference between the highest and lowest normal monthly precipitation and temperatures over the 30-year period. The variation variables measure the range of interannual variation.[3] We also measure the diurnal range (the difference between the average of the highest and lowest daily temperatures) for each of the four months. Altogether there are 12 variation measures in the study.

In order to link the agricultural data which is organized by county and the climate data which is organized by station, we conduct a spatial statistical analysis which examines the determinants of the climate of each county (see Mendelsohn *et al.*, 1994 for more details). The interpolation relies on a regression weighted by distance.

The next and crucial stage is to use the climate data to predict aggregate land values. Following Mendelsohn *et al.* (1996), we define the dependent variable as the aggregate value of farmland in each county rather than the farmland value per acre. This aggregate measure takes into account how the climate affects which land can be used for agriculture as well as how climate affects the value of the farmland that remains. In order to determine the marginal impact of each climate variable, we regress aggregate farm values on climate, soil, and economic variables. The soil and economic variables control for unwanted variation so that the climate variables are less likely to reflect correlated omitted variables. For example, the economic variables control for the effect of nearby local markets and speculative future land uses.

Alternative control variables in the theoretical model such as interest rates and farm input prices are not included in the empirical model because they are assumed to be the same for all counties. In a cross-sectional analysis, the capital market will equate interest rate expectations across parcels, so that this effect will be the same for all observations. Competitive market forces should also equate farm input prices for energy, labor, and equipment.[4]

In previous analyses, it was demonstrated that both precipitation and temperature

---

[3] An alternative formulation would have been to use the variance in monthly normals over the 30-year period. Our decision to rely upon the range is partially motivated by the availability of this measure and partially by a general concern about extreme events.

[4] To the extent that farm input and output prices vary because of proximity to an urban area, the urban variables used in the analysis would control for this effect.

have quadratic relationships with farm value (see Mendelsohn *et al.*, 1994, 1996). This same specification is used in this analysis:

$$V = a_o + \sum_1^I a_i E_i + \sum_1^I b_i E_i^2 + \sum_1^{12} c_i Q_i + \sum_1^n d_i Z_i + e, \tag{3.12}$$

where E_i represent the precipitation and temperature normals, Q_i represent the climate variation terms, Z_i represent the control variables and e is the error term. The climate variables have been de-meaned. The coefficients a_i can therefore be interpreted as the marginal effect of E_i on land values evaluated at the sample mean for the United States. The coefficients b_i measure the impact of the quadratic terms, the coefficients c_i measure the impact of the climate variation terms, and the coefficients d_i capture the impact of the control variables.

3.3   Empirical results

Following Mendelsohn *et al.* (1994) the empirical models are weighted regressions using either percent cropland or total crop revenue in the county. Weighting counties by total crop revenue makes sense if the focus of the study is on aggregate agricultural production since the counties with the highest valued production are more important. Weighting by percent cropland is justified if the focus is understanding what is happening to cropland. See Appendix A3 for a complete list of the variables used in the models and their definitions.

Columns 3 and 4 in Table 3.2 present the climate model without variation terms included. Some results are consistent across both weighting schemes. Higher average temperatures in January and July are harmful to farm values whereas higher temperatures in April and especially October increase values. Increased precipitation in July and especially October reduces farm values but more precipitation in January and April increases farm values. The coefficients of all the control variables exhibit consistent effects across the models (although magnitudes vary) with the exception of soil permeability.

The year-to-year and diurnal climate variation variables are introduced in the models in columns 1 and 2 of Table 3.2. F-tests of the variation terms as a group indicate that the variation coefficients are significantly different from zero in all regressions. All the individual coefficients of interannual climatic variation are significantly different from zero. The coefficients are all negative implying that increasing interannual variation reduces farm values with the exception of April temperatures and January precipitation. An increase in interannual variation in January precipitation and April

Table 3.2. *Regression models with and without climate variation*[a]

Variation	Climate variation		No climate variation	
Independent variables	Percent cropland	Crop revenue	Percent cropland	Crop revenue
January temp.	−120.0	−145.0	−86.7	−108.0
	(15.53)	(19.92)	(10.50)	(13.94)
January temp. sq.	−2.02	−2.71	−0.83	−0.93
	(11.11)	(16.87)	(4.04)	(5.25)
April temp.	21.9	49.6	59.6	65.0
	(2.20)	(6.20)	(4.83)	(6.06)
April temp sq.	−3.69	−3.76	−3.27	−1.36
	(6.46)	(8.97)	(5.65)	(3.09)
July temp.	−189.0	−182.	−117.0	−141.0
	(20.46)	(23.85)	(11.45)	(18.83)
July temp sq.	−5.63	−5.78	−1.90	−3.07
	(10.32)	(17.40)	(3.38)	(8.08)
October temp.	235.0	266.0	152.0	233.0
	(15.66)	(18.50)	(8.68)	(14.40)
October temp. sq.	7.81	10.1	3.00	2.87
	(9.65)	(17.24)	(3.50)	(4.49)
January rain	43.3	88.0	−131.0	−123.0
	(2.28)	(5.11)	(5.34)	(5.14)
January rain sq.	1.78	0.23	12.5	13.3
	(0.71)	(0.13)	(4.96)	(7.39)
April rain	110.0	36.3	117.0	99.0
	(5.40)	(1.74)	(4.24)	(3.31)
April rain sq.	−28.7	−14.5	−25.8	−40.6
	(3.92)	(2.19)	(3.59)	(5.87)
July rain	−53.4	−26.5	60.0	75.2
	(4.69)	(2.35)	(3.24)	(3.86)
July rain sq.	34.3	17.8	18.2	−6.7
	(7.18)	(4.33)	(3.50)	(1.46)
October rain	−188.0	−129.	−74.5	−7.9
	(9.70)	(6.66)	(2.40)	(0.26)
October rain sq.	−21.1	6.8	−24.5	−16.4
	(1.79)	(0.96)	(2.02)	(2.17)

Table 3.2. (*cont.*)

Variation	Climate variation		No climate variation	
Independent variables	Percent cropland	Crop revenue	Percent cropland	Crop revenue
Year-to-year variation				
January temp. Y-var.	−19.3	−19.1		
	(4.84)	(4.80)		
April temp. Y-var.	19.9	11.2		
	(3.01)	(1.76)		
July temp. Y-var.	−57.1	−63.9		
	(8.53)	(8.56)		
October temp. Y-var.	−27.6	−29.3		
	(4.61)	(4.46)		
January rain Y-var.	21.9	24.9		
	(2.89)	(4.05)		
April rain Y-var.	−19.5	−26.8		
	(2.85)	(3.42)		
July rain Y-var.	−29.1	−25.4		
	(6.39)	(4.74)		
October rain Y-var.	−13.2	−31.7		
	(2.09)	(5.15)		
Daily variation				
January daily var.	−61.7	−100.0		
	(7.19)	(13.13)		
April daily var.	−66.7	−4.4		
	(5.83)	(0.42)		
July daily var.	−10.6	−11.8		
	(1.00)	(1.43)		
October daily var.	59.8	73.8		
	(5.81)	(7.03)		

65

Table 3.2. (*cont.*)

Variation	Climate variation		No climate variation	
Independent variables	Percent cropland	Crop revenue	Percent cropland	Crop revenue
Control variables				
Constant	957.0	1060.0	870.0	945.0
	(49.96)	(57.31)	(37.44)	(47.08)
Income per	57.4	27.0	52.7	32.9
capita	(14.20)	(6.00)	(13.70)	(7.74)
Density	86.3	64.2	15.2	8.6
	(1.41)	(1.29)	(0.26)	(0.18)
Density sq.	−101.0	−68.6	−76.8	−44.7
	(3.60)	(4.24)	(2.92)	(3.00)
Solar	−84.4	−38.6	−41.7	2.9
radiation	(6.57)	(3.37)	(2.74)	(0.22)
Altitude	−121.0	92.9	54.9	90.4
	(5.10)	(4.53)	(2.19)	(4.20)
Salinity	−843.0	−467.0	−725.0	−715.0
	(4.74)	(3.51)	(4.30)	(5.73)
Flood prone	−136.0	−102.0	−185.0	−125.0
	(3.27)	(2.25)	(4.57)	(2.83)
Wetland	−509.0	−784.0	−656.0	−832.0
	(4.79)	(7.67)	(6.43)	(8.68)
Soil erosion	−799.0	−1480.0	−1050.0	−1420.0
	(4.54)	(7.64)	(6.14)	(7.65)
Slope length	24.6	73.1	29.0	64.1
	(4.76)	(14.29)	(5.94)	(13.42)
Sand	−86.1	−81.2	−21.2	−74.5
	(1.94)	(1.99)	(0.51)	(1.97)
Clay	91.6	21.5	86.6	49.4
	(5.00)	(1.03)	(4.97)	(2.52)
Water capacity	0.51	0.34	0.41	0.30
	(14.96)	(10.75)	(12.72)	(9.98)
Permeability	-0.70×10^{-3}	-598×10^{-3}	-0.37×10^{-3}	-8.42×10^{-3}
	(0.35)	(4.43)	(0.20)	(6.62)
Adjusted R^2	0.793	0.843	0.800	0.869
Number of observations	2938	2938	2938	2938

Notes:
[a] Dependent variable is aggregate farm value. All observations are weighted. Values in parenthesis are t-statistics.

temperatures may be beneficial because at least spring-planting farmers can adjust for the realized values before planting, thus permitting good years to outweigh bad years.

Increases in diurnal variation in January and April are harmful to farming. Increases in diurnal variation during the summer seem to have no effect on farm values. However, increases in diurnal variation in the autumn appear beneficial, possibly serving as a useful signal to plants to begin maturing and ripening fruit.

In order to understand the spatial implications of the climate model in Table 3.2, the climate coefficients from the regression using crop revenues as the weight (the second column) are used to predict the impact of current climate on the distribution of farm values in the United States. For each county, the deviation between that county's climate and the US mean climate is calculated. This deviation is then multiplied by the climate coefficient in column 3 of Table 3.2 and the effect is summed across the climate variables. The predicted effect of the range of climates observed in the United States on farm values is shown in Figure 3.3. All the climatic variables taken as a group predict that four areas of the country have climates which yield above average agricultural land values: the Gulf coast, the southern New England coast, the Pacific coast, and the Mississippi river valley. Climates which lead to below average land values include northern Maine, the western plains, and the Rocky Mountains.

This same process can isolate the spatial contribution of only the climatic variation. The parts of the country with the most stable climates include the Pacific coast, the southern Mississippi delta, and coastal New England. The part of the country most sensitive to climate variation lies near the dust bowl in Kansas, Missouri, and Oklahoma. Note that these were the states most devastated by the dust bowl in the 1930s. The range of values produced by climate variation across the United States is surprisingly large. The current spatial distribution of interannual and diurnal variation is quite important to crops.

Introducing climatic variation into the Ricardian model has important effects on the seasonal pattern of *mean* temperature and precipitation. Adding the climatic variation variables decreases the harmful effect of a warmer January or July, increases the benefits of a warmer April, and reduces the benefit of a warmer October. Adding the variation terms also alters the seasonal importance of increases in precipitation. January precipitation becomes harmful, July precipitation becomes beneficial, and October precipitation becomes less harmful. Overall, warmer temperatures and increased precipitation become more beneficial with the variation terms included in the model.

Adding the climate variation terms also affects two other control variables in the model. The effect of solar radiation is reduced with the variation terms in place. Altitude goes from being harmful to being beneficial. It is possible that the damaging influence of higher altitude is due to the increase in diurnal variation.

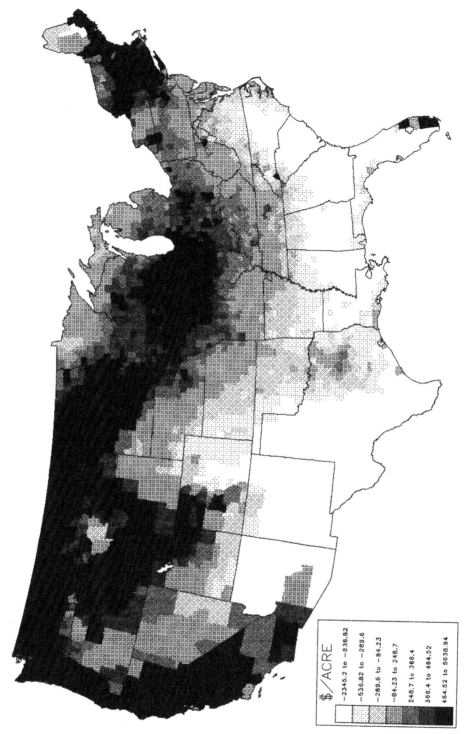

Figure 3.3 Farm values from current climate with crop revenue per county.

$/ACRE
-2345.2 to -536.82
-536.82 to -289.6
-289.6 to -84.23
-84.23 to 248.7
248.7 to 366.4
366.4 to 464.52
464.52 to 5838.94

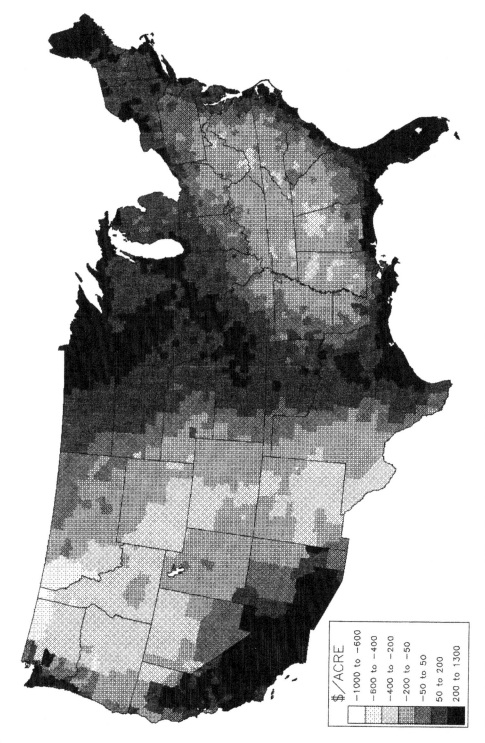

Figure 3.4 Change in value due to 5 °C uniform increase weighted by crop revenue/county.

Table 3.3. *Net agricultural effect of climate change without climate variation*[a]

Model: percent cropland – no climate variation

Precipitation change (%)	Temperature change (°C)		
	1.5	2.5	5.0
0	−11.9	−20.8	−39.1
7	−12.6	−21.3	−39.5
15	−13.4	−21.9	−39.8

Model: crop revenue – no climate variation

Precipitation change (%)	Temperature change (°C)		
	1.5	2.5	5.0
0	−2.6	−9.2	−15.7
7	−2.7	−5.7	−15.7
15	−2.7	−5.7	−15.6

Note:
[a] Change in annual net value to US agriculture in billions of dollars.

3.4   Climate simulations

In order to test what implications these models have for greenhouse warming, we simulate nine scenarios for each model. Following the protocol described in Chapter 1, we examine uniform temperature increases of 1.5, 2.5, and 5.0 °C for the entire United States under three precipitation scenarios of 0 percent, 7 percent, and 15 percent increases. Four impact models are explored: cropland weighted models and crop revenue weighted models with and without climate variation terms. The models without climate variations in columns 3 and 4 of Table 3.2 produce the results shown in Table 3.3. With both the cropland and crop revenue models, warming is increasingly harmful as one moves from 1.5 to 5 °C increases. Increased precipitation is also mildly harmful according to the cropland model and inconsequential according to the crop revenue model. Adding the climate variation terms (columns 1 and 2) changes these results dramatically, producing the results in Table 3.4. Gentle warming is strictly beneficial. As warming approaches 5 °C, however, the cropland model predicts that warming becomes harmful whereas the crop revenue model predicts even larger benefits. The regional impacts are shown in Figure 3.4.

Table 3.4. *Net agricultural effect of climate change with climate variation*[a]

Model: percent cropland – with climate variation			
	Temperature change (°C)		
Precipitation change (%)	1.5	2.5	5.0
0	2.7	2.9	−3.7
7	3.3	3.1	−3.1
15	3.9	3.7	−2.5
Model: crop revenue – with climate variation			
	Temperature change (°C)		
Precipitation change (%)	1.5	2.5	5.0
0	11.5	16.7	26.2
7	12.9	18.8	27.5
15	13.8	19.7	28.4

Note:
[a] Change in net annual income to US agriculture in billions of dollars.

3.5   Conclusion

This analysis examines the impact of climate variation on US farm values using the Ricardian approach developed by Mendelsohn *et al.* (1994). Three important results are developed. First, the assumption of constant output prices in the Ricardian model is shown to underestimate the damages and overestimate the benefits of climate change. However, these biases are very small, indicating the technique yields accurate estimates of welfare loss.

Second, climate variation (both diurnal and interannual) has important effects on farm values. In general, greater interannual variation is harmful to farm values. Variation in the beginning of the year, however, is less harmful than variation at the end of the year because farmers can more readily adjust to weather which occurs in winter and spring. Increases in diurnal variation are also important, generally reducing farm values in winter, spring, and summer. However, diurnal variation in the autumn appears to be beneficial, possibly because it serves as a useful signal to plants to begin ripening before dangerous frosts arrive.

Third, including climate variation in an empirical model is important because it is

71

correlated with mean temperatures. Increases in mean temperatures can be harmful if climate variation terms are omitted from a model. However, when climate variation terms are included, increases in mean temperatures are strictly beneficial.

The marginal effect of temperature variation is large. If the interannual variation of temperature increases by 25 percent in every month, average farm values would fall by about one-third.[5] Similarly, if the diurnal range of temperature decreased in every month by 25 percent, farm values would double. In contrast, if the interannual variation of precipitation in every month increased by 25 percent, farm values would fall just 6 percent. What farmers should fear, apparently, is years with unusual temperatures, not years with unusual precipitation levels.

After estimating the effect of diurnal and interannual variations in temperature and precipitation on agricultural land values, we tested the implications of these models. Impacts from a total of nine climate scenarios were estimated using the four different impact models. The models that include climate variation variables in the estimate yield quite different results to those from the models which omit these variables. Including climate variation suggests that small amounts of warming are beneficial. Only when the temperature increase is above 2.5 °C does the cropland with the climate variation model suggest that increased warming is harmful.

An alternative perspective on the four models can be obtained by examining the overall response function of the four models. Each predicts a quadratic relationship with an optimal average temperature (given US seasonal variation). The cropland and crop revenue models without variation terms predict the optimal average temperature for agriculture is 4 and 1 °C, respectively, less than the US average. The cropland and revenue models with variation terms included predict that the optimal agricultural temperature is 1 and 6 °C, respectively, warmer than the US average. Thus, the results are generally in agreement among all four models and they suggest as a group that modest warming will have either a mildly harmful or mildly beneficial effect. The model predictions, however, diverge with more severe climate scenarios.

There are a number of improvements which could strengthen our understanding of climatic impacts on agriculture. The direct effect of carbon dioxide must also be included for an accurate assessment. According to the model presented in Chapter 2, including carbon fertilization effects could add another $50 billion of benefits, making global warming clearly beneficial. The analysis also needs to be extended to other countries, especially in subtropical and tropical settings. Finally, this chapter demonstrates that changes in climatic variation are important to agriculture. More precise

---

[5] Summing across months, the product of the coefficient in Table 3.2 multiplied by a change in that variable yields an estimate of the net effect of that change.

climate work quantifying changes in diurnal and interannual variation will be important to final damage estimates.

References

Adams, R., Glyer, D. and McCarl, B. 1989. The Economic Effects of Climate Change in US Agriculture: A Preliminary Assessment. In: *The Potential Effects of Global Climate Change on the United States: Report to Congress*, Smith, J. and Tirpak, D. (eds.). EPA-230-05-89-050. Washington, DC: US Environmental Protection Agency.

Adams, R., Rosenzweig, C., Pearl, R., Ritchie, J., McCarl, B., Glyer, J., Curry, R., Jones, J., Boote, K. and Allen, L. 1990. Global Climate Change and US Agriculture. *Nature* 345: 219–24.

Adams, R., Fleming, R., Chang, C., McCarl, B. and Rosenzweig, C. 1995. A Reassessment of the Economic Effects of Global Climate Change in US Agriculture. *Climatic Change* 30: 146–67.

Crosson, P. and Katz, L. 1991 *Report IIA: Agricultural Production and Resource Use in The MINK Region With and Without Climate Change*. DOE/RL/01830T-H7. Washington, DC: US Dept. of Energy.

Freeman, M. 1979. *The Benefits of Environmental Improvement*. Baltimore: Johns Hopkins University Press.

IPCC. 1996. *Climate Change 1995: The Science of Climate Change*, Houghton, J.T., Filho, L.G., Callander, B.A., Harris, N., Kattenberg, A. and Maskell, K. (eds.). Cambridge: Cambridge University Press.

Johnson, S.R. and Haigh, P.A. 1970. Agricultural Land Price Differentials and Their Relationship to Potentially Modifiable Aspects of Climate. *The Review of Economics and Statistics* 52: 173–81.

Kane, S., Reilly, T. and Tobey, J. 1992. An Empirical Study of the Economic Effects of Climate Change on World Agriculture. *Climatic Change*. 21: 17–35.

Mendelsohn, R. and Nordhaus, W. 1996. The Impact of Global Warming on Agriculture: Reply. *American Economic Review* 86: 1312–15.

Mendelsohn, R., Nordhaus, W. and Shaw, D. 1994. The Impact of Global Warming on Agriculture: A Ricardian Analysis. *American Economic Review* 84: 753–71.

Mendelsohn, R., Nordhaus, W. and Shaw, D. 1996. Climate Impacts on Aggregate Farm Values: Accounting for Adaptation. *Agriculture and Forest Meteorology* 80: 55–67.

Rosenzweig, C. and Parry, M. 1994. Potential Impact of Climate Change on World Food Supply. *Nature* 367: 133–8.

Appendix A3. *Definition of major variables and terms used in this study*

Variable	Definition
Normal	As applied to temperature and precipitation refers to the value of that particular element averaged over the period from 1951–1980.
Temp	Normal daily mean temperature in the month, Fahrenheit. Computed as being the temperature one-half way between the normal daily maximum and normal daily minimum temperatures for the month.
Temp sq.	Temp for a month, squared.
Rain	Normal precipitation for the month, inches.
Rain sq.	Rain for a month, squared.
Daily var.	The difference between normal daily maximum and daily minimum temperatures in the month (diurnal cycle).
Temp y-var.	The range between the year with the highest and the year with the lowest mean monthly temperature over a 30-year period.
Rain y-var.	The range between the year with the greatest and the year with the least monthly precipitation over a 30-year period.
Income per capita	Annual personal income per person in $1000, 1984.
Density	Number of thousands of people per square mile, 1980.
Density sq.	Density, squared.
Solar radiation	Latitude measured in degrees from southern-most point in US.
Altitude	Height from sea level in feet.
Salinity	Percent of land which needs special treatment because of salt/alkaline in the soils.
Flood prone	Percent of cropland which is prone to flooding.
Irrigated	Percent of cropland with irrigation.
Water capacity	Ability of soil to hold water.
Permeability	Ability of water to pass through soil.
Wetland	Percent of land considered wetland.
Soil erosion	K factor–soil erodibility factor in hundredths of inches.
Slope length	Number of feet length of slope (not steepness).
Farm value	Estimate of the current market value of farmland including buildings for the county expressed in dollars per acre, 1982.
Sand	Mean surface layer texture of cropland from loamy sand to coarse sand.
Clay	Mean surface layer texture of cropland from sandy clay loam to clay.

4   Climate change and agriculture: the role of farmer adaptation

KATHLEEN SEGERSON AND BRUCE L. DIXON[1]

There has been considerable debate about the potential effect of emissions of "greenhouse gases" on climate change or "global warming" and its impact on economic and ecological systems (see Helms *et al.*, 1996). One sector thought to be sensitive to climate effects is the agricultural sector. The impact of global warming on the US agricultural sector has been studied by a number of previous authors (e.g. Adams *et al.*, 1988; Dudek, 1988; Adams, 1989; Crosson, 1993; Kaiser *et al.*, 1993; Mendelsohn *et al.*, 1994; Rosenzweig and Parry, 1994). However, most of these studies do not allow for the full range of adaptations that farmers could employ in response to climate change, such as changes in the crop/enterprise mix, input mix, and the timing of operations (with the exception of Mendelsohn *et al.*, 1994 which includes, but does not explicitly model adaptation). Those studies that do explicitly incorporate adaptation (e.g. Crosson, 1993; Kaiser *et al.*, 1993) base their estimates on simulated effects rather than actual evidence of adaptation that has occurred. Failure to reflect the full range of adaptation possibilities in estimates of impacts is likely to result in over-estimation of damages from climate change.

In order to assess the full range of adaptation possibilities, a study of the extent of farmer adaptations based on empirical adaptation data was undertaken. This chapter reports the results of that study. Some of the results reported here (specifically, the results from the estimated yield equations) were used in conjunction with other information on adaptation to generate "best guess" parameter adjustments for the Agricultural Sector Model (ASM). The ASM was then re-run with these adjustments to determine the effect of adaptation on the predicted aggregate welfare effects of climate change. Details regarding the parameter adjustments that were made and the resulting welfare impacts are reported in Chapter 2.

[1] We acknowledge the valuable research assistance of Susan Helms and Lih-Chyi Wen, as well as useful comments by Richard Adams, Robert Mendelsohn, James Neumann, and two reviewers.

4.1   Methodology

The basic approach used in this study was to estimate the adaptation possibilities by examining how farmers have responded to existing differences in climate across regions in the United States. The theoretical foundation for the approach is neoclassical duality theory. Duality theory suggests that farm-level production decisions depend on exogenous factors such as output prices, input prices, technological constraints, and environmental factors (Varian, 1992). Since environmental factors generally vary across regions, cross-sectional data can be used to estimate how production decisions (and the associated costs, revenues, and profits) have varied with these environmental factors. These estimated relationships reflect the adaptation possibilities, since in making the actual production decisions, farmers have taken advantage of all the mitigation or adaptation possibilities available to them. From the estimated relationships, we can then calculate how farm-level profits, for example, would change if an exogenous change in an environmental factor occurred and farmers adapted to that change.[2] The above approach could be applied in a number of different environmental contexts. For example, Garcia et al., (1986) used cross-sectional data for farms in Illinois to estimate the impact of ground-level ozone changes on farm profitability.

The duality-based approach is related to the Ricardian approach used by Mendelsohn et al., (1994, 1996) to estimate the impacts of global climate change (see also Chapter 3). Under the Ricardian approach, climate variables are assumed to affect farm-level profitability, which (among other things) determines land values. However, other factors affecting profitability, such as output prices, are not included. In addition, their methodology does not allow the estimation of yield changes that can be compared to yield change estimates based on crop simulation models to estimate the extent to which farmer adaptation can offset any negative impacts of climate change.

In this study, we take a two-pronged approach to estimating adaptation possibilities using duality theory. First, we directly estimate per-acre yield functions for corn, winter and spring wheat, and soybeans (the major field crops in the Midwest) that incorporate farmer adaptation to climate (temperature and precipitation). The yield equations are then used to predict the impact of alternative climate change scenarios on crop yields. Comparing these yield change estimates with the estimates obtained from crop simulation models that incorporate only modest adaptation allows the potential for adaptation to mitigate the yield losses to be measured. The yield effects,

---

[2] This approach does not incorporate input or output price changes that could occur if aggregate farmer responses are large. To incorporate price adjustments, the farm-level responses must be used in a market-level model in which prices are endogenous, such as the ASM model used in Chapter 2.

in turn, can be introduced to a market equilibrium model (such as the ASM in Chapter 2) to determine the effect on crop mix.

Alternatively, crop mix can be made endogenous by estimating a profit function rather than crop-specific yield response equations. The second component of the study estimates a per-acre profit function with endogenous crop choice. The estimated function is then used to simulate the effect of alternative climate scenarios on per-acre profitability. The methodology does not allow for movements of land in and out of agriculture, i.e. the impacts depend on the land remaining in agriculture after the climate change. In principle, one could derive supply functions from profit functions in order to predict land use changes, but collinearity problems in the data prevented us from being able to exploit this potential.

4.2   Data

Cross-sectional data for the 12 Midwest states of the Corn Belt, Lake States, and Northern Plains regions were used in both the yield and profit empirical studies. We restricted our analysis to this 12-state region (rather than including the entire United States, as in Mendelsohn *et al.*, (1994)) to maintain a relatively homogeneous production region. The duality theory underlying our methodology is only completely valid if the production technology is the same for all observations.

The yield equation estimates were based on a sample of counties for 1987. This year was chosen since most of the production data (as well as the revenue and cost data used in the profit function estimation) were taken from the Census of Agriculture, and the 1987 census was the most recent census available at the time of the study. In addition, 1987 appears to be a fairly typical or representative year, with yields slightly above the long-term trends. The profit function, on the other hand, was estimated from a time-series, cross-sectional sample of 975 counties for the years 1978, 1982, and 1987 (the three most recent census years available at the time of this study). Use of the panel data set for the profit function allowed us to exploit the variability in output price over time, thereby increasing the efficiency of our estimates. In addition, it allowed us to "smooth" out some of the effects of weather in a given year, since aggregate production does shift from year to year depending on the weather outcomes. A panel approach could also have been used for the yield equations but time and resource limitations prevented collection of the yield data for other census years.

Our initial sample consisted of 988 counties in the 12-state region. These represented all counties for which at least 20 percent of the land was in agriculture, which eliminated urban counties such as Cook County from the sample. However, in estimating the yield equations, some counties had to be dropped because they did not produce a crop (corn, winter wheat, spring wheat, or soybeans) or produced very little (less than 1000 acres). The resulting sample sizes were 854 counties for corn, 521 counties for winter wheat, 167 counties for spring wheat, and 727 counties for soybeans. Winter and spring wheat are distinguished because they are planted and grown at different times of the year and so are expected to have a different relationship with climate.

For estimation of the profit function, we started with the same initial 988 county sample. However, we were forced to drop seven counties because of incomplete data on the site characteristics. In addition, six other observations had to be dropped because of insufficient data on acreage (see below for details). This left a sample of 975 counties over 3 years, for a total of 2925 observations.

The production, cost, and revenue data were taken directly from the Census of Agriculture. Yields were calculated as the ratio of production (measured in bushels) to harvested acreage for farms with at least $10 000 in sales. Farms with less than $10 000 in sales are likely to be part-time farms with yields that are not necessarily representative. Profit (or more accurately net revenue) per acre was defined as the market value of the agricultural products sold (excluding government payments) minus the sum of the variable farm production expenses[3] and the interest payments for machinery and equipment, divided by the sum of cropland and pastureland for all farms. Nominal values were converted into real values by dividing by the GNP deflator for that year.

Because the data were not subdivided by irrigated and non-irrigated status, we had no way in which to distinguish yields and profits on irrigated vs. non-irrigated acreage. However, since there is relatively little irrigation in our study region (only Nebraska irrigates more than 10 percent of its land), failure to account for irrigated acreage is not likely to introduce a significant error into the estimated coefficients. We note, however, that because we do not explicitly consider irrigation, we are unable to account for any increase in irrigated acreage that might occur as a result of climate change.

For estimation of the yield equations, county-level output price indices were calculated as the ratio of value of sales of a crop to total production. While this provides variability in prices across counties, it does not account for on-farm consumption of crops for feed or other unpriced uses. Thus, in estimates of the profit function, we

---

[3] Farm production expenses consist of livestock and poultry purchased, feed and seed, fertilizers and chemicals, petroleum products and electricity, hired and contract labor, custom work, and machine hire. Production expenses, such as insurance, water, animal health costs, grazing fees, marketing charges, and miscellaneous farm supplies, could not be included.

simply used state-level prices from agricultural statistics (USDA, 1989) for each county within a state. Nominal output prices were converted into real terms by dividing by the GNP deflator for the corresponding year.

Since county and state data on input prices were not available, the effect of any regional input price variation was captured by using regional dummy variables. It would clearly have been preferable to include cross-sectional input price data (had they been available). However, for moderate variations in input prices, the comparative advantages and biological imperatives of various crops are likely to outweigh the effect of input price fluctuations. Thus, the results reported here should be valid for the input prices that were in effect during the study period.

All of the climate data (seasonal temperature and precipitation) as well as the data on soil and site characteristics come from Mendelsohn et al., (1994). The climate data represent 30-year normals for January, April, July, and October as described in detail in Chapter 3. Summary statistics for all of the variables used in the profit function regressions are given in Table 4.1.

4.3   Yield equations

As noted above, yield equations were estimated for corn (for grain or seed), winter wheat, spring wheat, and soybeans. The dependent variable is the logarithm of yields measured in bushels per acre. The explanatory variables include both economic and physical variables. The economic variables are: price of corn/price of soybeans, price of wheat, and the percentage of farm operators who are full-time farmers. The prices for all three crops are included in each of the yield equations to reflect the substitution possibilities among the crops. Because corn and soybeans are frequently substitute crops for each other, the ratio of their prices more directly reflects their relative profitability than entering them into the regressions linearly (see Dixon et al., 1994). Thus, we include the corn and soybean prices in ratio form. Own-price effects are expected to be positive, while cross-price effects are expected to be negative. Thus, the price ratio of corn to soybeans is expected to have a positive sign in the corn equation and a negative sign in the soybean equation. We hypothesize that the effect of higher crop prices causing more intensive input use and higher yields is stronger than the effect of higher prices causing less productive land to be drawn into production. Similarly, the price of wheat is expected to have a positive sign in the wheat equations but a negative sign in the corn and soybean equations. Full-time farmers act as a proxy for managerial input. This variable is therefore expected to have a positive sign in all three equations.

79

Table 4.1 *Descriptive statistics of profit function variables*

Variable		Mean	Minimum	Maximum
Market value		55 428.250	1 510.000	381 532.000
Production expenses		35 148.500	1 025.000	338 310.000
Machinery and equipment		44 921.620	2 456.000	189 731.000
Real profit		105.487	−14.213	562.443
Cropland		183 005.700	8 393.000	917 398.000
Pastureland		74 760.720	183.000	3 502 685.000
January temperature	(°F)	21.192	−0.553	35.956
April temperature	(°F)	50.023	38.413	60.189
July temperature	(°F)	74.561	66.354	82.478
October temperature	(°F)	53.224	43.193	61.079
January precipitation	(in)	1.352	0.254	3.717
April precipitation	(in)	3.051	1.104	4.732
July precipitation	(in)	3.634	1.845	4.937
October precipitation	(in)	2.197	0.690	3.769
Erosion	(K-factor)	0.281	0.090	0.390
Water-holding capacity	(in/lb)	2.926	0.643	11.915
Permeability	(in/hour)	4 194.990	85.250	27 782.350
Slope length	(ft)	2.245	0.420	13.520
Percent wetlands	(%)	0.040	0.003	0.393
Salinity	(%)	0.013	0.000	0.314
Latitude (° from reference point)		15.857	10.576	23.198
Altitude	(ft)	1 184.030	324.570	4 832.600
Price of corn	($/bushel)	2.396	1.550	3.116
Price of soybeans	($/bushel)	6.579	4.489	9.349
Price of wheat	($/bushel)	3.176	1.985	4.640
Price of cattle and cows	($/head)	387.593	290.859	690.000
Price of hogs and pigs	($/head)	90.291	59.625	123.346
Price of milk	($/100 lbs)	12.711	9.455	14.875

The physical variables are: seasonal temperatures and precipitation levels (January, April, July, October), the water-holding capacity of the soil, and a measure of soil erodibility. The temperature and precipitation variables are obviously included to reflect the role of climate in determining yields. Water-holding capacity and erodibility indicate soil quality and type. In general, we expect increased erodibility to decrease yields. The effect of increased water-holding capacity could be positive or negative. Some ability to hold water is beneficial but too much can lead to wet conditions that impair planting and other farm operations.

All variables are transformed by taking their natural logarithms (1° was added to the temperature variables to assure positive values). This transformation allows an easy interpretation of the coefficients and gives diminishing returns for the linear terms when the coefficients are less than one. Since the log function is asymptotic as the data approach zero, it can produce extreme results at very low values of the variables, for example with very low temperatures. However, most of the observations are not at very low levels (the mean of January temperatures is 21 °F). Squared terms were included for the climate variables. This allows climate change to have a non-monotonic effect on yield, i.e. increasing yield in cool and dry places and possibly decreasing yield in warm and wet locations.

It should be emphasized that the climate variables here are measured differently than in conventional yield response studies. The variables are 30-year averages, not the levels of these variables observed for the 1987 crop season (which were not available). The climate variables have more of a long-term character than is typical in most yield response studies. To the extent that the weather in a particular county deviated from its long-term average in 1987, the climate variables did not reflect this deviation. However, the 30-year averages used here are appropriate for the analysis of the effects of climate changes on yields.

Finally, note that the variables reflecting the major farm programs (such as deficiency payments or acreage enrolled in the Conservation Reserve Program) were not included either here or in the profit function estimation below. These programs have clearly had an impact on acreage and the 1996 farm legislation will certainly affect future acreage. However, both the yield equations and the profit function are estimated on a per-acre basis. We do not estimate changes in total acreage. In addition, our analysis of the impacts of climate change are *ceteris paribus* with respect to the general policy regime that was in effect during 1978–87. Any future loosening of acreage controls will allow farmers to adapt even more to climate change. This suggests that the degree of adaptability estimated in this study is probably conservative.

Since the equations are estimated using cross-sectional data, the major econometric concerns are heteroscedasticity, multicollinearity, and the stability of parameters across counties. In general, all four of the equations estimated have these problems. Heteroscedasticity has been noted in yield response models as discussed in Dixon *et al.* (1994) and in Yang *et al.* (1992). Consequently, six different tests for heteroscedasticity were examined for each equation and all of the equations except spring wheat clearly indicated the presence of heteroscedasticity. However, it is not clear what pattern generates the observed heteroscedasticity. Because least squares estimators of the coefficients in heteroscedastic models are consistent and the sample sizes are all in excess of 100 (and in excess of 700 for corn and soybeans), a search for a pattern of heteroscedasticity consistent with the data was not undertaken. Instead the regression coefficients were estimated by least squares and their standard errors were estimated by White's consistent estimator of the least squares covariance matrix. Thus, the t-ratios of the reported regressions should be interpreted as having a standard normal distribution asymptotically.

The multicollinearity problems are severe as measured by standard regression diagnostics (Belsley *et al.*, 1980). Condition indices for each regression are as large as 700 000. The highest levels of collinearity arise from the climate variables, and particularly from the inclusion of the squared terms. If these are removed, the collinearity is substantially reduced although not eliminated. However, inclusion of the squared terms is important in that it allows for non-monotonicity. The clear implication of collinear climate variables is that the estimated coefficients of any particular climate variable should be interpreted carefully. Most importantly, the result that a particular coefficient is insignificant at some standard level of significance should not be taken as compelling evidence that a particular variable is an irrelevant regressor. Because of these high levels of collinearity, the model is most appropriately used to estimate the impact of climate changes where all eight of the temperature variables or all eight of the precipitation variables are changed by some common amount or percentage. Under such an approach, any inaccuracies of a particular coefficient are more likely to be offset by the coefficients of other collinear variables, and their joint impact is likely to be estimated more accurately. Collinearity also suggests, though, that the estimated impacts of a change in temperature and/or precipitation in all seasons (as in the scenarios considered in this analysis) are not likely to change much even if an individual temperature or precipitation variable were omitted from the model.

The final econometric problem is the stability or homogeneity of the coefficients across counties. The region used in estimation of the equations is large and covers a diverse area of climate and soil conditions. To allow for regional variation in the specification, regional dummy variables were included in the estimated equations for corn, soybeans, and winter wheat (but not for spring wheat which was grown in only three states). Three regions were identified: Lake States (MI, MN, WI), Northern Plains (KS, NE, ND, SD) and Corn Belt (IL, IA, IN, OH, MO). These correspond to the United States Department of Agriculture farm production regions.

Parameter estimates are presented in Table 4.2 for all four equations. The overall fit of the corn equation is good for cross-sectional data with an R^2 of 0.56. The price ratio of corn to soybeans is positive and significant, as anticipated, while the price of wheat is negative (although not significant). The managerial input is positive and significant. Furthermore, increased erodibility reduces yields, although the coefficient is not significant. The coefficient of the regional dummy variable for the Northern Plains

Table 4.2. *Parameter estimates for yield equations[a]*

Variable	Corn	Soybeans	Spring wheat	Winter wheat
January temp.	0.575[c]	0.021	−0.212[c]	29.039[c]
	(3.38)	(0.56)	(−4.35)	(7.28)
April temp.	28.740	111.330[c]	117.490	−31.540
	(0.90)	(3.72)	(1.29)	(−0.92)
July temp.	175.030	147.60	−387.920	324.020[c]
	(1.43)	(1.33)	(−1.16)	(3.32)
October temp.	−62.060	−115.26[b]	116.890	−146.240[b]
	(−1.27)	(−2.15)	(0.83)	(−2.06)
Jan. temp squared	−0.068	0.014	0.106[c]	−4.508[c]
	(−1.54)	(0.66)	(4.03)	(−7.19)
April temp squared	−3.742	−14.340[c]	−16.370	3.724
	(−0.91)	(−3.75)	(−1.38)	(0.85)
July temp squared	−20.260	−17.410	45.210	−37.320[c]
	(−1.42)	(−1.35)	(−1.16)	(−3.32)
Oct. temp squared	7.830	14.790[b]	−14.100	18.540[b]
	(1.28)	(2.20)	(−0.78)	(2.10)
Jan. precipitation	−0.074	−0.167[c]	0.062	0.047
	(−1.35)	(−5.07)	(0.22)	(0.91)
April precipitation	−1.540[c]	−0.960[c]	−1.056[b]	1.320[c]
	(−6.00)	(−3.77)	(−2.03)	(−4.77)
July precipitation	2.450[c]	−0.207	−0.629	1.814[b]
	(3.37)	(−0.32)	(−0.80)	(2.35)
October precipitation	0.098	0.657[c]	1.173[c]	0.349[b]
	(0.52)	(3.85)	(5.57)	(2.15)
Jan. precip squared	−0.077[b]	−0.062[b]	0.057	0.023
	(−2.26)	(−2.17)	(0.31)	(0.70)
April precip squared	0.838[c]	0.531[c]	0.578	0.764[c]
	(7.10)	(4.79)	(1.39)	(6.25)
July precip squared	−0.830[c]	0.160	0.328	−0.728[b]
	(−2.91)	(0.66)	(0.84)	(−2.34)
Oct. precip squared	−0.320[c]	−0.465[c]	−1.209[c]	−0.500[c]
	(−3.23)	(−5.04)	(−3.55)	(−6.13)
Corn/soybean price	0.019[c]	0.003	0.011[c]	0.002
	(8.08)	(0.87)	(4.29)	(0.46)
Wheat price	−0.001	−0.001	−0.320	0.019
	(−0.46)	(−0.76)	(−1.69)	(0.20)
Full-time farmers	0.232[c]	0.136	0.237	0.314[c]
	(3.25)	(1.75)	(0.90)	(4.84)
Erodibility (K-factor)	−0.039	−0.023	0.170	−0.005
	(−1.40)	(−0.79)	(1.29)	(−0.13)

83

Table 4.2. (*cont.*)

Variable	Corn	Soybeans	Spring wheat	Winter wheat
Water-holding capacity	0.019	0.008	0.016	0.016
	(1.55)	(0.85)	(0.85)	(1.50)
Lake State dummy	0.021	0.000		0.008
	(−0.95)	(0.01)		(0.24)
Northern Plains dummy	−0.104^c	−0.070^c		−0.285^c
	(−3.42)	(−2.95)		(−8.41)
Constant	−308.610	−301.28	381.930	−393.070^b
	(−1.39)	(−1.55)	(0.75)	(−2.42)
R^2	0.56	0.58	0.66	0.76
N	854	727	167	521
Mean yield (bu.)	111	35.7	31.1	43.8

Notes:
All variables except the regional dummy variables are in natural logs.
^{a} Figures in parentheses are asymptotic "*t*" ratios.
^{b} significant at the 5% level, ^{c} significant at the 1% level.

indicates that yields are lower in the Northern Plains. Given the double log form of the model, this coefficient indicates that the 1987 yields were approximately 10 percent lower in the Northern Plains than in the Corn Belt, after adjusting for climatologic and economic factors.

With the exception of the January temperature, the individual coefficients on the temperature variables are not significant. The lack of significance among these variables reflects the fact that all four of them are strongly linearly related as indicated by the collinearity diagnostics. The combined effect of the coefficients will determine the overall impact of a temperature increase for all seasons.

All four of the precipitation variables are significant in either linear or quadratic terms or both. The signs on the April precipitation coefficients imply that initial increases in precipitation in April are yield-decreasing. For July and October, when precipitation levels are low, the increases in precipitation during these seasons are initially yield-increasing. However, "too much" precipitation during these seasons will decrease yields. This could reflect the fact that high precipitation levels in April could delay planting and therefore decrease yields.

The soybean equation also fits well with an R^2 of 0.58. The price of wheat is again negative (as expected) but insignificant. However, the coefficient of the corn to soybean price ratio variable is not negative as expected (although it is insignificant). This may reflect the fact that soybeans are grown as a substitute for corn in some

Midwest regions and as the second crop in a winter wheat and soybean double-crop regime in other regions. Thus, the price effects are likely to be different in different parts of the Midwest. As with corn, the managerial input and the erodibility variables have the signs expected, although neither of them is significant. Again, the regional dummy indicates that yields in the Northern Plains are below those in the Corn Belt, in this case by an estimated 7 percent, indicating the regional superiority of the southerly regions of the Midwest in growing soybeans.

The climate variables are individually more significant in the soybean equation than in the corn equation. Warmer temperatures in April and July increase soybean yields whereas warmer temperatures in October decrease these yields. These effects are modified in the opposite direction by the squared terms, implying the impacts reverse with sufficiently high temperatures. As noted above, given the collinearity in the data, the model is likely to be more accurate in predicting the effects of changes that occur across all seasons than in predicting individual season effects.

As with the corn and soybean equations, the explanatory power of the two wheat equations is surprisingly good, with R^2 values of 0.66 (spring wheat) and 0.76 (winter wheat). The average yields for the two crops are markedly different. Spring wheat has a mean county-level yield of 31 bushels while winter wheat has a mean yield of 44 bushels. In addition, the regional dummy in the winter wheat equation indicates that yields in the Northern Plains were again significantly lower than yields in the Corn Belt, in this case by 28.5 percent. Overall, the price variables are not very important in determining wheat yields. Three of the four price variables in the two equations are insignificant, and the one that is significant (the corn/soybean price ratio for spring wheat) has an elasticity of only 0.01. The managerial input has the expected sign in both equations, although it is only significant in the winter wheat equation. Erodibility has the expected sign for winter wheat but not for spring wheat although both coefficients are insignificant.

In the winter wheat equation, all of the temperature variables except those for April are significant. Similarly, all of the precipitation variables except those for January are significant. The only temperature that is significant in the spring wheat equation is the January temperature. In terms of precipitation effects, both April and October precipitation levels show a significant impact on yields.

The estimated yield equations were used to simulate the effects of three different climate change scenarios, with two alternative temperature increases (2.5 °C and 5.0 °C) and one change in precipitation (7 percent). We assumed that the temperature and precipitation changes were uniform across all four seasons and locations. Five separate locations were chosen to evaluate each crop in order to illustrate the importance of initial temperature and precipitation. Two of the sites are relatively "wet":

Table 4.3. *Comparison of predicted yield changes: corn*

	Agronomic	Empirical yield
Scenario: DT = 2.5°C, DP = 0		
Des Moines, IA	−14.4	−4.4
North Platte, NE	−7.5	−8.8
Scenario: DT = 5.0 °C, DP = 0		
Des Moines, IA	−30.4	−16.4
North Platte, NE	−19.1	−23.1
Scenario: DT = 2.5 °C, DP = +7%		
Des Moines, IA	−10.7	−2.9
North Platte, NE	−7.5	−6.7

Indianapolis and Des Moines. The other three sites are relatively "dry", but vary in temperature: Fargo (cold), North Platte (warm), and Dodge City (warmer still). Indianapolis and Des Moines currently grow soybeans and corn, Dodge City grows wheat, North Platte grows some corn (but is mainly in pasture), and Fargo grows wheat. In each county, we contrast the yield change predicted by Cynthia Rosenzweig using an agronomic model (CERES) with little adaptation and the yield change predicted by the estimated yield equations (Table 4.2). The results are presented in Tables 4.3 to 4.5.

The comparison between the empirical yield results and the agronomic results is complex. For two of the crops, the two methods predicted roughly the same effects from climate change at one site whereas the empirical method predicted much lower damages to yields at the other site. In the comparison of corn yields (Table 4.3), the two sets of damage estimates were virtually the same for North Platte. In Des Moines, however, the empirical model predicted damages that are one-half to one-quarter the size of those predicted by the agronomic model. Similar results were found for wheat in Table 4.4 where the Dodge City results were similar but the empirical model predicted benefits rather than damages in Fargo. Averaging these results suggests that adaptation might offset nearly half of the yield loss effect of a 2.5 °C warming but might only be able to offset a smaller percentage of the effect of a 5.0 °C warming. The evidence suggests that farmers can adapt to temperature increases and thereby reduce yield losses, but that this mitigation does not offset all the damages. The results also suggest that warming is going to have very different impacts depending upon whether temperatures are currently relatively cool or warm.

The results for soybeans (Table 4.5) do not show much potential for adaptation. In fact, the empirical yield equations predict larger yield losses than the agronomic projec-

Table 4.4 *Comparison of predicted yield changes: wheat*[a]

	Agronomic	Empirical yield
Scenario: DT = 2.5 °C, DP = 0		
Dodge City, KS	−24.3	−26.7
Fargo, ND	−26.6	+0.2
Scenario: DT = 5.0 °C, DP = 0		
Dodge City, KS	−44.1	−50.1
Fargo, ND	−40.9	+14.4
Scenario: DT = 2.5 °C, DP = +7%		
Dodge City, KS	−15.7	−27.8
Fargo, ND	−19.4	+2.1

Note:
[a] Dodge City results based on winter wheat and Fargo results based on spring wheat.

Table 4.5. *Comparison of predicted yield changes: soybeans*

	Agronomic	Empirical yield
Scenario: DT = 2.5 °C, DP = 0		
Indianapolis, IN	−11.6	−14.1
Des Moines, IA	−5.7	−10.4
Scenario: DT = 5.0 °C, DP = 0		
Indianapolis, IN	−32.7	−33.8
Des Moines, IA	−24.9	−28.8
Scenario: DT = 2.5 °C, DP = +7%		
Indianapolis, IN	−5.6	−12.8
Des Moines, IA	−2.8	−8.9

tions. One explanation for these results is that the soybean results are being affected by complex cropping patterns. As temperatures warm, farmers can grow an extra crop of soybeans as a double-crop. Although the warming might make the soybeans more productive, the fact that warmer areas are double-cropped can result in lower yields per acre.

4.4 Profit function estimates

In this section, we estimate the effects of climate change using a profit function, which allows for endogenous changes in crop mix and other related farm-level decisions (e.g. double-cropping and crop rotation). The dependent variable in the profit function was net revenue per acre. A semi-logarithmic functional form was used, where all the independent variables (except the regional dummy variables) were transformed by taking their logs. Although other forms were explored, the semi-logarithmic functional form provided the best fit.

The price variables included are the market prices received for the major field crops (corn, wheat, and soybeans) and the mean *per capita* value of the inventory for the major livestock animals (cattle and swine). The price data come from various issues of *Agricultural Statistics*. Since profits increase with output prices, the coefficients of all of the price variables are expected to be positive. Input prices were not included since county-level (or even state-level) data on input prices were not available (see discussion above). To capture the impact of the Dairy Termination Program in 1987, we included the price of milk in 1987. While the herd liquidations were carried out over a period of time to minimize the impact on beef markets, this program had the effect of lowering milk supplies abruptly compared with 1978 and 1982. It is likely that this gave a boost to milk prices, thereby increasing the profits of those farmers who remained in dairying. Hence, the parameter on this variable is expected to be positive.

Because of severe collinearity, only two of the four seasons (January and July) were included. In addition, quadratic terms were not included since the resulting collinearity made the regression results unreliable. In addition to the climate variables, the following variables were included to capture the impact of site characteristics: erosion, soil water-holding capacity, soil permeability, slope length, percentage of wetlands, latitude, altitude, and salinity. Finally, dummy variables for the Lake States (LS) and Northern Plains (NP) were included to capture regional differences unrelated to climate or other included site characteristics (such as differences in input prices or market access).

The results from the estimation of the profit function are reported in Table 4.6. Four of the five output price variables are significant at the 5 percent level and have positive coefficients, as predicted. The one price with a negative coefficient (wheat) is insignificant. In addition, the dairy buy-out program in 1987 appears to have had a significant impact on profits in that year. During that year, in counties where the milk price was high, the per-acre profit was higher. Of the eight physical site characteristics, seven are significant. As expected, profit per acre is increased by increases in permeability and slope length, and is reduced by increases in erodibility and altitude. In addition, profits are significantly lower in the Northern Plains and the Lake States than in the Corn Belt. Unexpectedly, higher latitudes are also associated with higher profits which suggests that latitude represents something other than just the length of the growing season.

Table 4.6. *Parameter estimates for profit function*

Variable (in logs)	Coefficient	t-ratio
January temperature	42.943[b]	12.872
July temperature	−355.50[b]	−4.379
January precipitation	−47.827[b]	−9.470
July precipitation	99.731[b]	13.138
Corn price	117.90[b]	6.358
Soybean price	34.398[b]	2.513
Wheat price	−6.4662	−0.483
Cow and cattle price	66.614[b]	5.982
Hog and pig price	100.36[b]	5.156
Milk price – 1987	16.168[b]	8.972
Erodibility	−29.318[b]	−6.106
Water-holding capacity	−3.4926[a]	−1.975
Permeability	8.5526[b]	7.647
Slope length	23.098[b]	11.889
Wetlands	−1.5774[a]	−2.174
Salinity	−1.4926[b]	−3.326
Latitude	45.529[a]	2.019
Altitude	−41.840[b]	−7.964
Lake State dummy	−11.625[b]	−2.749
Northern Plains dummy	−55.977[b]	−14.777
Constant	474.495	1.047

Notes:
$R^2 = 0.625, N = 2925$
[a] significant at the 5 % level, [b] significant at the 1% level.

All of the climate variables are significant. The results suggest that increases in January temperatures will increase profitability, while increases in July temperatures decrease profit per acre. In contrast, increased precipitation is beneficial in July but detrimental in January. Again, this may reflect the fact that increased winter precipitation can delay the spring planting if the soil becomes too wet.

The estimated profit function was used to simulate the impacts of climate change on profits per acre. The simulations were done for the nine climate scenarios used in the simulation of yield changes. The predicted changes in aggregate profits are reported in Table 4.7 for these scenarios. The results are reported in both percentage terms and total dollars (net gain or loss measured in 1982 dollars) for the sample region. These simulated changes are based on 1987 output prices and do not reflect

Table 4.7. *Simulated effects of climate change: predicted changes in aggregate profit*

Scenario		Change in aggregate profit	
ΔT(°C)	ΔP%	($ billion)	%
1.5	0	−1.26	−6.50
1.5	+7	−0.43	−2.23
1.5	+15	+0.45	+2.32
2.5	0	−2.26	−11.60
2.5	+7	−1.43	−7.33
2.5	+15	−0.54	−2.78
5.0	0	−4.93	−25.37
5.0	+7	−4.10	−21.10
5.0	+15	−3.22	−16.55

any changes in prices that might result from aggregate changes in output. The results suggest that temperature increases reduce aggregate profit per acre, or reduce acreage cultivated, or some combination of these two effects. With no change in precipitation, the predicted percentage reduction is 6.50 percent for a 1.5 °C increase in all seasons, 11.60 percent for a 2.5 °C increase, and 25.37 percent for a 5 °C increase.[4] However, since precipitation increases improve profits, changes in precipitation can mitigate the negative effect of temperature increases (and in some cases more than offset it). For example, the 11.60 percentage reduction in profits under a 2.5 °C warming with no change in precipitation is reduced to a 7.33 percent reduction if the temperature rise is accompanied by a 7 percent increase in precipitation. If the increase in precipitation is 15 percent, the overall impact of the climate change is reduced to 2.78 percent.

The simulation results from the profit function can be compared to the results of Mendelsohn *et al.* (1994) who report results based on a scenario with a 2 °C temperature increase accompanied by an 8 percent increase in precipitation. This is close to our scenario of a 2.5 °C temperature increase and 7 percent precipitation increase. For this scenario, the predicted change in aggregate profit per acre based on the profit function is −7.33 percent. Using cropland weights, which emphasize the vast grain growing territory of the Midwest, Mendelsohn *et al.* predict annual damages for the

---

[4] These predicted changes do not reflect the possibility that profit reductions will be sufficiently large to cause land to move out of agriculture to other uses that would be more profitable under a given climate scenario. In addition, they do not reflect the price effects of climate change. Large decreases in output (for example, in response to a 5 °C warming) would be likely to lead to significant price changes.

United States of -4.5 percent. Although not exactly comparable, the results of the two studies are consistent with each other.

The predicted changes reported in Table 4.7 are aggregated across the region. There is, however, considerable variation in the predicted changes for individual counties within the region. For example, while a $2.5\,°C$ warming with a 7 percent increase in precipitation yields aggregate losses of about 7.5 percent, some counties in the study region actually gain from this climate change. The cooler northern counties tend to gain slightly from warming whereas the warmer southern counties and dryer western counties are the biggest losers.

4.5   Conclusion

This chapter has examined farmer adaptation to climate change. The study uses a cross-sectional empirical analysis of Midwestern counties to measure the sensitivity of yields and profits to climate. These empirical relationships were then used to forecast how yields and profits would change in different climate scenarios. The results were compared to agronomic models which did not include adaptation.

A model of yields was estimated using economic variables, site characteristics, and climate variables (seasonal temperatures and precipitation levels). The estimated yield equations were then used to simulate the yield impacts of alternative climate change scenarios. The predicted yield changes were compared to the results of crop simulation models to provide an indication of the extent to which adaptation (as embodied in the estimated yield response equations) might reduce yield losses from climate change. Limitations of the available data and differences in the methodologies make a precise estimate of adaptation impossible. Nonetheless, the results support the hypothesis that adaptation does reduce predicted yield losses. The "best guess" estimate is that the losses in corn and wheat yield from a moderate temperature increase might be reduced by about 50 percent through adaptation. This suggests that climate change analyses that do not allow for adaptation are likely to overestimate damages by a considerable amount.

The second component of the research uses a combination of cross-sectional and time-series data to estimate a profit function. The profit function relates per-acre profits to economic variables, site characteristics, and climate variables. Unlike the yield equations, the profit function allows for changes in crop mix in response to climate changes. The profit function was used to simulate the effect of alternative climate change scenarios on per-acre profits as a measure of economic impacts. The results suggest that, for the aggregate region, temperature increases will reduce profit per acre while precipitation increases will increase aggregate profits per acre. For a moderate climate change scenario (2.5 °C increase in temperature and 7 percent increase in precipitation), aggregate profit per acre is reduced by 7.3 percent or $1.4 billion over the sample region. This estimate is roughly comparable to the damage estimate obtained by Mendelsohn *et al.* (1994) using a Ricardian approach. However, our aggregate damage estimates mask considerable differences within the regions, which indicate that cooler counties may benefit from mild warming whereas dry and warm counties are likely to be more severely damaged.

While the estimates reported here incorporate the ability of farmers to adapt to changes in climate by varying their crop/enterprise mix as well as the input mix and timing of operations, there are other adjustments that are not incorporated. Specifically, the estimates do not reflect potential changes in the land base, price changes, or new technology. A complete investigation of the role of adaptation to climate change should account for these adjustments as well.

References

Adams, R.M. 1989. Global Climate Change and Agriculture: An Economic Perspective. *American Journal of Agricultural Economics* **71**: 1272–9.

Adams, R.M., McCarl, B.A., Dudek, J.W. and Glyer, J.D. 1988. Implications of Global Climate Change for Western Agriculture. *Western Journal of Agricultural Economics* **68**: 886–94.

Agricultural Finance Databook, Division of Research and Statistics, Board of Governors of the Federal Reserve System, Washington, DC, various issues.

Belsley, D.A., Kuh, E. and Welsch, R.E . 1980. *Regression Diagnostics: Identifying Influential Data and Sources of Collinearity.* New York City, New York: John Wiley & Sons, Inc.

Crosson, P. 1993. Impacts of Climate Change on the Agricultural Economy of the Missouri, Iowa, Nebraska and Kansas (MINK) region. In: *Agricultural Dimensions of Global Climate Change*, Kaiser, H. and Drennen, T. (eds.). Delray Beach, FL: St. Lucie Press.

Dixon, B.L., Hollinger, S.E., Garcia, P. and Tirupattur, V. 1994. Estimating Corn Yield Response Models to Predict Impacts of Climate Change. *Journal of Agricultural and Resource Economics* **19**: 58–68.

Dudek, D.J. 1988. Climate Change Impacts Upon Agriculture and Resources: A Case Study of California. In: *The Potential Effects of Global Climate Change in the United States.* Smith, J.B. and Tirpak, D.A. (eds.). Washington DC: US Environmental Protection Agency.

Garcia, P., Dixon, B.L., Mjelde, J.W. and Adams, R.M. 1986. Measuring the Benefits of Environmental Change Using a Duality Approach: The Case of Ozone and Illinois Cash Grain Farms. *Journal of Environmental Economics and Management* **13**: 69–80.

Helms, S., Mendelsohn, R. and Neumann, J. 1996. The Impact of Climate Change on Agriculture. *Climatic Change* **33**: 1–6.

Kaiser, H., Riha, S., Wilkes, D. and Sampath, R. 1993. Adaptation to Global Climate Change at the Farm Level. In: *Agricultural Dimensions of Global Climate Change.* Kaiser, H. and Drennen, T. (eds.). Delray Beach, FL: St. Lucie Press.

Mendelsohn, R., Nordhaus, W.D. and Shaw, D. 1994. The Impact of Global Warming on Agriculture: A Ricardian Approach. *American Economic Review* **84**: 753–71.

Mendelsohn, R., Nordhaus, W.D. and Shaw, D. 1996 Climate Impacts on Aggregate Farm Values: Accounting for Adaptation. *Agriculture and Forest Meteorology* **80**:55–67.

Rosenzweig, C. and Parry, M. 1994. Potential Impacts of Climate Change on World Agriculture. *Nature* **367**: 133–8.

Varian, H.R. 1992. *Microeconomic Analysis*, 3rd edn. New York, NY: W.W. Norton & Company, Inc.

USDA/ERS. (1994). *Feed Situation and Outlook.* FDS–330. Washington, DC: Economic Research Service.

USDA. (1989). Agricultural Statistics. Washington DC: Government Printing Office.

Yang, S.R., Koo, W.W. and Wilson, W.W. (1992). Heteroskedasticity in Crop Yield Models. *Journal of Agricultural and Resource Economics* **17**:103–9.

5   The impacts of climate change on the US timber market

BRENT L. SOHNGEN AND ROBERT MENDELSOHN[1]

In this chapter, ecosystem and economic models are integrated to estimate the impact of climate change on the US timber industry. Beginning with alternative General Circulation Model (GCM) climate scenarios, the steady-state response of ecosystems in the United States to climate change is predicted. We then develop dynamic scenarios of ecosystem change from these steady-state predictions. Using a dynamic economic model, harvests and endogenous prices are calculated over time, which adjust to these gradual changes in forest distribution and productivity. Model results vary depending on the combination of models used. For example, combining a dynamic ecological scenario with a dynamic economic one results in net benefits because this allows for salvage and other human management and adaptation.

Changes in temperature and precipitation can affect the natural ecosystems by altering the growing season, available soil moisture, or nutrient balances. The combined impact of these changes may affect the growth rates of plants or alter the competitive balance among species (Solomon and West, 1985). Ultimately, this will lead to a change in species composition on the land, and/or a change in land productivity (Neilson and Marks, 1994; Melillo et al., 1993; McGuire et al., 1993; Neilson et al., 1992; Prentice et al., 1992; IPCC, 1996b).

Although substantial work has been done on agricultural impacts, there are few studies of the impact of climate change on timber (Binkley and Van Kooten, 1994). Past work was handicapped by limited modeling of the link between climate change and forest response. For example, although Binkley (1988) used a global trade model of timber markets, the ecological model he used considered only the effect of temperature on forest growth. Other ecological models have examined only particular subnational regions. Solomon (1986), for example, considered only the eastern United States. Consequently, the US EPA report on climate impacts (Smith and Tirpak, 1989) focused on the impact of warming only on Northern hardwoods. These regional studies are not sufficient, however, to model the timber market impacts in the United States.

[1] The authors would like to thank Richard Haynes, Roger Sedjo, A. Myrick Freeman, III, Joel Smith, and others for their comments on earlier drafts of this chapter.

Callaway *et al.* (1994) attempted to overcome these barriers by piecing together the forecasts from several regional gap models. The gap models they used predict substantial forest decline from warming. Further, they assumed that all the ecological damage will occur in 1990 instead of over the next century despite the fact that warming will gradually occur over many decades. These assumptions lead to large estimated damages with a present value of $294 billion. In contrast, the US Forest Service (Joyce, 1995) estimated climate change impacts using four GCM scenarios and the Terrestrial Ecosystem Model (TEM; Melillo *et al.*, 1993). The TEM model generally predicts that climate change will increase forest productivity. The economic model used in that study, the Timber Assessment Market Model (TAMM; Adams and Haynes, 1980), predicts that these changes would increase supplies in every region in the United States and decrease domestic timber prices by 10–40 percent. Although a specific figure is not calculated, these results suggest that warming will produce benefits in timber markets. Perez-Garcia *et al.* (1997) also utilized the TEM model for predictions, but they estimated impacts globally. Their predictions suggest that US timber markets will benefit from the increased forest productivity predicted by TEM.

In this study, we carefully address both the theoretical and empirical modeling limitations of past studies. First, we build a dynamic forward looking economic model which endogenously adjusts both prices and harvest quantities over time in response to future conditions. Although this effort does not explicitly model random events in the future, it does capture expected changes in future conditions. Second, we develop a dynamic climate change scenario by combining several GCM doubling scenarios with predictions of linear change over time. This allows for a dynamic and spatially detailed prediction of climate change. Note that although the underlying climate change and ecological response is a linear path, the economic response to this path is not a linear projection. Third, we rely on a massive ecological effort, the Vegetation/Ecosystem Modeling and Analysis Project (VEMAP, 1995), which includes a broad set of biogeographic and biogeochemical cycle models.[2] This set of studies provides a broad range of results from an ecological perspective. The outcome is a research project that tightly coordinates the use of a sophisticated economic model and advanced natural science models.

This study integrates climate change scenarios, ecosystem model predictions, and a sophisticated dynamic economic model in order to estimate the economic response of timber markets in the United States to climate change. Figure 5.1 illustrates our

---

[2] The Vegetation/Ecosystem Modeling and Analysis Project was a joint international ecological modeling effort funded by the National Air and Space Administration, the US Forest Service, and the Electric Power Research Institute. Three biogeochemical models and three biogeographical models predicted ecological outcomes from three GCM doubling scenarios for the United States. An overview of the project and some results are discussed in VEMAP (1995).

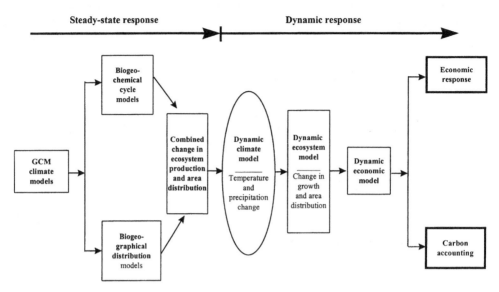

Steady-state response   **Dynamic response**

Figure 5.1 Integrated model of dynamic forest ecosystems with climate change.

modeling system, which flows from GCM model temperature and precipitation predictions, to the steady-state ecosystem response, to a fully integrated, dynamic economic response. The emphasis we place on paying careful attention to the ecological impacts and to how we implement those impacts in our economic model distinguishes this work.

The response of ecosystems to climate change has been characterized in terms of a steady-state change in ecosystem distribution (Neilson and Marks, 1994; Neilson *et al.*, 1992; Prentice *et al.*, 1992) and a steady-state change in ecosystem production (Melillo *et al.*, 1993). The likely time path of ecological change, however, will be driven by changes in temperature and precipitation (among other variables), which are projected to occur gradually as atmospheric carbon increases (IPCC, 1996a). It is therefore necessary to incorporate a dynamic ecosystem response into our integrated model.

What is true of ecological systems, certainly can be said of economic systems; people will adjust and adapt to change. We thus integrate the ecological changes with a dynamic model of timber markets that captures both the harvesting adjustment to evolving stock levels and the replanting adjustment to changing land productivity. This allows us to capture the impact of future changes in growing conditions because the model views timber as a capital asset. Adaptation through planting decisions is tied to future changes in land productivity in this way as well. If actors in the timber market act rationally, and are well-informed, timber markets will adjust efficiently to climate change.

Welfare changes in the timber market are defined as the change in net surplus, or consumer surplus minus costs. Net surplus in the baseline case scenario is compared with net surplus where ecosystems are affected by changes in temperature and precipitation. Because these changes are spread across time, we rely upon the change in the present value of net surplus as our measure of net economic value, or welfare. Our measure of net surplus could be further broken into consumer's and producer's surplus, although in this study we limit our analysis to the total impact. A change in consumer's surplus would represent the benefit or damage to timber consumers, while the producer's surplus would represent the benefit or damage to timberland owners.

This chapter is broken into four sections. The first two sections present the model, following Figure 5.1. In Section 5.1, we present the steady-state model and in Section 5.2, we present the dynamic model. Section 5.3 presents the results, while Section 5.4 contains our conclusions.

5.1   Steady-state response

In this analysis, we rely on a set of ecological models which were compared by the Vegetation/Ecosystem Modeling and Analysis Project, a multiple party effort to assess the sensitivity of terrestrial ecosystems and vegetative processes to climate change (VEMAP, 1995). The underlying strategy of this project was to combine a set of different ecological models to get a sense of the range of ecosystem predictions which are possible for a given climate scenario. The analysis in this paper builds on this important foundation.[3]

The steady-state response of ecosystems to climate change is driven by GCMs. These models produce predictions of steady-state temperature, precipitation, wind speed, and humidity, among other variables, at the earth's surface. Ecosystem models are linked to the predictions of the GCMs in order to determine changes in ecosystem distribution and productivity. The VEMAP project used the results of three GCMs provided by the United Kingdom Meteorological Office (UKMO; Wilson and Mitchell, 1987), Oregon State University (OSU; Schlesinger and Zhao, 1989), and the Geophysical Fluid Dynamic Laboratory (GFDL–R30; Manabe and Wetherald, 1987). Table 5.1 shows the mean temperature and precipitation changes predicted by these three models for the coterminous United States. The UKMO model produces

---

[3] There are many people to thank who helped us obtain and utilize this data. Tim Kittel, Nan Rosenbloom, and Hank Fisher at the University Corporation for Atmospheric Research, Boulder, CO, made the vegetation distribution data and the ecological production data available. Paul Barten and Rob Fraser at Yale allowed us to use GIS computing resources, and provided much needed information about the GRASS GIS system.

Table 5.1. *Change in temperature and precipitation under the three GCM models used in the analysis*

	Change in temperature (°C)	Change in precipitation (%)
OSU	3.01	3
GFDL–R30	4.33	23
UKMO	6.73	15

the highest temperature change, 6.73 °C, while the GFDL–R30 predicts the greatest increase in precipitation, 23 percent. The OSU model predicts the least precipitation and temperature change overall.

Two types of ecological models are used in this study, biogeographic distribution models (geographic models) and biogeochemical cycle models (ecosystem production models). The geographic models predict the distribution of ecosystem types based on climate conditions. The ecosystem production models predict steady-state production for a given ecosystem type and climatic condition. Together, they describe the steady-state response of ecosystems to climate change through the area redistribution of ecosystems and changes in ecosystem production. A summary of all the steady-state models used is presented in Table 5.2.

Biogeographical distribution models

The biogeographical distribution models include DOLY (Woodward *et al.*, 1995), BIOME2 (Prentice *et al.*, 1992), and MAPSS (Neilson *et al.*, 1992).[4] Figure 5.2 shows the distribution of ecosystems under current CO_2 climate conditions in the coterminous United States for the BIOME2 model and the UKMO climate model.

Each of these models is based on the mechanistic relationship among driving forces, such as water availability and temperature, and the ecosystem most suitable for that climate. As plants compete for inputs to growth (water, light, nutrients, etc.), these models draw boundaries between ecosystems, denoting areas where certain combinations, or types, of plants can grow. These combinations are called biomes or ecosystem types, descriptions which will be used interchangeably throughout the text.

---

[4] We thank VEMAP participants I. Colin Prentice of the University of Lund, Sweden, Ron Neilson of the USDA Forest Service, Ian Woodward of the University of Sheffield, and Tom Smith of the University of Virginia for letting us use their model results. Note that the model results actually used in the analysis do not come from the referenced journal articles. Instead, they come from updated models developed and used by the VEMAP process.

Table 5.2. *Steady-state models used*

Global circulation models (GCM)	UKMO
	OSU
	GFDL R30
Biogeographical distribution models	MAPSS
	BIOME2
	DOLY
Biogeochemical models (ecosystem production)	TEM
	BIOME-BGC
	Century

DOLY bases the predicted vegetation type on physiological constraints as well as climate gradients. Maximum leaf area index is calculated, based on long-term carbon and hydrogen budgets, so that the soil – water balance is maintained. This information is combined with other physiological considerations, such as net primary productivity (NPP), evapo-transpiration, and potential evapo-transpiration, and climatic variables to determined biome types. Biome types are developed from a predetermined classification system.

The BIOME2 model predicts global vegetation patterns from driving variables such as mean coldest-month temperature, annual accumulated temperature over 5 °C, a drought index, and available water capacity. A two-step procedure is used to determine the type of biome present on a plot of land. First, the model predicts the types of plants that potentially can exist at a given location. Second, the model determines the dominant set of vegetation that will maximize net primary production for the entire ecosystem. Combinations of plant types determine biomes.

The MAPSS model depends on two types of parameters, temperature thresholds that define the seasons, and precipitation thresholds that determine ecosystem types based on the temperatures. The basic ecological assumption of MAPSS is that vegetation stature, rooting depth, and leaf area mediate a steady-state balance between precipitation inputs and evapo-transpiration withdrawals from the system. Thus, the dominant characterization of biomes will depend on the availability of water. Above a certain threshold, forests will dominate and below this, other ecosystem types will be present. Temperature and precipitation then interact to determine the specific type of forested or nonforested ecosystem.

Ecosystem distribution is shown for the current climate in Figure 5.2. Although each model draws the particular boundaries somewhat differently, there is general

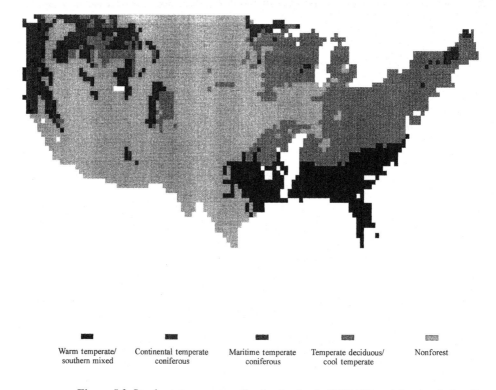

| Warm temperate/
southern mixed | Continental temperate
coniferous | Maritime temperate
coniferous | Temperate deciduous/
cool temperate | Nonforest |

Figure 5.2 Steady-state ecosystem distribution for the BIOME2 model under the baseline climate scenario. (Adapted from VEMAP members, 1995.)

agreement among them as to the distribution of ecosystem types. For example, they agree for the most part on the boundaries of forested ecosystems in the East and the Pacific Northwest. Differences, however, occur in determining the boundary between the Eastern deciduous forests and the plains, and in determining the distribution of conifer forests in the mountains of the West. Although larger differences occur in determining the exact distribution of nonforest types, we limit our consideration mainly to the forested ecosystems, and therefore shade all nonforested types the same.

Using BIOME2, Figure 5.3 shows the predicted ecosystem distribution under a doubled CO_2 scenario for the UKMO climate. The UKMO model is characterized by relatively large increases in both temperature and precipitation on average for the United States. Looking across the model results for this climate experiment, there are some similarities and some differences. For instance, there is broad agreement among them that the warm temperature/southern mixed forests (WTSMF) of the Southeastern United States will migrate northward and replace temperate deciduous

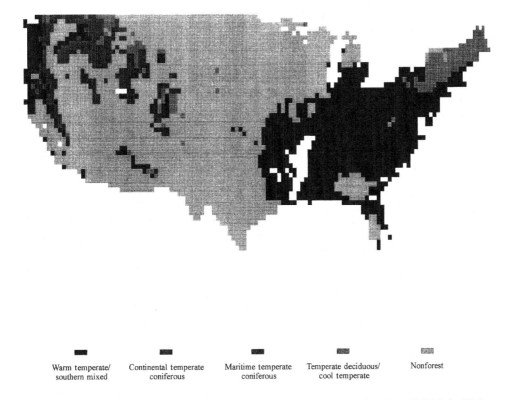

Warm temperate/   Continental temperate   Maritime temperate   Temperate deciduous/   Nonforest
southern mixed   coniferous   coniferous   cool temperate

Figure 5.3 Steady-state total ecosystem biomass for BIOME2 and the UKMO doubled CO_2 climate. (Adapted from VEMAP members, 1995.)

and cool temperate forests (TDCTF). The southern range of TDCTF include the oak–hickory types that currently dominate from the Appalachian Mountains northward through Ohio, Indiana, and Illinois and eastward to the Atlantic Ocean. Large differences occur in determining the relative size of future continental temperate coniferous forest (CTCF) in the western mountain region and the maritime temperate coniferous forests (MTCF) in the Pacific Northwest.

The BIOME2 model is characterized both by a fairly substantial expansion of forestland and by a fairly large area of dieback. The area of WTSMF expands into the Northeast and into the Pacific Northwest, at the expense of the current dominant forest type. Despite this prediction that WTSMF move into the Pacific Northwest, discussions with the BIOME2 modeling group suggest that the current MTCF would still be able to thrive in that region, so we decided to maintain that area in the MTCF type. The implications of this restriction are explored further in sensitivity analysis where we allow the adjustment of WTSMF into the Pacific Northwest. Although

101

WTSMF and MTCF types expand, the WTSMF contain pockets of dieback, and the TDCTF and CTCF experience widespread dieback. In the Northeast, however, areas of TDCTF dieback convert to more productive WTSMF, while in the West many areas of forest dieback convert to grassland.

DOLY suggests a fairly large expansion of the WTSMF type to the north, and a limited expansion of the MTCF, whereas CTCF die back throughout the West, with most of the dieback converting to nonforest types (grasslands and arid woodlands). Arid woodlands are dry forests with low production, such as the current pinyon–juniper type. Some areas of CTCF, however, do convert to more productive MTCF in western Montana and Idaho.

MAPSS appears to be the most sensitive to changes in soil moisture availability, as drying occurs in the Southeastern United States, and eastern forests convert to grassland and savanna. MAPSS does predict that the CTCF will increase somewhat, but this is tempered by the large decline in MTCF. The influence of changes in soil moisture can be seen by the relatively large increase in nonforest in the West and by the presumed expansion of CTCF into the Great Plains.

Biochemical cycle models

The biogeochemical models include TEM (Melillo *et al.*, 1993), BIOME-BGC (Running and Gower, 1991), and Century (Parton *et al.*, 1988).[5,6] The total biomass accumulation in grams carbon per meter^2 for ecosystems within the lower 48 states of the United States is calculated for each model under current conditions, doubled CO_2 with the UKMO climate change parameters, and doubled CO_2 with the UKMO climate change parameters and the MAPSS redistribution scenario. The change in ecosystem production predicted by the combined models shown in the final column provides input for our economic model.

Ecological production models attempt to measure the productivity of a particular ecosystem type for a given amount of time. Ecosystem productivity in any time period

---

[5] We thank VEMAP participants Jerry Melillo of the Marine Biology Laboratory at Woods Hole, MA, Steve Running of the University of Montana, and Denis Ojima at Colorado State University for allowing us to use their model results. The models used in this analysis have been updated from the version(s) referenced here (VEMAP 1995).

[6] Setting the initial conditions is a problem for all three ecosystem production models because the initial standing biomass is not known for all grid cells in the country. Although this problem should not affect the direction of the changes predicted by these models, it may affect the relative size of the changes. Some examples of how well the BIOME-BGC model predicts ecosystem production when these initial conditions can be set accurately are Running (1994), Running and Gower (1991), Hunt *et al.* (1991), Korol *et al.* (1991), and Running and Coughland (1988).

is measured with net primary productivity (NPP), which is the net amount of carbon fixed by terrestrial plants within that time period. NPP measures how much of the carbon fixed by plants in a given time period can be used in biomass accumulation. Biomass accumulation can be summed for each time period to obtain total biomass. Thus, NPP is a flow variable, while total biomass is a stock.

The TEM (Melillo *et al.*, 1993) is a process-based model describing how photosynthesis respiration, decomposition, and nutrient cycling, among other important ecosystem processes, interact to determine productivity. TEM uses spatial information on climate, elevation, soil, vegetation, and water availability to make monthly estimates of important carbon and nitrogen fluxes and pool sizes (McGuire *et al.*, 1992). From this information, TEM predicts net primary productivity and biomass accumulation in a particular ecosystem type for whatever CO_2 and climate conditions exist.

Whereas TEM was developed specifically to accommodate large spatially referenced data sets (Melillo *et al.*, 1993) over a variety of sites and ecosystem types, BIOME-BGC was developed originally as a site-specific tool to determine productivity of forested ecosystems (Running and Coughland, 1988; Running and Gower, 1991). The driving forces are canopy interception and evapo-transpiration, transpiration, photosynthesis, growth and maintenance respiration, carbon allocation, litterfall, decomposition, and nitrogen mineralization. Given meteorological inputs, including air temperature, radiation, precipitation, and humidity, as well as the ecosystem type, productivity is determined by mechanism equations relating the carbon, nitrogen, and other significant ecosystem processes.

The Century model (Parton *et al.*, 1988) is similar to BIOME-BGC in that it was developed initially for site-specific ecosystem production measurements, but it differs from BIOME-BGC in that it was utilized initially in assessing ecosystem production in grassland. This model simulates the dynamics of carbon, nitrogen, phosphorous, and sulfur in ecosystem soils in order to predict productivity. After determining the available pools of these nutrients in the soil and in live and dead above- and belowground plant parts, the maximum potential growth of plants is estimated as a function of annual precipitation. Century has been validated for large areas of grassland in the Midwest (Parton *et al.*, 1990).

There are important differences between the models even with current climate. Overall, Century predicts the greatest level of ecosystem biomass. This is particularly true in the Southeastern United States, where Century predicts fairly high levels of total biomass. BIOME-BGC predicts the least amount of total biomass, for both total land and forested land.

On the other hand, the models are very similar in how they treat the vegetation types. For example, a fairly sharp line is drawn where the eastern forestlands give way to the plains, and the models easily distinguish forestland and agricultural land north of the Mississippi valley in Illinois and southern Wisconsin. Further north, total biomass increases as forestlands dominate northern Wisconsin and Minnesota. In the West, the pattern of forestland to grassland is pervasive throughout the models.

The models predict productivity would be different under climatic change even if biomasses do not shift. These differences are striking, particularly in the East. The area of relative drying in the Southeast evident in the UKMO model is shown in all three models, although it is most apparent in Century and BIOME-BGC. Biomass increases in more northern areas of the East. There is relatively less difference in the Western United States.

This research examines how total biomass changes when ecosystems have redistributed according to the biogeography models. With area redistribution, larger changes in biomass occur for each model. For example, the area of southern pine dieback determined by the MAPSS model is easily identified for all three cases. The loss of forestland along the Pacific Northwestern coast and the shift in land types throughout the rest of the West can also be seen.

These models are limited in that they show conditions only in a steady state. Steady state refers to ecological systems that have had a chance to stabilize after CO_2 has doubled and temperature and precipitation have reached an equilibrium. While this provides a template for the future of ecosystems, it does not answer the important question of how these ecosystems will change through time. We discuss dynamic change in the next section.

5.2   A dynamic model

The dynamic model begins where the steady-state response left off. Two things must happen. The first is to develop the framework for a dynamic model of ecosystem change. The second is to develop an economic model that allows us to capture the dynamic response, as well as the various mechanisms for adaptation that timber markets are sure to undertake during climate change.

Dynamic climate model

The first step is to define the "dynamic" climatic response to a gradual doubling of CO_2 in the atmosphere. This means proposing a time path for temperature and precipitation change. The Intergovernmental Panel on Climate Change

104

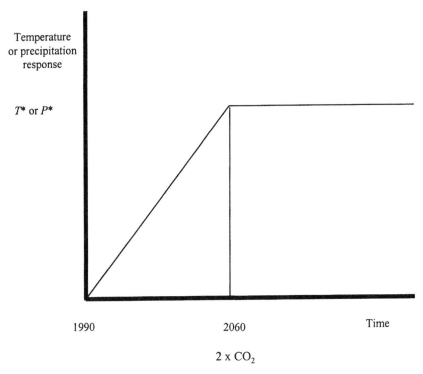

Figure 5.4 Dynamic response of temperature and precipitation.

(IPCC, 1996a) projects the uncontrolled carbon emissions will result in a linear increase in temperature from now until 2060, the time by which greenhouse gases will have doubled. This type of change is shown in Figure 5.4 for a 70-year adjustment to additional atmospheric carbon. We limit ourselves to linear paths such as this, and we explore the implications of the IPCC projection, assuming that temperature and precipitation will stabilize in 2060.

Dynamic model of ecosystem change

The second step is to develop a dynamic model of ecosystem change. This begins by relating ecological change directly to temperature and precipitation: we assume that ecosystems change proportionally to changes in temperature, precipitation, or both. As climate variables are assumed to increase linearly over a given number of years, the ecological adjustment, described by area and productivity change, occurs linearly over that time period as well. While the assumption of a proportional adjustment is a strong assumption, it follows directly from several ecological studies on tran-

105

sient ecosystem adjustment during climate change (King and Neilson, 1992; Neilson, 1993; Smith and Shugart, 1993).

Dynamic area change

Introducing a dynamic shift into the redistribution of timber types is difficult because transient models of biogeographic change generally have not been developed (several ecological modeling groups are presently working on this problem, however). We therefore develop two models of biogeographical distribution that are intended to capture a range of possible transient behaviors.

The first model of biogeographical redistribution is dieback. Some modelers have suggested that existing trees will die as the conditions under which they grow change (Neilson *et al.*, 1992; Shugart *et al.*, 1986; Solomon, 1986). These modelers imply that the changing climate conditions will stress the plants, and cause the standing stock to die prematurely due to insect infestations, fires, or other natural disturbances that may otherwise have no impact on trees. Once climate change has caused the stock to die back on a given parcel of land, species that are adapted to the new climate there must migrate into the region.

The second model is called limited regeneration. For this model, we assume that the standing stock is not itself affected by a changing climate. Instead, we assume that climate change has its biggest impact during regeneration. When a parcel of land is undergoing climatic change, the existing trees are able to continue growing, but they are unable to successfully regenerate. If left to natural forces, as new species migrate into the area, they will begin to compete with the old stock, and slowly replace it. This scenario is generally consistent with silvicultural evidence, which suggests that many tree species are able to live under widely differing climates. Red oaks, for example, range from Maine all the way to Louisiana. Since many tree species are found to live in such a broad spectrum of climates now, there is little reason to believe that the standing trees will die back as a result of climate change.

Both dynamic models of biogeographical change must incorporate some assumptions about tree migration. Tree migration involves both seed dispersal rates and land management activities. While natural rates of migration are relatively slow, the economic model assumes that humans react quickly by replanting the most healthy and profitable timber types within a region. We account for both natural and human processes by assuming that there are two types of land: high intensity and low intensity managed land. High intensity managed lands are forest plantations, and are replanted with the correct species for the new climate immediately after either dieback or harvest. Low intensity managed lands are held for multiple management reasons, and will have lags before they are restocked fully.

106

While ecosystem distribution and ecological production models do not consider competition among tree species directly, by allowing for high and low intensity management, we account for the impacts of competition during climate change. On high intensity managed land, competition is assumed away, because managers will replant the right species and will suppress the growth of competing trees or shrubs. On low intensity managed land, lags to regenerate are included so that competition may play a role in limiting the ability of species to compete effectively when they are invading a new area.

Dynamic ecosystem productivity change

Like the redistribution of timber types, changes in yield will occur proportionally to temperature changes. The main issue here is how to relate the production of biomass, or "biomass accumulation", available from the ecosystem production models to a change in yield functions. Biomass accumulation measures the total vegetation carbon stored in ecosystems at a given time. For those familiar with accounting, it can be linked to a balance sheet, where biomass accumulation is an asset.

A yield function, while it relates the total amount of biomass in terms of timber yield, actually contains much more information about the growth of trees than does the amount of biomass accumulation. It relates information not only about current yield, but also about the growth rate of those trees at different times during their life cycle. Since biomass accumulation from the ecosystem production models is a steady-state concept, it cannot be used to infer information about the growth rates of those trees. We define V_i as the yield of species i and $\dot{V}_i$ as the growth rate of species i.

We assume that changes in growth are proportional to changes in biomass accumulation at every age. Thus, if biomass accumulation is 20 percent greater after climate change, growth increases 20 percent. At any moment, the increase in growth for trees under climate change, $\dot{V}_{i,cl}$, is equal to the product of the original growth rate times the proportional change, α_i:

$$\dot{V}_{i,cl}(a_i) = \alpha_i \dot{V}_i(a_i),$$ (5.1)

where a_i is the age of the trees, $\dot{V}_i(a_i)$ is the base line yearly growth and $\dot{V}_{i,cl}(a_i)$ is the climate enhanced yearly growth. In a steady-state situation, growth is increased or decreased by the proportion α_i.

The proportion, α_i, adjusts dynamically with temperature or precipitation change. As these climate variables move linearly over time, α_i will also change linearly over time (t):

$$\alpha_i(t) = 1 + \gamma_i t,$$ (5.2)

107

where γ_i is the yearly increment in biomass accumulation. The total amount of change is limited by the number of years over which this shift can occur. Once a steady-state climate and ecosystem is attained, $\alpha_i(t)$ is held constant.

The stock or size of trees is consequently a function of historic growth rates. If there is an instantaneous change in biomass accumulation at time 0, the standing stock of trees at time t after time 0 is equal to the standing stock at time 0 plus all the growth that occurs in each of the n years from time 0 to time t. Mathematically, this can be expressed as

$$V_{i,cl}\big(a_i(t),t\big) = \hat{V}_i\big(a_i(0)\big) + \int_0^t \big\{\alpha_i(n)\,\dot{\hat{V}}_i\big(a_i(n)\big)\big\}\,dn, \tag{5.3}$$

where $\hat{V}_i\big(a_i(t)\big)$ is the base yield function of timber $a_i(t)$ years old at time t, $V_{i,cl}\big(a_i(t),t\big)$ is the climate adjusted yield function at time t. Climate begins changing at time 0, and affects all future growth. This adjustment accounts for the age of the trees when the climate shock begins, as well as the amount of time the trees have had to grow at the new, changed rates before they are harvested.

Dynamic economic model

Utilizing a dynamic economic model allows us to account for the future impacts of climate change on current harvesting and replanting decisions. Assuming that timber consumers and producers rationally predict future changes in ecological production and economic conditions (such as income and population), they will adapt to climate change. Given that climate change is predicted to occur gradually over time, there is little reason to expect that producers will be caught off guard. We describe the economic model first, and then discuss adaptation more thoroughly.

The economic model

Although we discuss many of the most important aspects of the timber market model that is used in this valuation effort, the reader is referred to Sohngen (1996) for further explanation of both the model and the parameter values chosen. All parameter values used in this model were either estimated from data or taken from published empirical work. The timber market model is a dynamic, partial equilibrium model that maximizes the net present value of timber market welfare in the United States over an infinite horizon. We assume that timber markets are competitive throughout.

In reality, timber markets in the United States are linked with the worldwide market for timber through imports and exports. While modeling world timber markets is

108

beyond the scope of this project, we assume that global prices have the same behavior as predicted in our model. Failing to incorporate global trade flows explicitly, therefore, would affect our results only if climate change had a different impact in other regions of the world. If other regions became more (less) productive, for example, production in those regions would increase (decrease) and relative prices would decrease (increase). The balance of trade consequently would change. Assuming that climate change affects other regions similarly to the United States, production in other regions of the world would also change in a similar manner to the United States. Research under way at the moment (Sohngen et al., 1996) will address how biogeographical and biogeochemical changes will affect timber production in all regions of the globe.

Our description of the economic model begins with an inverse demand function, D, for timber:

$$P_s(t) = D(Q(\cdot), t), \qquad (5.4)$$

where P_s is stumpage price, which is the price at the mill net of harvesting and delivery costs, $Q(\cdot) = Q\left\{ \sum_i \theta_i H_i(t) V_i(a_i(t)) \right\}$ and represents timber output in each period, θ_i is a quality adjustment factor, $H_i(t)$ is the number of acres of timber of type i harvested, and $V_i(a_i(t))$ is the yield function for timber of type i. The quantity harvested in each type is quality adjusted in this national market. Quality adjustments capture the large differences in characteristic output of regional species. For example, consumers do not purchase a board foot of timber per se, but instead they purchase strength, or fiber length, or some other desired set of characteristics. Historical price differences between regional species are assumed to reflect the different level and value of the characteristic outputs these species provide. Using this historical price information, we develop a set of factors that adjust actual output in each region to a standard set of characteristic output. This concept is discussed further in Sohngen (1996).

Given the quality adjusted supply of timber in any period, the inverse demand function predicts the marginal stumpage value or price for that timber. Time is an argument in the demand function because we expect that many of the determinants of demand such as population and income will grow over time. In each period, prices are assumed to adjust so that the quantity demanded equals the quantity supplied.

Net surplus in a timber market is measured as total consumer surplus minus all costs, or alternatively, consumer's plus producer's surplus. This is the integral of the area under the demand function minus costs up to the quantity consumed each period, Q^*:

$$S(Q) = \int_0^{Q^*} D\big(Q(H_i)\big)\, dQ - C(\cdot), \qquad (5.5)$$

where $C(\cdot)$ is the set of costs associated with regenerating and maintaining timberland. We are considering a stumpage market, so that the price for stumpage reflects the costs associated with harvesting and transporting logs from a logging site to a mill. This is analogous to assuming a constant marginal harvesting cost per unit of timber harvested. We do not suspect that this assumption will impact our welfare measures significantly, given that timber markets represent only a small fraction of national capital and labor markets. Note we are using a Marshallian measure of welfare.[7]

In addition to the costs associated with harvesting, the replanting and capital costs must also be considered in the social planner's problem. They are real costs associated with keeping land in timber. Costs for replanting are allowed to vary by region because there are such large differences between productivity and replanting success from area to area. Land rent describes the capital cost of maintaining land in timber. It is included explicitly because landowners in reality have many choices over what they can do with their land. Because competition for land exists between forestry and other market sectors, and because land use is predicted to change drastically during climate change, land rent is an important component that must be incorporated into our measure of welfare. If land is to remain in timber, the net present value of future harvests of that land must be enough to pay the rent that could otherwise be earned on the land.

As well known in economies, a market will yield efficient results as long as agents are competitive and all goods and services are priced. A competitive timber market will thus make the same decisions as a benevolent social planner. That is, the market will maximize the present value of net surplus:

$$\underset{H_i(\cdot,t)}{\text{Max}} \int_t^\infty e^{-rt} \left\{ \int_0^{Q^*} D\big(Q(H_i(t)),t\big) \, dQ - \sum_i \beta_i G_i(t) - \sum_i R_i(t) X_i(t) \right\} dt, \qquad (5.6)$$

where β_i is the cost of replanting, G_i is the number of acres to be replanted, R_i is the land rent, and X_i is the number of acres of timber type i in the forest. This expression defines a dynamic optimization problem which determines a time varying harvest schedule that maximizes the net present value of net surplus in the timber market. Note, this equation maximizes net surplus and not net revenue so that no monopoly behavior is introduced.

There are several constraints to this problem. A general stock constraint must be placed on the system so that

$$\dot{X}_i(t) = -H_i(t) + G_i(t). \qquad (5.7)$$

---

[7] As developed in detail by Willig, (1976), Marshallian consumer surplus measures are reasonably accurate measures of net willingness to pay under the circumstances explored in this model.

Additionally, there are other constraints that imply

$$H_i \geq 0; \; G_i \geq 0; \; P_s \; \text{all} \geq 0, \tag{5.8}$$

and starting conditions for each variable and parameter.

Under these conditions, plus the assumption that marginal costs are constant across every acre harvested and a transversality condition, this problem can be solved for the following set of first order conditions:

$$\dot{P}_s V_i\big(a_i(t)\big) + P_s(t)\dot{V}i = rP_s(t)V_i\big(a_i(t)\big) + R_i(t), \tag{5.9}$$

where $\dot{P}_s(t)$ is the change in stumpage price. It is assumed to be the same for each timber type, although quantities are quality adjusted. Equation (5.9) must hold in each time period for each timber type, i. The left-hand side of Equation (5.9) is the marginal benefit of waiting to harvest, while the right-hand side is the marginal cost of waiting. Marginal benefits are defined by the value added by waiting an extra moment to harvest. This is the price growth multiplied by the volume, the first term, plus volume growth multiplied by price, the second term. Marginal costs are the sum of the net opportunity cost of not harvesting that acre, the first term on the right-hand side, and the land rent, the second term.

This basic condition has been derived by others (Brazee and Mendelsohn, 1990; Hardie et al., 1984). Over time and with a given price schedule defined by a competitive market, timberland owners must harvest along the path defined by this equation. If prices were constant, this equation would resolve to the more familiar, steady-state Faustmann formula. A good discussion of the Faustmann formula, and the debate over optimal forest rotations, is found in Samuelson (1976). We assume that demand for timber products continues to grow indefinitely, and land supply is fairly inelastic. Under these assumptions, prices will continue to grow indefinitely, the Equation (5.9) represents timber harvesting behavior.

The value of the timber market is captured by the net present value of consumer's plus producer's surplus, as shown in Equation (5.6). In any particular year, consumer surplus is the area under the demand curve and above the price line, while producer's surplus is the total revenue from harvesting timber less regeneration and land rental costs. The value of the welfare benefits or damages that arise from climate change, then, is the difference between Equation (5.6) in the baseline case and Equation (5.6) in the climate change case.

For the remainder of this section and the next section, we discuss several adaptations to the theoretical model above that are incorporated into our empirical analysis. We discuss these in order to give the reader a sense of the complexities involved with modeling timber markets. We do not attempt to show formally how they are incorporated, nor do we show the relevant parameter values and elasticities that are used. For parameter values and for further explanation of the empirical model itself, the reader is referred to Sohngen (1996).

In our model, not all land is assumed to be managed strictly for timber purposes (Powell *et al.*, 1993). Many authors deal with land management in terms of ownership types, classifying private land either as industrial or as nonindustrial (see Newman and Wear, 1993, for an example). This classification, however, does not appear to capture the full range of possible management schemes because current inventory levels suggest that diverse management strategies exist on both industrial land and nonindustrial land. For example, where industrial landowners may be expected to harvest timber around year 30 in the South, they have in cases held timber longer. Likewise, nonindustrial owners appear to harvest some timber in Faustmann rotations, and appear to hold other timber for much longer time periods.

We thus allow for two classifications of land, high intensity and low intensity managed land, either of which may contain industrial and nonindustrial private forestland. Assuming that timberland which exceeds the Faustmann rotation age by more than 10 years is not intensively managed, we distinguish our two management types by age. In general, stands below the Faustmann rotation age plus 10 years are included in high intensity land classes, and stands above the Faustmann rotation age plus 10 years are included in low intensity land classes.

For each timber type, high intensity land will be harvested according to Equation (5.9). The remaining land will be harvested according to a price responsive supply function. Over time, some low intensity land will convert to high intensity status due to changes in management intensity. The proportion of land in each intensity class will vary with the stumpage price over time. In addition, we allow yield on high intensity land to vary with price. As price increases, managers will adopt practices that increase the effective yield from timberland, such as planting genetically improved stocks, increasing stocking density, or thinning control.

By accounting for the type of land as well as management intensity, this economic model presents a more realistic picture of timber markets than would a simple Faustmann model. We have attempted to initialize all parameters based on current conditions in the timber market. Our base year is 1990, allowing us to use the most recent forest inventory data (Powell *et al.*, 1993), as well as the most recent economic data.[8]

---

[8] We would like to thank Richard Haynes and John Mills at the USDA, Forest Service, Pacific Northwest Research Laboratory in Portland, OR for providing up-to-date information on timber harvest, price, and inventory levels.

Adaptation

In this section, we discuss how timber markets are likely to adapt to climate change. As discussed above, we assume throughout that timber markets evolve based on rational expectations. Further assumptions about long-term shifts in land owner-ship, as well as differences between dieback and regeneration are discussed below.

Two broad land-related issues complicate the discussion: whether or not land is owned by government or private individuals and companies; and the relative quanti-tites of agricultural, forest, residential, or urban land. In order to deal with the first issue, we assume that the timber supply from government-owned land is constant over time, which suggests it is not price responsive. While the structure of many govern-ment timber sales allows firms to adjust timber harvests to short run price fluctuations, the long run supply of timber from government lands, particularly federal lands, is essentially fixed. Fixing long-term government supply is consistent with the empirical timbermodel because the model captures only long-term timber market adjustments. It is not intended to model short-term adjustments associated with business cycle activities.

As for land use, we consider mainly agriculture because it is a large, competing land use. We deal with it in two ways. First, areas with high quality farmland, such as the states of the Great Plains, are masked from the analysis because we do not expect them to move away from agriculture during climate change. Much of this land is classified ecologically as grassland, and even if farming were abandoned, it would not turn into forest (Figure 5.2).

Other areas with a lower proportion of farmland, such as the Southeast, Northeast, and Rocky Mountains have not been excluded because they potentially could contrib-ute to timber markets, depending on economic and climatic conditions. These types of land, often referred to as nonindustrial private forest lands (NIPF), account for nearly 59 percent of all timberland acres in the United States. This group represents an important component of yearly timber supply, so we maintain most of these areas in our analysis (Powell *et al.*, 1993).

Shifting from nonforestland types to forestland types is expected to be an issue mainly in the West, where large land areas may be shifted between agricultural uses and timber. The rate at which this can occur is tied to the current proportion of timberland area, however. Land that becomes suited for timber because of climate change is assumed to become timberland in direct proportion to the original ratio of timberland to nontimberland in the region. This proportion is given in Table 5.3 as the proportion the actual area is of the predicted area.

In the East, most land is suited to forestry initially, even though it may be used by agriculture. We assume that the proportion of forestland to agricultural land will

Table 5.3. *Baseline area, predicted area, and the proportion the baseline area is of the predicted area (actual/predicted) for softwood forested ecosystem types in the DOLY and MAPSS biogeographical ecosystem models and the middle climate scenario*

Timber type	Baseline area (Sq. km)	DOLY		MAPSS	
		Predicted area (Sq. km)	Actual/ predicted	Predicted area (Sq. km)	Actual/ predicted
Southern pine	224 163	887 038	0.25	1 004 812	0.22
PNWW conifer	33 835	182 946	0.18	180 711	0.19
Northern softwood	103 360	664 809	0.16	603 249	0.17
Western conifer	107 918	344 296	0.31	407 248	0.26

Note:
PNWW is the portion of the Pacific Northwest west of the Cascade Mountain range.

remain the same. Climate change will cause some shifting of land use, however, by changing ecosystem types.

Finally, since other types of land use are ultimately associated with the relative value of the land, we relate future shifts in timberland area to timber prices. While timberland area in general has decreased over the past 40 years (Haynes, 1990), much of this may be explained by increased population pressures and the increased relative value of land for nontimber purposes. As population growth tails off in the future, however, and as agriculture continues to become more productive, it is entirely possible that more land will flow into forestry. This is particularly true if timber prices are increasing. Thus, if prices increase, we allow more land to flow into timber; if prices decrease, we allow some additional acres to lie fallow after harvest. We account for regional differences by using separate elasticities for each region which are shown in Sohngen (1996).

We turn now to considering adaptation within climate change. First, we have assumed that die back occurs when climatic conditions have changed enough so that they stress timber to the point where pathogens, fire, or some other disturbance kills the existing trees. We assume that timberland owners will adapt by salvage harvesting the remaining stock. One study on salvage from southern pine beetle damage indicated that the value of material salvaged ranged between 21 percent and 75 percent of the material that was affected (de Steiguer *et al.*, 1987). We assume that landowners are able to salvage, on average, 75 percent of the value of affected high intensity stocks, and we assume that they salvage 50 percent of the affected low intensity stocks. We consider both lower and higher values in the sensitivity analysis.

114

The second way in which timberland managers will adapt to dieback is by harvesting lands at a faster rate if they are more susceptible to dieback. The effect of dieback on harvest decisions is identical to increasing the interest rate by the rate of dieback for the time period in which dieback occurs (Reed, 1984). Harvesting from the original stock will continue to occur optimally during climatic change in this scenario. Timberland managers will adjust harvest on forests which have no dieback only in response to changes in growth rates.

In both the dieback and the regeneration scenarios, when an acre shifts from one ecosystem type to another, it may enter the new timber type either as high intensity or low intensity land. These shifting acres enter in exactly the same proportion as the acres that were just harvested, but that did not shift. For the acres that enter into the high intensity managed land, they are replanted instantly. For the acres that enter low intensity managed land they regenerate only after a lag.

Replanting behavior, like harvest behavior, will vary based on climatic influences. Because timberland owners are able to adapt to the slowly changing climate, they will replant the correct species on their high intensity type of land. Only acres that shift into low intensity status have a lag to regenerate. We keep track of these separate stocks of land within the economic model itself.

5.3   Results

Thus far, we have discussed the basic modeling framework as described in Figure 5.1. This section follows the same general flow. We begin by describing the temperature and precipitation scenarios that will be analyzed. We then go through each step in the model process for the two specific ecological scenarios chosen for the middle climate scenario. This provides a detailed account of how we move from the steady-state response to a dynamic ecosystem response to a dynamic economic response. Finally, we present the full range of economic responses for all nine combinations of ecosystem models.

Temperature and precipitation scenarios

In order to make this study consistent with the other impact studies in this book, we analyze a specific set of temperature and precipitation scenarios. The different combinations that we consider are shown in Table 5.4. Temperature changes range from 1.5 °C average increases to 5.0 °C average increases over the entire United States, while precipitation changes are limited to 7 percent and 15 percent average increases. Unfortunately, there are no GCM model runs, and consequently no

Table 5.4. *Matrix of temperature and precipitation scenarios used in the dynamic analysis*

Precipitation change (%)	Temperature change		
	+1.5	+2.5	+5.0
+7	GFDL (0.325)	UKMO (0.5)	OSU
+15	N/A	GFDL (0.65)	UKMO

ecological results that correspond exactly to those changes (Table 5.1). In the VEMAP process, only three GCMs were used to calculate the full set of ecosystem model results.

In order to obtain information on temperature and precipitation changes that are not explicitly described by the GCM models, we interpolated the results of the ecological models. For example, no GCM provides an experiment for a 2.5 °C temperature change and 7 percent precipitation increase. The UKMO model, however, has an average temperature increase of 6.73 °C and a 15 percent precipitation increase. Scaling the ecological results by one-half, we can approximate a scenario with a 2.5 °C average temperature change and 7 percent average precipitation increase. Scaling factors, where used, are shown in parenthesis in Table 5.4. Note that actual temperature and precipitation changes will vary across the grid cells, depending upon the GCM being used.

Model results for two ecological scenarios and the middle climate scenario

Steady-state ecosystem response

Here, we show how the results of the steady-state ecosystem models can be adapted for use in the dynamic economic model. We utilize the combined results of biogeographical distribution models and ecosystem production models to define the ecosystem change. Two particular cases are presented in detail here because their ecological and economic results contrast. One case is described by combining the DOLY and TEM models (DOLY–TEM), while the other is found with MAPSS and BIOME-BGC models (MAPPS–BIOME-BGC).

We begin with ecosystem area redistribution. Changes in the geographical distribution of ecosystems is shown in Figure 5.2 for the three biogeographical redistribution models under the UKMO model. With a geographic information system (GIS), we can translate these maps into the amount of land in each timber type initially and the amount of land in each type after climate change. The total quantity of land that shifts can be calculated as the difference between these two.

Tables 5.5 and 5.6 show the area that shifts with the DOLY and MAPSS models under the middle climate scenario. The initial area of land potentially in each type is given in the first column. The types used in the economic model are shown above the types described by the ecological models. The potential area includes all land that is suited to the particular ecosystem type. Because humans have managed land in the United States, the actual area will be different from this. Recall that Table 5.3 shows the actual and predicted area for each timber type and each model.

Looking across Tables 5.5 and 5.6, we can see where the original land will shift. The last two columns show how much of the original land remains in forest types and how much shifts over to "other" types, or to nonforest. Reading down the table shows us the total amount in a particular type after climate change. It also tells us from what type the land was derived. Net gains (losses) are shown in the second to last row, and the total area after climate change is given in the final row.

This information is then summarized by two ecosystem parameters that can be used in the dynamic economic model. First, we need to know the relative size of each forest type before and after climate change. This is shown in Table 5.7. If additional land shifts the forest type and not much shifts out, then the type area will increase, and this value will be greater than 1. If some land converts over to other nonforested ecosystem types and not much land shifts into the type, then this value will be less than 1. The DOLY model predicts the largest increase in area for southern pine and Western Pacific Northwest (PNWW) conifers, while the MAPSS model predicts the largest increase in area for the northern softwoods and western conifers. Changes in southern pines and the PNWW conifers turn out to be especially significant given the economic importance of these two regions.

Because some ecosystem changes entail movement from one forest type to another, some of the costs of climate change are related to change, not just the eventual size of forests. The actual proportion of area that shifts is shown also in Table 5.7 as the "shift proportion". This proportion is assumed to die in the dieback scenarios and to shift after harvest in the regeneration scenarios. DOLY predicts a lower level of shifting in southern pines and PNWW conifers, whereas MAPSS predicts a lower amount of shifting for the western conifers.

Table 5.7 also presents the steady-state change in yield expected to occur with climate change and the given ecosystem model combinations. Recall that this change is proportional to the change in per-acre, total ecosystem biomass predicted by those models. Distinct differences can be seen. TEM predicts increases for each type, while BIOME-BGC predicts all decreases. The drying that occurs in the South and North

117

Table 5.5. *Steady–state area change in forested ecosystems with the DOLY geographic ecosystem model in the middle climate scenario*

| | Economic type | | | | Final area | | | | | |
| | | | Northern softwoods | | | Southern pine | | Western conifer | PNWW conifer | | Total |
Ecosystem type	Initial area (Sq. km)	BCF	CTF	TDF	WT SMF	TEF	CTCF	MTCF	Nonforest	forest
Northern softwoods										
Boreal conifer	226 640	143 503	13 066	4 266	0	0	23 348	1 072	41 368	185 254
Cool temperate	438 169	0	230 150	84 358	76 825	0	0	2 143	44 694	393 475
Temperate deciduous	698 039	0	0	349 020	345 627	0	0	0	3 393	694 647
Southern pine										
Warm temp./subtropical mixed	887 038	0	0	0	871 324	0	0	0	15 714	871 324
Tropical evergreen mixed	0	0	0	0	0	0	0	0	0	0
Western Pine										
Continental temperate conifer	344 296	1 164	10 938	13 989	0	0	192 962	37 199	90 046	254 251
PNWW conifer										
Maritime temperate conifer	182 946	0	0	3 287	0	0	0	178 607	1 052	181 894
Nonforest	2 401 439	7 206	1 042	2 243	26 800	0	21 069	7 668	2 335 412	66 028
Net area gain (loss)		(74 768)	(182 974)	(240 878)	433 538	0	(108 917)	43 742	130 257	(130 257)
Final steady-state area		151 873	255 195	457 195	1 320 576	0	235 379	226 688	2 531 696	2 646 872

Table 5.6. *Steady-state area change in forested ecosystems with the MAPSS geographic ecosystem model in the middle climate scenario*

Economic type / Ecosystem type	Initial area (Sq. km)	Northern softwoods			Southern pine			Western conifer	PNWW conifer	Nonforest	Total forest
		BCF	CTF	TDF	WT	SMF	TEF	CTCF	MTCF		
Northern softwoods											
Boreal conifer	143 942	71 962	14 314	0	0	0	0	27 177	4 516	29 956	117 969
Cool temperate	459 325	0	232 033	92 395	54 294	0	0	4 169	3 107	83 328	375 997
Temperate deciduous	723 043	0	0	363 722	339 902	0	0	0	0	19 419	703 624
Southern softwoods											
Warm temp./subtropical mixed	1 004 812	0	0	0	846 973	0	3 973	0	0	153 867	850 945
Tropical evergreen mixed	0	0	0	0	0	0	0	0	0	0	0
Western softwoods											
Continental temperate conifer	407 208	0	20 897	0	3 448	0	0	353 557	1 081	28 265	378 983
PNWW conifer											
Maritime temperate conifer	180 711	0	26 646	0	1 033	0	0	1 223	130 173	21 638	159 074
Nonforest	2 453 241	0	14 795	2 200	39 354	0	0	64 437	1 231	2 331 225	122 017
Net area gain (loss)		(71 962)	(150 640)	(247 726)	280 191		3 973	43 314	(40 640)	210 456	(210 456)
Final steady-state area		71 962	308 685	448 317	1 285 003		3 973	450 562	140 107	2 663 697	2 708 608

Table 5.7. *Predicted change in softwood forest types*

| Ecosystem change | Model combination | | | | | | | |
| | DOLY–TEM | | | | MAPSS–BIOME-BGC | | | |
	South	PNWW	North	West	South	PNWW	North	West
Relative size	1.49	1.24	0.61	0.69	1.27	0.78	0.63	1.08
Shift proportion	0.02	0.02	0.42	0.45	0.16	0.28	0.47	0.13
Yield change, gain (loss)	0.07	0.17	0.15	0.25	(0.08)	(0.02)	(0.18)	(0.06)
Economic change	(Welfare gains, net present value over 150 years)							
Dieback	$16.2 billion				$2.2 billion			
Regeneration case	$17.3 billion				$4.9 billion			

Note:
South is southern pine, PNWW is the western portion of the Pacific Northwest, North is northern softwoods, West is western conifer.

under the UKMO model is evident through the relatively large decreases in yield in BIOME-BGC.

Dynamic ecological and economic model results

The steady-state shifts described in Table 5.7 above are assumed to occur proportionally to the 2.5 °C temperature change and the + 7 percent precipitation change. Over 70 years, the full adjustment to climate change occurs; after 2060, CO_2, temperature, and precipitation stabilize. Timber markets continue to adjust dynamically, however, to the change that has occurred.

The dynamic economic model is a free end-point problem in that prices will continue to change in response to a continuously increasing demand function. A 5 percent rate of discounting is used throughout. Demand is driven by population and GNP growth, both of which are assumed to increase throughout the simulation and beyond. In addition to demand factors, large amounts of land shift from ecosystem type to type, as well as from low to high management intensity. In climate change, it will take many years before a full adjustment actually occurs.

Two issues arise when the long-term nature of this problem is considered, setting terminal conditions and calculating welfare effects. While demand will continue to rise indefinitely, we assume that the amount of land in forestry ultimately becomes fixed. After an initial adjustment period, whether the adjustment is caused only by increasing demand or whether it is caused by climate change, prices will increase at the rate of

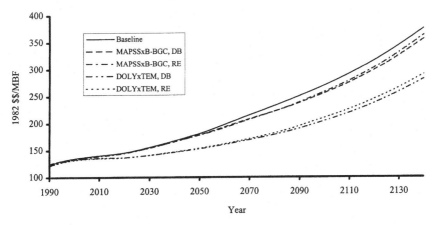

Figure 5.5 Price paths for alternative climate-ecological outcomes.
(Middle Climate Scenario. "DB" is Dieback and "RE" is Regeneration)

growth of demand (Berck, 1981). One aspect of the terminal conditions involves ensuring that prices ultimately rise at the rate of demand increase, with stable land supply. Another aspect is a transversatility condition which ensures that we remain on this path far into the future. Experience with the model suggests that after 150 years, continued stock adjustments have very little impact on the value of the objective function in earlier periods.

It is precisely these earlier periods which are important in measuring welfare, because we used the discounted difference between the stream of net surplus in the climate case and net surplus in the baseline case. Discounting ensures that future adjustments have only small impacts on earlier welfare measures. For calculating welfare, we need only calculate the first 150 years of welfare effects, because changes farther in the future have little impact on the measure.

Turning to the economic results, the price paths for each model combination under the middle climate scenario (a 2.5 °C and 7 percent precipitation increase), as well as the base case, are shown in Figure 5.5. The DOLY–TEM combination predicts a significant decrease in the trend of prices relative to the baseline. Large amounts of land convert to both southern pine and PNWW conifers, and biomass increases for both types. Interestingly, even under the MAPSS–BIOME-BGC combination, where dieback ranges from 13 percent of the land to 47 percent, and where ecosystem production declines from 2 percent to 18 percent, prices are lower. Despite an 8 percent decrease in biomass in the South, more timber is available for markets because the size of the southern pine forest increases 27 percent.

Under the middle climate scenario, the DOLY–TEM combination implies a $16.2

billion or $17.3 billion increase in the net present value of welfare over the first 150 years, depending on whether dieback or regeneration occurs. Welfare in MAPSS–BIOME-BGC likewise increases, $2.2 billion or $4.9 billion for dieback and regeneration respectively. One interesting result is that there appears to be fairly little difference between the dieback and regeneration scenarios. This runs counter-intuitive to the generally held belief that dieback means that timber markets will be irreparably damaged. There are two reasons for this. First, salvaged timber makes up some of the difference. The other reason has to do with the complex dynamic adjustment of harvests from regions where there is relatively less dieback to regions where there is more dieback during the early stages of climate change. This limits the overall damage caused by timber that is lost due to dieback.

Another interesting point is that the area changes outweigh adjustments in productivity. As shown in Table 5.7, the BIOME-BGC model predicts fairly large decreases in productivity in all regions, but the MAPSS–BIOME-BGC model predicts lower prices and increased market welfare. This results mainly from the large increase in area of southern pines as northern types die back or are replaced. Long rotation, low-valued, northern types are replaced with shorter rotation, higher-valued southern types.

Aggregate results of empirical model

Table 5.8 presents the mean net present value of net surplus from the nine model combinations, as well as the range of potential results. Increases in total surplus range from $1.13 billion in the 5 °C and +15 percent precipitation climate change case with dieback to $35.26 billion in the 5 °C and +7 percent precipitation climate change case with regeneration. Averaging the result is helpful for understanding the general direction of the implied change in welfare, but it masks potentially important underlying differences in the ecological response.

For example, the 2.5 °C and 7 percent precipitation increase and the 5 °C and 15 percent precipitation increase scenarios are weighted heavily by the low benefits predicted by the MAPSS–BIOME-BGC ecological change scenarios. These ecological scenarios predict a high level of dieback and low level of forest expansion in the South and Pacific Northwest. This result is driven by how MAPSS–BIOME-BGC reacts to the UKMO-GCM model (recall that the UKMO-GCM model provided the basis for these two climate change scenarios). Given the large range of results within the middle climate scenario, most of the variability in climate change impacts in timber markets is explained by differences between the ecological models themselves.

The values in Table 5.8 describe the present value of the benefits and damages which occur over a 150-year dynamic period. In order to understand what happens in

122

Table 5.8. *Change in net present value across ecological model combinations for each climate change scenario (% change in parenthesis)*

Precipitation change (%)		Temperature change (°C)		
		1.5	2.5	5
7a: Dieback				
7	Average ($)	12.11 (4%)	11.28 (4%)	19.97 (6%)
	Range ($)	*7.41–15.91*	*2.22–16.18*	*11.86–31.77*
15	Average ($)	N/A	17.16 (5%)	15.96 (5%)
	Range ($)	N/A	*10.79–26.13*	*1.13–30.05*
7b: Regeneration				
7	Average ($)	14.13 (4%)	13.88 (4%)	23.28 (7%)
	Range ($)	*9.95–19.12*	*4.90–17.98*	*15.06–35.26*
15	Average ($)	N/A	20.53 (6%)	18.90 (6%)
	Range ($)	N/A	*14.51–28.19*	*3.87–32.58*

Note:
The estimates given are in billions of 1982 US $. Net present values calculated from 1990 to 2140.

any given year, we also present Figure 5.6. The net annual impact from four different climate model combinations are presented. All four combinations reflect a series of benefits over time which are quite small for two decades and then gradually increase. The projected benefits increase more rapidly under the DOLY–TEM model combination than under the MAPSS–BIOME-BGC model combination. Interestingly, the net benefits are not particularly sensitive to whether or not dieback is part of the ecological scenario. These annual estimates are used to generate the impacts cited in Chapter 12 for 2060.

Sensitivity analysis

Additional analysis was performed over several key assumptions to see how sensitive the results are to changes in some of the underlying assumptions. The sensitivity analysis was performed over the two particular climatic and ecological scenarios presented in the results section. These scenarios are the DOLY–TEM ecological model combination and the MAPSS–BIOME-BGC ecological model combination, both under the middle climate scenario (recall that the middle climate scenario is assumed to be 50 percent of the change predicted by UKMO).

The following sensitivity analyses were considered:

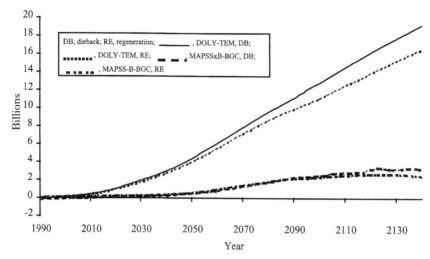

Figure 5.6 Net welfare over time.

(1) Long-term climate adjustment. We changed the dynamic climate impact from a 70-year adjustment to a new steady-state climate to a 150-year adjustment.

(2) Low intensity extended lag. We allowed for a longer period after climate change between dieback or harvest, and the subsequent natural regeneration of low intensity land. Recall that in the regeneration scenario, when a parcel of land converts from one type to another, there is no dieback. The trees live out their natural lives, but once they have naturally died, or they have been harvested, natural re-establishment will occur only after the lag period. Lags in both the dieback and regeneration scenarios were increased to 30 years in southern pines, 50 years in the PNWW conifers, and 70 years in northern and western pines.

(3) High and low salvage. The high salvage case allows 90 percent of the value of the stock affected by dieback to be salvaged. The low salvage case allows only 25 percent of the value of the stock to be salvaged.

(4) High intensity land lags. This analysis tests the possibility that foresters are unable to correctly foresee the future with climate change. We assume that they do not correctly plant the new types on the land when climate has changed enough for them to do so. To incorporate this, we introduce lags into the high intensity timberland, and we assume that foresters replant once with the incorrect species that cannot grow, and then replant correctly later. The lags before they make the correct regeneration decisions are 10

124

Table 5.9. *Sensitivity analysis*

| Sensitivity analysis | Ecological scenario | | | |
| | DOLY–TEM | | MAPSS–B-BGC | |
	Dieback	Regen.	Dieback	Regen.
Initial value of change	16.18	17.29	2.22	4.90
1. Long-term climate adjustment	9.98	10.61	4.30	1.38
2. Low intensity extended lag	14.78	15.72	1.66	3.30
3. Salvage				
High	17.47	17.29	4.47	4.90
Low	15.46	17.29	1.90	4.90
4. High intensity land lag	11.59	13.85	(4.29)	(0.40)
5. Alternative baselines				
High demand	29.08	34.39	7.43	9.77
Low demand	8.97	9.57	2.54	8.06

Note:
The estimates are given in billions of 1982 US $. All values in the table represent the change in the net present value of net surplus from the baseline. In case 5, alternative baselines are used. Each of these cases assumes the middle climate change scenario.

years in the southern pines, 20 years in the PNWW conifers, and 30 years in the northern and western pines.

(5)   High and low demand. These sensitivity analyses consider a high demand case where demand increases at 2.7 percent annually until the year 2030 and then remains constant. The low demand case involves holding demand constant at the 1990 levels forever.

The results of these sensitivity analyses are shown in Table 5.9 which presents the change in net present value of net surplus from the baseline for each case. In the long-term climate change example, benefits are reduced in both of the DOLY–TEM scenarios because the benefits of increasing the area of southern pine production are held off until future, lower value periods. This same explanation also causes the decline in benefits in the MAPSS–BIOME-BGC regeneration scenario. The increase in benefits in the MAPSS–BIOME-BGC dieback case, however, occurs because dieback in southern pines and PNWW conifers is pushed into future periods. The results, therefore, are fairly sensitive to the path of climatic and ecological change when the change has significant stock dieback effects in the most valuable timber producing regions.

Extended low intensity lags have a small impact on the welfare estimates, although they reduce the benefits in each case. This reduction results from small transitional inventories on the low intensity lands, which reduces harvests and increases prices slightly. Increasing (decreasing) salvage increases (reduces) benefits in the dieback cases. Salvage has no impact on the regeneration cases because timber is not subject to dieback influences. Even when only 25 percent of the value of stands is salvaged, there is a minimal effect on welfare. By optimally adjusting harvests from region to region in a forward looking manner, the social planner minimizes the impact of dieback events.

Introducing lags on high intensity land has a fairly significant, negative impact on welfare estimates. This results partially from reductions in the available stocks during the transition period, but it also results from the increased regeneration costs associated with trying twice to negotiate the correct species when climate change occurs. Finally, the alternative baseline scenarios consider high and low demand cases. In both cases, the estimated welfare impacts have the same sign as the initial values in the first line of Table 5.9.

One other sensitivity analysis was considered. In the original set of scenarios, we did not allow southern pines to move into the Pacific Northwest in the BIOME2 model under the UKMO-GCM scenario (shown in Figure 5.2), despite the fact that the ecological model predicted it. Our reasoning was based on the suggestion of the ecological modeling group that produced those results. Given that the economic model allows timberland owners to adjust to these biogeographical shifts through replanting, we tested the impact of this restriction by allowing this type of conversion to occur. For this analysis, we considered the BIOME2 and TEM model combination under the middle climate scenario. Under this analysis, the difference between the net present value of net surplus in the ecological scenarios and the baseline case is found to be $14.79 billion for the dieback case, and $19.09 billion for the regeneration case. The welfare benefits in the BIOME2 and TEM combination under the middle climate scenario decrease by $0.57 billion under dieback if we allow southern pines to move into the Pacific Northwest, but they increase by $1.11 billion under regeneration. The decrease in welfare benefits in the dieback case occurs because conversion to southern pines implies that a large proportion (50 percent) of the stock of PNWW conifers dies back. While other ecological scenarios include larger overall dieback effects, those impacts are limited to western and northern species with smaller initial standing stocks. Nevertheless, welfare estimates do not appear to be significantly altered by our original assumption.

In general, the results are driven more by changes in area distribution than by changes in productivity. This is particularly true of the conversion of northern softwoods to the southern pines. Under MAPSS–BIOME-BGC, for example, large decreases occur in southern ecosystem productivity and 16 percent of the initial timber stock dies back. Despite this, two factors mediate any potentially negative impacts. The first is harvest adjustment. Early on during climate change, timberland managers adjust by utilizing existing stocks efficiently so as to minimize damages. In the dieback case, managers adapt by harvesting more quickly in southern pines and by utilizing the stock that dies back in salvage.

The second adjustment occurs because of replanting efforts. Timberland managers replant both the original land that remains in southern pine and the land that shifts into southern pine according to their perception of future economic and climatic conditions. With rational expectations, they choose good places to plant, and they anticipate future conditions. Given short rotations, southern pines planted early in the simulation can be used within 30 years.

In addition to the importance of southern pines, the significance of expansion in western types, particularly the PNWW conifers, cannot be ignored. DOLY implies the greatest expansion of this type, and this results in lower long-term prices. Although the planting of additional land with this type does not have an immediate pay-off, it will have long-term benefits. This is apparent in Figure 5.5, where prices continue to rise more slowly than the baseline case in the latter parts of the simulation.

5.4 Conclusion

In this chapter, we have described a process for combining the ecological impacts of climate change with a dynamic, economic model of the timber industry (Figure 5.1). We began by looking at GCM models and showing how those are linked to steady-state ecosystem change models via the VEMAP process. The steady-state ecosystem change models capture the spatial shift of ecosystem, as well as changes in ecosystem production (Figures 5.2 and 5.3).

From the steady-state climate and ecosystem response, we developed a dynamic framework. The first step involved developing a linear path of climate change over 70 years, followed by a stabilization of the global climate. Stabilization occurs because experiments were limited to doubling of the atmospheric CO_2. Ecosystem change was then assumed to occur proportionally to the change in climate.

The dynamic, ecological changes were then linked to a dynamic, economic model. The economic model is dynamic in that it captures the influence of future changes in productivity and land use in harvesting and replanting decisions. This is important in a situation like climate change where most changes will occur gradually over time. In addition to a dynamic optimization model, we incorporate several assumptions about the ability of timber markets to adapt to climate change. The model is driven by rational expectations.

This work captures some aspects of uncertainty by using a broad range of climate and ecosystem model predictions, reflected in Table 5.8. By examining this range of results, one can sense both the consistency and reliability of the findings. Importantly, although a large range of changes in the net present value of net consumer's surplus exists, all changes are positive.

This study relies on GCM predictions of climate change and thus is inconsistent with the uniform climate change assumption utilized in the other impact studies in this book. In Table 5.8, for example, we classify each climate scenario in terms of annual average changes in temperature and precipitation in order to determine where the results from each GCM scenario will fit into our matrix. The geographical and seasonal distribution of climate change inherent in the GCM predictions, however, are captured in the results. Some of the oddities of comparing results across GCMs is actually due to the different distributions of climate change variables within those models.

Our model of ecological change is simple. Unfortunately, neither the GCM models nor the ecosystem models have incorporated dynamics explicitly. As modeling of dynamics improves, we can obtain more accurate predictions of the adjustment. The economic model can then assess the implications of alternative ecological predictions.

Another assumption underlying this analysis concerns global timber markets. We assume that the rest of the world has the same climate sensitivity as the United States. Forest distribution and production is assumed to change proportionally across the globe, leading to the price changes predicted in this chapter. This is indeed possible as other temperate regions in the world are likely to be affected similarly. Increases in the tropical and boreal forests are expected by the global ecological models (Melillo *et al.*, 1993). If global forest production increases by more (less) than US production, US benefits will increase (decrease). These benefits are largely enjoyed by consumers, as prices decline. US producers benefit mainly from higher prices so that higher (lower) global timber production may decrease (increase) their economic well being.

Overall, the timber market is likely to adapt to climate change, thereby ameliorating the potential problems associated with ecological change. This work shows how harvest schedules will adjust from region to region and from moment to moment so as to use timber stocks efficiently during the transition period. These adjustments occur regardless of the specific climatic and ecological scenarios. This chapter also shows how timberland owners will adjust their replanting behavior by responding to future ecological and economic conditions. Despite the apparent severity of some ecological effects, market behavior offsets the potential damages through adaptation.

128

References

Adams, D.M. and Haynes, R.W. 1980. The 1980 Timber Asset Market Model: Structure, Projections, and Policy Simulations. *Forest Science* **26**(3): Monograph 22.

Berck, P. 1981. Optimal Management of Renewable Resources with Growing Demand and Stock Externalities. *Journal of Environmental Economics and Management* **8**: 105–17.

Binkley, C.S. 1988. A Case Study of the Effects of CO_2-induced Climatic Warming on Forest Growth and the Forest Sector: B. Economic Effects on the World's Forest Sector. In: *The Impact of Climatic Variations on Agriculture*, Parry, M.L., Carter, T.R. and Konjin, N.T. (eds.) Dordrecht: Kluwer.

Binkley, C.S. and Van Kooten, G.C. 1994. Integrating Climatic Change and Forests: Economic and Ecologic Assessments. *Climatic Change* **28**: 91–110.

Brazee, R. and Mendelsohn, R. 1990. A Dynamic Model of Timber Markets. *Forest Science* **36**: 255–64.

Callaway, M., Smith, J. and Keefe, S. 1994. *The Economic Effects of Climate Change For US Forests*: Final Report. Washington: US EPA; Adaptation Branch, Climate Change Division, Office of Policy, Planning, and Evaluation.

de Steiguer, J.E., Hedden, R.L. and Pye, J.M. 1987. *Optimal Level of Expenditure to Control the Southern Pine Beetle*. Research Paper SE-263. Asheville: US Department of Agriculture, Forest Service. Rocky Mountain Research Station.

Hardie, I.W., Daberkow, J.N. and McConnell, K.E. 1984. A Timber Harvesting Model with Variable Rotation Lengths. *Forest Science* **30**(2): 511–23.

Haynes, R.W. 1990. *An Analysis of the Timber Situation in the United States*. General Technical Report RM-199. Ft Collins: US Department of Agriculture, Forest Service. Rocky Mountain Research Station.

Hunt, E.R. Jr, Martin, F.C. and Running, S.W. 1991. Simulating the Effect of Climatic Variation on Stem Carbon Accumulation of a Ponderosa Pine Stand: Comparison with Annual Growth Increment Data. *Tree Physiology* **9**: 161–72.

IPCC. 1996a. *Climate Change 1995: The Science of Climate Change*. Houghton, J.T., Filho, L.G., Callander, B.A., Harris, N., Kattenberg, A. and Maskell, K. (eds.). Cambridge: Cambridge University Press.

IPCC. 1996b. *Climate Change 1995: Impacts, Adaptations, and Mitigation of Climate Change: Science-Technical Analyses*. Watson, R., Zinyowera, M., Moss, R., Dokken, D. (eds.). Cambridge: Cambridge University Press.

Joyce, L. 1995. *Productivity of America's Forests and Climate Change*. General Technical Report RM-271. Ft Collins: US Department of Agriculture, Forest Service. Rocky Mountain Research Station.

King, G.A. and Neilson, R.P. 1992. The Transient Response of Vegetation to Climate Change: A Potential Source of CO_2 to the Atmosphere. *Water, Air and Soil Pollution* **94**: 365–83.

Korol, R.L., Running, S.W., Milner, K.S. and Hunt, E.R., Jr 1991. Testing a Mechanistic Carbon Balance Model Against Observed Tree Growth. *Canadian Journal of Forest Resources* **21**: 1098–105.

Kuchler, A.W. 1975. *Potential Natural Vegetation of the United States* (2nd Edition). New York: American Geographical Society.

Manabe, S. and Wetherald, R.T. 1987. Large-Scale Changes in Soil Wetness Induced by an Increase in Carbon Dioxide. *Journal of Atmospheric Science* **44**: 1211–35.

McGuire, A.D., Melillo, J.M., Joyce, L.A., Kicklighter, D.W., Grace, A.L., Moore, B., III, and Vorosmarty, C.J. 1992. Interactions Between Carbon and Nitrogen Dynamics in Estimating Net Primary Productivity for Potential Vegetation in North America. *Global Biogeochemical Cycles* **6**: 101–24.

McGuire, A.D., Joyce, L.A., Kicklighter, D.W., Melillo, J.M., Esser, G. and Vorosmarty, C.J. 1993. Productivity Response of Climax Temperate Forests to Elevated Temperature and Carbon Dioxide: A North American Comparison Between Two Global Models. *Climatic Change* **24**: 287–310.

Melillo, J., McGuire, D., Kicklighter, D., Moore, B., III, Vorosmarty, C. and Schloss, A. 1993. Global Climate Change and Terrestrial Net Primary Production. *Nature* **363**: 234–40.

Monserud, R.A. and Leemans, R. 1992. Comparing Global Vegetation Maps with the Kappa Statistic. *Ecological Modeling* **62**: 275–93.

Neilson, R.P. 1993. Vegetation Redistribution: A Possible Biosphere Source of CO_2 During Climatic Change. *Water, Air, and Soil Pollution* **70**: 659–73.

Neilson, R.P. and Marks, D. 1994. A Global Perspective of Regional Vegetation and Hydrologic Sensitivities from Climatic Change. *Journal of Vegetation Science* **5**: 715–30.

Neilson, R., King, G. and Koerper, G. 1992. Toward a Rule-Based Biome Model. *Landscape Ecology* **7**: 27–43.

Newman, D.H. and Wear, D.N. 1993. Production Economics of Private Forestry: A Comparison of Industrial and Nonindustrial Forest Owners. *American Journal of Agricultural Economics* **75**: 674–84.

Overpeck, J.T., Bartlein, P.J. and Webb, T,. III. 1991. Potential Magnitude of Future Vegetation Change in Eastern North America: Comparison with the Past. *Science* **254**: 692–5.

Parton, W.J., Stewart, J.W.B. and Cole, C.V. 1988. Dynamics of C, N, P and S in Grassland Soils: A Model. *Biogeochemistry* **5**: 109–31.

Parton, W.J., Cole, C.V., Stewart, J.W.B., Ojima, D. and Schimel, D.S. 1990. Simulating Regional Patterns of C, N, and P in the US Central Grassland Soils, In: *The Ecology of Arable Lands*, Clarholm, M. and Bergstrom, L. (eds.). Dordrecht: Kluwer Academic Publishers, pp. 99–108.

Perez-Garcia, J., Joyce, L.A., McGuire, A.D. and Binkley, C.S. 1997. Economic Impact of Climatic Change on the Global Forest Sector: An Integrated Ecological/Economic Assessment. In Press. *Critical Reviews in Environmental Science and Technology*.

Powell, D.S., Faulkner, J.L., Darr, D.R., Zhu, Z. and MacCleery, D.W. 1993. *Forest Resources of the United States, 1992*. General Technical Report RM-234. Ft. Collins: US Department of Agriculture, Forest Service. Rocky Mountain Research Station.

Prentice, C., Cramer, W., Harison, S., Leemans, R., Monserud, R. and Solomon, A. 1992.

A Global Biome Model Based on Plant Physiology and Dominance, Soil Properties, and Climate. *Journal of Biogeography* **19**: 117–34.

Reed, W.J. 1984. The Effects of the Risk of Fire on the Optimal Rotation of a Forest. *Journal of Environmental Economics and Management* **11**: 180–90.

Running, S.W. 1994. Testing BIOME-BGC Ecosystem Process Simulations Across a Climatic Gradient in Oregon. *Ecological Applications* **4**: 238–47.

Running, S.W. and Coughland, J.C. 1988. A General Model of Forest Ecosystem Processes For Regional Applications. I. Hydrologic Balance, Canopy Gas Exchange and Primary Productivity Processes. *Ecological Modeling* **42**: 125–54.

Running, S.W. and Gower, S.T. 1991. FOREST BGC, A General Model of Forest Ecosystem Processes for Regional Applications. II. Dynamic Carbon Allocation and Nitrogen Budgets. *Tree Physiology* **99**: 147–60.

Samuelson, P. 1976. Economics of Forestry in an Evolving Society. *Economic Inquiry* **14**: 466–72.

Schlesinger, M.E. and Zhao, Z.C. 1989. Seasonal Climate Changes Induced by Doubled CO_2 as Simulated by the OSU Atmospheric GCM-Mixed Layer Ocean Model. *Journal of Climate* **2**: 459–95.

Shugart, H.H., Antonovsky, M. Ya, Jarvis, P.G. and Sandford, A.P. 1986. CO_2, Climate Change and Forest Ecosystems. In Bolin, B., Doos, B.R., Jager, J. and Warrick, R.A. (eds.). *The Greenhouse Effect, Climate Change, and Ecosystems*, Chichester: Wiley, 1986. 475–521.

Smith, T.M. and Shugart, H.H. 1993. The Transient Response of Terrestrial Carbon Storage to a Perturbed Climate, *Nature* **371**: 523–6.

Smith, J.B. and Tirpak, D. 1989. *The Potential Effects of Global Climate Change on the United States*. Washington DC: US Environmental Protection Agency. EPA-230-05-89-050.

Sohngen, B.L. 1996. *Integrating Ecology and Economics: The Economic Impact of Climate Change on Timber Markets in the United States*. Doctoral Dissertation. New Haven: Yale University.

Sohngen, B.L., Sedjo, R.A., Mendelsohn, R. and Lyon, K. 1996. *Analyzing the Economic Impact of Climate Change in Global Timber Markets*. Resources For the Future, Discussion Paper 96-08, Washington: Resources For the Future.

Solomon, A.M. 1986. Transient Response of Forest to CO_2-induced Climate Change: Simulation Modeling Experiments in Eastern North America. *Oecologia* **68**: 567–79.

Solomon, A.M. and West, D.C. 1985. Potential Responses of Forests to CO_2-Induced Climate Change. In: *Characterization of Information Requirements for Studies of CO_2 Effects: Water Resources, Agriculture, Fisheries, Forests and Human Health*, White, M.R. (ed.). Rep. DOE/ER-0236. pp. 145–69.

VEMAP. 1995. Vegetation/Ecosystem Modeling and Analysis Project: Comparing Biogeography and Biogeochemistry Models in a Continental-Scale Study of Terrestrial Ecosystem Responses to Climate Change and CO_2 Doubling. *Global Biogeochemical Cycles* **9**(4): 407–37.

Willig, R. 1976. Consumer's Surplus Without Apology. *American Economic Review*. **66**: 589–97.

Wilson, C.A. and Mitchell, J.F.B. 1987. A Doubled CO_2 Climate Sensitivity Experiment with a Global Climate Model Including a Simple Ocean. *Journal of Geophysical Research* **D11**-11/20, 13, 315–13, 343.

Woodward, F.I., Smith, T.M. and Emanuel, W.R. 1995. A Global Land Primary Productivity and Phytogeography Model. *Global Biogeochemical Cycles* 9(4): 471–90.

6 Economic effects of climate change on US water resources

BRIAN HURD, MAC CALLAWAY, JOEL B. SMITH,
AND PAUL KIRSHEN[1]

Water is a critical resource in many activities, including domestic use (e.g. cooking, cleaning, and drinking), food production, power generation, transportation, many commercial and manufacturing processes, pollution assimilation, recreation, and many biological and ecological processes. Changes in the spatial and temporal distributions of runoff, and in the quality of water, can have profound social and economic consequences. Such changes are projected by some climate researchers as a result of increased atmospheric concentrations of greenhouse gases (IPCC, 1996). The symptoms of climate change, including sustained changes in temperatures, precipitation patterns, and the frequency and intensity of droughts and storms, may signal the need for changes in water-use patterns and other strategies to mitigate the impacts of climate change.

In a comprehensive assessment of possible climate change effects, it is important to consider both the physical and economic dimensions of the change. Existing assessments of climate change impacts on water resources have been largely based on the results from physical models, which have simulated changes in runoff and occasionally in water-use patterns. The value of these assessments, however, is limited by the absence of economic adjustment, specifically the response of water users to changes in water scarcity (i.e. prices). To describe more completely how the changes in water availability and climate affect social welfare, it is necessary to integrate models describing the physical effects (e.g. hydrologic changes) with models describing economic and institutional responses.

In this assessment of climate change impacts on US water resources, we have responded to the limitations of existing studies by developing methods that integrate models of physical change and economic response. This assessment consists of two parts. First, we construct spatial equilibrium (SE) models of four selected US river

[1] We are grateful to Jim Booker for lending his model of the Colorado River and helping to design the other basin models. We would also like to acknowledge: Rob Mendelsohn, Bruce McCarl, Howard Perlman, Eric Wood, Dennis Lettenmaier, Norm Rosenberg, Dan Epstein, Mark Leuffgren, Lynne Bennett, and Rich Adams.

basins: the Colorado, Missouri, Delaware, and Apalachicola–Flint–Chattahoochee (A–F–C). These models depict the physical movement of water and its economic use within a basin, and are used to analyze the optimal response of water users to changes in water availability and runoff. Second, we extrapolate from the river basin models to larger regions and then to the national level.

This chapter is organized into six sections. Section 6.1 summarizes the literature on water resources and climate change, and provides the context for our assessment. Relevant economic concepts and a description of our methods, data, and models are given in Section 6.2. Scenario and model assumptions are the subject of Section 6.3. Individual basin results are presented in Section 6.4, and Section 6.5 presents the national level results. Section 6.6 presents the conclusions.

6.1   Literature review

Studies of the effects of climate change on water supply and allocation have evolved from physical assessments of runoff changes to integrated assessments from runoff to water management and planning. Early studies of the effects of climate change (e.g. Stockton and Boggess, 1979; Revelle and Waggoner, 1983) were based on statistical models relating annual temperature and precipitation to annual runoff levels at the basin level. These studies, however, suffered from the inadequacy of statistical models to account for changes in underlying physical mechanisms. Improvements to this approach were made by Němec and Schaake (1982), who calibrated a rainfall and runoff model to the Pease River in Texas, and projected the effects of changes in daily temperature and precipitation on runoff. This effort was followed by a number of studies, the most important of which was Gleick's (1987) study of the Sacramento Basin, which found (using general circulation model – GCM – results) that winter runoff would increase and summer runoff would decline. The Lettenmaier *et al.* (1992) study of the American River in Washington pointed to similar effects in the Cascades, where shifts in the runoff peak would exacerbate the conflicts between power production, irrigation, and salmon protection. These studies and others like them (e.g. Frederick, 1993) helped to advance the state of the art in hydrologic model-ing, and raised interesting issues about how climate change might influence the exist-ing competition for water. However, these studies did not grapple directly with allocation issues in a quantitative fashion.

Existing studies of the response of water users and water institutions to runoff changes generally fall into three categories. The first integrates reservoir and system management models with rainfall/runoff models to determine how to best adapt

134

reservoir operation to climate change. For example, Lettenmaier and Sheer (1991) examined the implications of climate change for management of the Central Valley and State Water Projects in California using the California Water Planning Model. They combined results from the hydrologic modeling of GCM outputs with a water management model simulating water delivery requirements. They concluded that, even under scenarios of increased annual runoff, increased winter runoff would not be retained by California's reservoir system and would result in decreased water supplies during the rest of the year. Perhaps the most comprehensive study of this kind is Nash and Gleick's (1993) study of the Colorado River basin. They used the National Weather Service rainfall and runoff model in conjunction with the Colorado River Simulation System (CRSS) to evaluate the impacts of both hypothetical and GCM projected changes in temperature and precipitation on water withdrawals, power production, salinity, and storage. Their results indicated that projected climate changes for the region could have potentially severe impacts on all of the above indicators of system performance, especially if the runoff peak were displaced from April to May. Studies like these are important because they deal directly with the attempts of water resource managers to adapt to climate change; however, they are limited by the absence of explicit water demand schedules and economic responses. In particular, these studies do not account for endogenous adjustments to changes in the relative value of water, or for how these changes in value provide economic signals of a need to allocate water more efficiently in a market system.

The second type of study focuses on the issue of the economic valuation of water resources, but in a context that divorces the issue of valuation from that of economic response (e.g. reallocation of resources). Noteworthy among these efforts are the "back of the envelope" calculations of the economic value of climate change damages at the national level by Cline (1992) and Titus (1992). Cline assumed climate change would cause a 10 percent decrease in water supplies across the country. Using an average value of water of $250 per acre foot (af) for municipal and industrial uses, and $100 per af for irrigation, he estimated that the reduction in supplies would result in damages of $7 billion per year. Titus (1992) estimated changes in supply and demand for surface and groundwater state by state. In his analysis, he used estimates of changes in water availability based on Waggoner (1990) and changes in water demand based on Gibbons (1986), and incorporated some adjustments and adaptations in his analysis by estimating the cost of additional point source controls for water pollution and accounting for changes in hydropower production. He concluded that annual damages to water resources would be between $21 and $60 billion, with $15 to $52 billion of that due to the increased costs of water pollution controls. Although these studies have been useful in establishing some "first-order" estimates of the magnitude of the economic effects of climate change on water users, these estimates were not based on a consistent model of economic behavior.

In the third type of study, information about the physical effects of climate change is integrated with a model of resource allocation based on economic theory. Vaux and Howitt (1984) pioneered this approach in their application of spatial equilibrium (SE) models to water resources in California. Their SE model of the state joins the regional water supply functions and demand functions for specific uses with a linear representation of the water delivery system. The model maximizes the sum of producer and consumer surplus in all regions, subject to the constraints imposed by the water distribution system. This approach effectively simulates competition among water users everywhere in California. When water supplies in the model were reduced through reductions in system inflows, Vaux and Howitt showed how water everywhere would be allocated to more highly valued uses. In an unpublished paper, the authors showed that in the face of a 50 percent reduction in water supplies, it would be less costly to society to redistribute supplies based on economic principles than to construct additional storage capacity in the state.

To value the impacts of climate change on water resources, we have expanded on Vaux and Howitt's approach by integrating river basin SE models with hydrologic models within a multiregional framework. By using this physical/economic framework, we provide a more thorough and detailed analysis than either Cline or Titus, and we do it in a framework that is grounded in economic theory.

6.2   Methods and data

The components and information flow of our basin-level approach to modeling climate change impacts on water resources are shown in Figure 6.1. The first step in the methodology is characterizing a climate change scenario. The study adopts ten climate change scenarios including most of the scenarios discussed in Chapter 1. These climate change scenarios are then used to model changes in hydrology and runoff. Projections of the hydrologic impacts for the scenarios were developed by Lettenmaier and Wood (1994), using a variable infiltration capacity (VIC) model to translate the changes in monthly average precipitation and temperature into changes in monthly runoff.[2]

---

[2] The resolutions of the hydrologic models are 1° latitude and longitude for the Colorado and Missouri basins, and 0.5° for the Delaware and A–F–C basins. The monthly runoff data were then aggregated to match the basin models, both spatially and temporally.

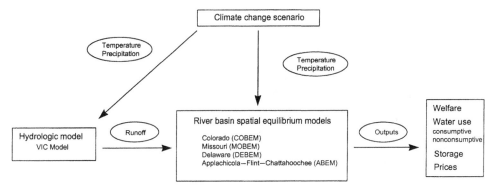

Figure 6.1 Information flow in the assessment of climate change effects on water resources.

Hydrologic model

The river basin models require hydrologic input data. These data must be matched to the inflow points of the river model, and must be consistent with the temporal scale of the model. These data represent the contribution of runoff from precipitation and snowmelt to the volume of water flowing in the river system, net of evapo-transpiration and groundwater losses. The hydrologic data used in the individual river basin models were produced by hydrologic models.

Hydrologic models simulate streamflow at varying spatial and temporal resolutions. They do this by translating the climatic events and factors, such as precipitation and temperature, into runoff while taking into account the dynamics of soil, vegetation, and atmospheric water transfers. Many of the complexities of river basin hydrology are difficult to capture. These complexities are reflected in the variations between actual and modeled streamflow. The calibration and validation of hydrologic models, therefore, depend on how well the model captures and mirrors variation in the observed streamflow data at various points along the river system.

The hydrologic input data are, therefore, important to the analysis of water resource impacts. The hydrologic data that we use derive from regional variants of the two-layer, Variable Infiltration Capacity (VIC-2L) model (Nijssen *et al.*, 1997; Liang *et al.*, 1994; Lettenmaier and Wood, 1994). The VIC-2L is a hydrologically based soil–vegetation–atmosphere transfer scheme designed to represent the land surface in climate and weather models. The model was designed to work in an integrated fashion with GCMs. It can, however, perform analyses off-line, as was done for these studies. In this case, the model simulated the incremental changes in precipitation and temperature that were prescribed in the study design.

137

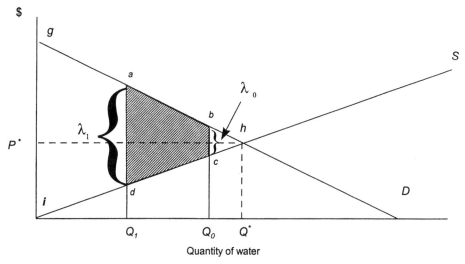

Figure 6.2 Efficient water use in a single water market.

Spatial equilibrium models

The river basin SE models are the primary assessment method. These models use the runoff data from the hydrologic models as model input. Model inputs also include demand parameters (for agriculture only) that are scenario/climate dependent. For example, the agricultural water demands are climate sensitive, typically increasing in response to greater temperatures. The basin models optimize welfare derived from water use and storage, and result in estimates of water use, price, storage, and sector welfare.

The river basin SE models are basically economic models that allocate water to different activities (both consumptive and nonconsumptive) over space and time. The solutions to these models are water allocations, storage levels, and regulated river flows that generate the maximum economic welfare across all water uses, i.e. the maximum consumer and producer surplus. In other words, redistributing water away from the modeled allocations would result in a net welfare loss to the system.

The economic principles at work in these models can be understood by considering Figure 6.2, which shows a supply (*S*) and demand (*D*) schedule for a typical water use. The demand schedule results from water users optimizing their use of water (e.g. to maximize utility or expected profits) in a productive activity, and describes how the marginal value (benefit) of water in this activity varies inversely with the quantity of water used. The total value of water in this activity (i.e. the total willingness to pay of the consumers) is measured by the area under the demand curve up to the quantity consumed. Consumer surplus is defined by this area less the total amount paid (price multiplied by quantity).

138

In a similar fashion, the supply schedule describes the marginal resource costs required to supply a given quantity of water. In general, the marginal costs of providing water (e.g. pumping, storing, distributing, and treating) vary directly with the level supplied (in particular, if the costs of developing new supplies are included). Total resource costs are measured by the area below the supply curve, and producer surplus is equal to the amount paid by the consumer less total resource costs. In the absence of competing uses and supply constraints, welfare is maximized at the intersection of supply and demand (Q^*), where marginal benefits equal marginal costs. This allocation is achieved in a market setting, with consumer surplus shown by the area ghP^* and producer surplus shown by the area iP^*h.

The introduction of water supply (runoff) constraints or competing uses (e.g. downstream users) alters the mechanism of efficient allocation slightly. When runoff is limited and insufficient to reach Q^*, marginal benefits (MB) may exceed marginal costs (MC). This difference, defined as $\lambda = MB - MC$, is the implicit marginal value or shadow price of water, and reflects the value of an additional unit of water to the system. This shadow price is a complex function of available water (runoff), the marginal value of its use (both consumptive and nonconsumptive), the costs of supplying water, and return flows (see Appendix A6 for a discussion of how return flows affect the analysis and determination of water value). A change in any one of these factors could change the shadow price, and affect the efficient allocation of water.

In Figure 6.2, assume initially that the available runoff results in a shadow price equal to λ_0. At this price, water use is Q_0 at the point where the shadow price equals the net marginal value of that use. A decrease in runoff, as some models of climate change predict, increases the shadow price of water to λ_1, and results in a lower level of water use at Q_1.[3] The change in economic welfare associated with this reduction in runoff is measured by the shaded area *abcd*, which is the change in consumer and producer surplus.

Several other important economic concepts can be conveyed by examining the case of two competing uses with different demand elasticities. Figure 6.3 shows two demand curves, D_1 and D_2, that compete for water at the same point in the system. The horizontal sum of these demands is the total demand curve (shown in bold), and water supply is shown as S. Initially, the shadow price of water in the system is equal to λ_0, and total water use is equal to Q_0. The share of this total that is allocated to each use is determined by equating the total demand price (P_0) with the demands associated with each respective use, Q_{10} and Q_{20}, respectively.

---

[3] Alternatively, we could show the change in runoff directly in Figure 6.2 as an inelastic supply curve that shifts exogenously (by the amount corresponding to the shift in λ). However, we want to emphasize the generalized nature of the economic response in these models to system-wide changes in either physical or economic dimension, such as the important effects of other competing users.

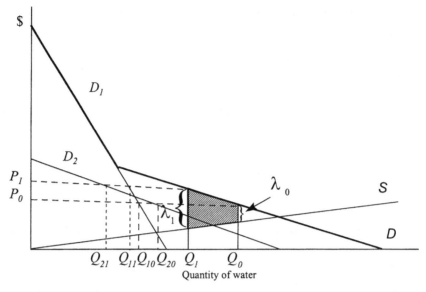

Figure 6.3 Efficient water allocation and welfare change in a two sector market.

When runoff is reduced, the shadow price increases to λ_1, total water use falls to Q_1, and the total demand price rises to P_1. Equating this new price with the demands in each use results in a greater reduction in the more elastic use, D_2, compared to the less elastic use, D_1. Compare the reduction by the former, $Q_{20}-Q_{21}$, with that of the latter, $Q_{10}-Q_{11}$. A greater share of water is retained in the use with the higher marginal values and lower elasticities. The shaded area in Figure 6.3 shows the net change in consumer and producer surplus associated with a reduction in runoff. These results suggest that in a competitive market for water, reductions in runoff are not shared equally across uses. A corollary is that if uniform reductions are imposed, as is often assumed by researchers using physical response models alone, then it stands to reason that welfare losses are greater than they would be if market adjustments are allowed. These economic principles characterize the fundamental nature of the allocation decisions made in the river basin SE models.

The river basin SE models are dynamic and nonlinear, patterned after the spatial equilibrium models first described by Samuelson (1952) and further developed by Takayama and Judge (1964). They account for important spatial features in resource supply and demand, and model the flow of resources between regions. SE models are widely used to characterize market behavior in natural resource sectors, and have also been used to model the agricultural, forestry, water, and energy sectors. Vaux and Howitt (1984), for example, developed an SE model of California's water resources to measure the potential benefits of relaxing restrictions on water markets.

We developed (or, for the Colorado river, modified) models for each of four river basins: the Colorado, Missouri, Delaware, and Apalachicola–Flint–Chattahoochee rivers. These models maximize the sum of consumer and producer surplus subject to physical, economic, and institutional constraints, and consist of the following two basic elements:

- a nonlinear *objective function* containing consumer and producer surplus value functions
- a set of *constraints* describing physical and institutional dimensions of water flow, distribution, storage, exports, and use within a basin.

The objective function consists of benefit and cost functions relating water use and river flow to economic welfare. The economic surplus of consumptive uses, for example, are modeled generally as quadratic functions derived from linear demand and supply schedules. Consumptive water uses include agriculture, municipal and industrial, and thermoelectric energy. The parameters that define these demand and supply functions were based on available data and information (including Gardner and Young, 1985; Gibbons, 1986; Ogg and Gollehon, 1989; Griffin and Chang, 1990, 1991; Booker and Young, 1991; Schneider and Whitlatch, 1991; US Bureau of Reclamation, 1991; Nieswiadomy, 1992; Gutwein and Lang, 1993; Solley *et al.*, 1993).

Nonconsumptive uses, which depend on river flows and runoff directly, were also modeled when and where the data were available. In general, the modeled non-consumptive uses included hydropower, navigation, flood damages, and three measures of value associated with water quality/pollution assimilation. The water quality measures include thermal waste heat from once-through cooling plants, secondary municipal wastewater treatment, and advanced municipal wastewater treatment. The Colorado model contains a different mix of economic sectors than the other models; for example, it includes flatwater recreation on reservoirs, instream recreation (rafting through the Grand Canyon), and salinity damages in the lower basin (see Booker and Young, 1991).

The functional form of the value functions for nonconsumptive water users varied across economic sectors, depending on the data and the assumed relationship between river flows and the generation of economic benefits or costs. For example, flood damages were approximated by an increasing quadratic function in the Missouri basin model, and by an increasing linear function in the A–F–C model.[4] Table 6.1 describes the nature of the nonconsumptive sectors and functions represented in the models.

Model constraints define the physical and institutional dimensions of the river

---

[4] Parameters for flood damage functions vary by location and were derived off-line based on available data on flood damages and associated flow rates.

Table 6.1. *Nonconsumptive economic sectors in the regional basin models*

Sector/basin model	Valuation method	Data description/source
Hydropower/all models	The value is modeled as the avoided cost of lost hydropower. This is modeled as a function of the release rate from reservoirs and average hydraulic head which is a quadratic function of reservoir storage.	US Army Corps of Engineers (1981, 1993a, 1994a, b); Gibbons (1986); US Bureau of Reclamation (1986).
Navigation/Missouri and A–F–C	The value is modeled as an S-shaped logistic function of the rate of flow at one or more specified reaches.	US Army Corps of Engineers (1993a, 1994b).
Flood damages/ Missouri, Delaware, and A–F–C	Linear or quadratic flood damage parameters are estimated for flows above the threshold flow. Estimates of flood damages primarily reflect urban flooding except in the Missouri which includes agricultural damages.	US Water Resources Council (1978) and US Army Corps of Engineers (1993b, 1994a, b).
Thermal waste heat/ Missouri, Delaware, and A–F–C	Costs from lost electricity production of thermal electric plants are modeled as an exponential function of reduced river flows.	Regional power authorities in each basin.
Secondary wastewater treatment/Missouri, Delaware, and A–F–C	The net benefits are modeled as a linear function of flow. They proxy the value of the river for diluting and assimilating biochemical oxygen-demanding (BOD) materials.	Gibbons (1986).
Advanced wastewater treatment/Missouri, Delaware, and A–F–C	The costs are modeled as a function of the volume of the return deficit below which water quality standards are satisfied.	Regional water authorities and from US EPA (1978a, b).
Flatwater recreation/ Colorado	The value is the product of visitation, which is a quadratic function of reservoir surface area, and $16 to $35 values per visitor day.	Booker and Young (1991).
Instream recreation/ Colorado	The value is a quadratic function of river flow.	Booker and Young (1991).
Salinity damages/ Colorado	Damages are modeled as an increasing function of salt concentrations, which are assumed to vary inversely with the flow of water.	Booker and Young (1991); Gardner and Young (1985); Lohman et al. (1988).

basin system so that the models are physically faithful to the spatial distribution of major tributaries, reservoirs, and points of water use. The spatial structure of the model is typically patterned from a schematic diagram of the basin showing major points: basin inflows (i.e. runoff), tributaries, major water diversions, and reservoirs. Physical continuity is maintained in the model by a system of mass balance equations that define both spatial and temporal water balances. Institutional constraints, which depict important basin-specific legal and regulatory provisions (e.g. the Colorado River Compact and the Mexican Treaty), can also be incorporated into the model.

Flow balance constraints define water flow and distribution to mimic the physical behavior of river systems. The flow balance acts like a network, connecting points where runoff enters the system to points where the water is used, stored, or passed to another region. Within each time period the flow balance is modeled as

$$F_n = F_{n-1} + I_n + r_n W_{n-1} + R_n - W_n,$$

where the flow from node n (F_n) is equal to the flow from the node $I-1$ (F_{n-1}) plus inflows from tributaries or runoff (I_n), plus return flows from previous uses (nW_{n-1}), plus net reservoir releases (R_n), less withdrawals (W_n). Evaporation from streamflow is not explicitly modeled. Evaporation losses are accounted for in the reservoir storage balances, and are assumed to reflect the overall evaporation losses in the system. Return flows are assumed to occur within the same period as the associated withdrawals.

Storage balance constraints maintain the physical continuity of reservoir storage levels across time periods. Storage decisions are made in each time period about the net volume of water to release downstream or store for future use. These decisions account for both inflows into the reservoir system and evaporative losses. Reservoir storage balances are maintained by the following equation:

$$S_{t1} = S_{t0} - R_{t1} - E_{t1},$$

where ending storage in period 1 (S_{t1}, with t a time step) must be equal to the ending storage from the previous period (S_{t0}), less any net reservoir releases in period 1 (R_{t1}), less net evaporation losses in period 1 (E_{t1}).

In addition to physical continuity, we model important institutional relationships that regulate the pattern of river flow and water use. For example, the Colorado basin model includes constraints representing the Colorado River Compact of 1922, which requires that the upper basin states (i.e. Wyoming, Colorado, Utah, and New Mexico) release a minimum of 75 million acre-feet (maf) of water during each consecutive 10-year period. This type of institutional requirement can be approximated by a set of minimum flow constraints.

143

Table 6.2. *Summary of river basin planning model components*

Sector/Component	Colorado	Missouri	Delaware	A–F–C
Number of consumptive uses by sector				
Agriculture	16	6	3	4
Municipal and industrial	5	6	4	4
Thermoelectric	4	6	2	4
Number of nonconsumptive uses by sector				
Hydropower	7	3	1	3
Navigation	not applicable	1	not applicable	1
Flood damage	not modeled	8	3	4
Pollution assimilation	4 salinity damages	2 M&I 3 thermal heat	2 M&I 3 thermal heat	2 M&I 2 thermal heat
Number of modeled reservoirs, inflow points, and river reaches				
Recreation	2 instream 7 flatwater	not modeled	not modeled	not modeled
Reservoirs	7	4	3	3
Inflow points	14	8	4	4
Mainstream reaches	21	13	7	9
Tributary reaches	6 Green	3 Platte, 3 Kansas, 2 Osage	3 Lehigh, 3 Schuykill	3 Flint

Regional basin models

In this section we briefly describe each of the river basin models: the Colorado Basin Economic Model, the Missouri Basin Economic Model, the Delaware Basin Economic Model, and the Apalachicola–Flint–Chattahoochee Basin Economic Model. The structure and composition of each basin model is summarized in Table 6.2.

Colorado basin model

The Colorado River is the dominant source of surface water in the arid Southwest, supplying a drainage area of more than 244 000 square miles. In addition to supplying water for consumptive water users, the Colorado River generated over 11 million megawatt-hours (MWh) of hydroelectric power in 1990 (Solley *et al.*, 1993). The model extends Booker and Young (1991) by lengthening the optimization period, separating municipal and industrial uses, and incorporating temperature sensitivity

144

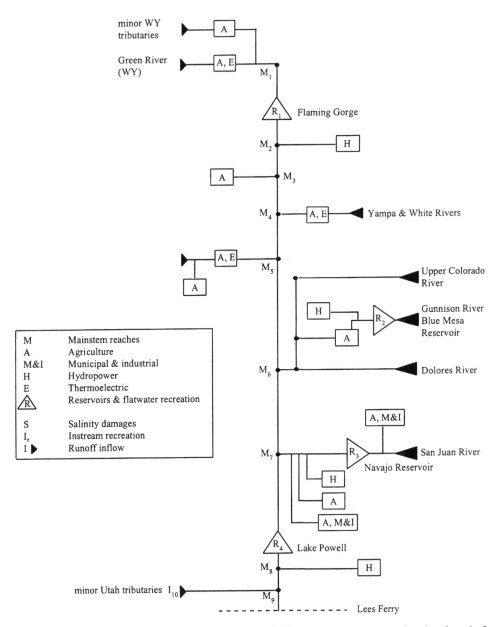

Figure 6.4 Colorado basin model schematic diagram.　　　(*continued overleaf*)

into the agricultural demand schedules. The spatial dimensions of inflows, water use, and storage facilities are shown in the schematic diagram in Figure 6.4.

The Colorado model (COBEM) is the most detailed basin model used in this study with more spatial and intertemporal detail than the other basin models. The demand functions are approximate Cobb–Douglas demand functions, which depict demand

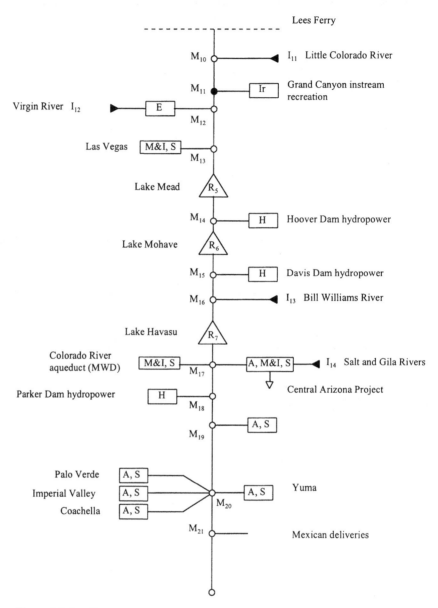

Figure 6.4 (*cont.*)

prices rising asymptotically with decreasing quantities. There are more non-consumptive sectors in this model including salinity and recreation. Salinity damages vary with salt concentration and have been documented for agricultural and municipal and industrial (M&I) users in the lower basin of the Colorado River (Gardner and Young, 1985; Lohman *et al.*, 1988; Booker and Young, 1991).

Institutional constraints are also important in the Colorado basin. The Colorado

River Compact and Mexican Treaty obligations apportion water between the upper and lower basins and Mexico, respectively. These institutions are modeled by defining the minimum annual flow rates at each of two specific reaches (i.e. Lees Ferry and the Mexican border). A penalty function is used to model these constraints and penalize the objective function for failure to meet the required flows.

Missouri basin model

The Missouri River and its tributaries are important to a large agricultural region of the United States. The river basin drains a region of more than 500 000 square miles of the Midwestern United States, a region that produces a variety of grain, oilseeds, and livestock products. The river is a primary source of drinking and industrial water for the region as well as an important transportation resource for moving raw agricultural and mineral products downstream. The Missouri River cools thermoelectric plants which generate more than 159 million MWh and powers 12 million MWh of hydroelectric power (Solley *et al.*, 1993). The river is also used as a sink for industrial and municipal wastewater.

The Missouri model, and also the Delaware and A–F–C models, rely upon linear demand functions for the consumptive sectors. Two nonconsumptive uses were added for navigation and flood damage. The model contains fewer nodes, inflow points, reservoirs, and diversion points than the Colorado model. This greater regional aggregation allows for greater temporal resolution and flexibility, particularly for many of the economic sectors with pronounced seasonal variations such as agriculture, flooding, and navigation. The Missouri model is illustrated in Figure 6.5.

Delaware basin model

The Delaware River flows from its headwaters in the Catskill Mountains of New York and along the border between Pennsylvania and New Jersey. The river drains nearly 15 000 square miles before it empties into Delaware Bay. The Delaware River is a primary source of municipal and industrial water in the mid-Atlantic region, serving water users in New York, New Jersey, Pennsylvania, and Delaware. Water use is governed by a multipurpose compact among the states. Agricultural use of the Delaware River is relatively small compared to that used in the arid Colorado and Missouri basins. Annual agricultural withdrawals in 1990 were estimated at 17.5 thousand acre feet per year (kaf/yr) (Solley *et al.*, 1993). The Delaware River assimilates wastes from many of the cities and towns that use its waters, and it provides water for thermoelectric generation and agriculture.

The Delaware River model is shown in Figure 6.6. There is no significant navigation on the Delaware River above Philadelphia, and navigation to the port of

147

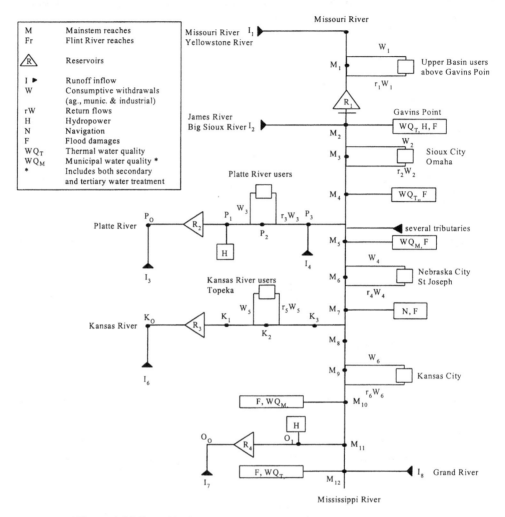

Figure 6.5 Missouri basin model schematic diagram.

Philadelphia is maintained by tidal flows. The consumptive water use in the Delaware basin is dominated by and approximately equally split between municipal and thermal energy uses. Together these two sectors account for over 99 percent of withdrawals for consumptive use (Solley *et al.*, 1993). Withdrawals from the Delaware River averaged 6.8 maf/yr whereas annual average runoff was over 13.5 maf/yr. The Delaware and other eastern basin systems may be less vulnerable to climate change and runoff reductions because they currently use only a fraction of the available runoff.

A–F–C basin model

The Apalachicola, Flint and Chattahoochee (A–F–C) rivers together form an important waterway system serving central and western Georgia, eastern Alabama,

148

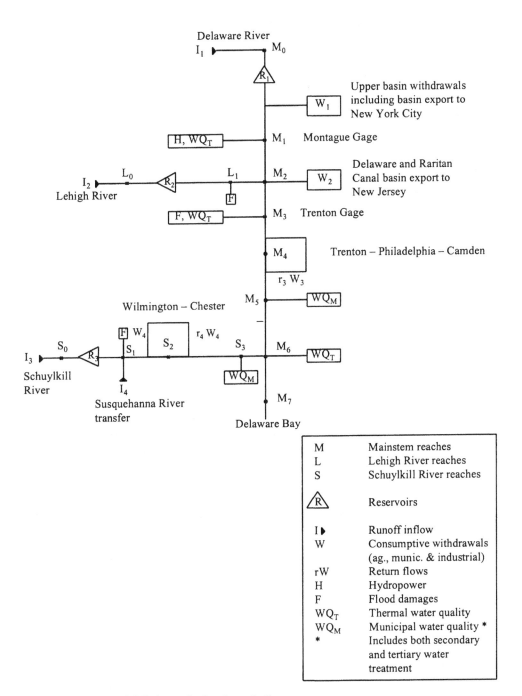

Figure 6.6 Delaware basin schematic diagram.

and the Florida Panhandle. The surface area drained by these rivers is approximately 20 000 square miles. These three rivers are used to transport minerals, timber, and agricultural products to the Gulf of Mexico, in addition to serving as a freshwater resource for municipal and industrial water users in cities such as Atlanta, Columbus, and Bainbridge, Georgia, and Phoenix City, Alabama. The rivers are also used to assimilate wastewater from cities and thermal plants, and to generate steam and hydro-electric power. These rivers are also susceptible to major flooding events, as recently seen along the Flint River in southwestern Georgia. This area may also be more susceptible to droughts because the A–F–C river basin, unlike the basins of the western United States, does not have sufficient reservoir storage capacity to mitigate against prolonged drought conditions.

The A–F–C model includes both consumptive and nonconsumptive water use, and is similar in composition and scale to the Missouri and Delaware models. The physical structure and network of inflows, reservoirs, and water uses are different, as illustrated in Figure 6.7.

National models

National estimates of the economic impacts of climate changes were developed for both consumptive and nonconsumptive water uses. These estimates extrapolate the impacts derived from the four regional models to other regions and to the whole of the United States. Each of the four modeled regions is paired with a set of similar basins from the remaining US water regions (pictured in Figure 6.8), as given in Table 6.3.

The extrapolation method for consumptive uses relies on the modeled regional estimates of economic impacts, and on data from all regions regarding water-use characteristics and changes in runoff.[5]

The change in national welfare for consumptive uses is equal to the sum of net changes in consumer and producer surplus across sectors and regions, and is given as:

$$\Delta \text{ National consumptive} - \text{use welfare} = \sum_j \sum_i \Delta R_{ij}, \tag{6.1}$$

where ΔR_{ij} is the change in consumer and producer surplus in sector i and region j. This surplus change is defined as

$$\Delta R_{ij} = (\overline{\$}_{ij_0} \times \Delta \overline{W}_{ij}) + \tfrac{1}{2} (\Delta \overline{\$}_{ij_0} \times \Delta \overline{W}_{ij}), \tag{6.2}$$

where $\overline{\$}_{ij_0}$ is the estimated net marginal value of water under baseline conditions for the reference model, and $\overline{W}_{ij}$ is the baseline annual surface water withdrawal for sector

---

[5] Hydrologic data for the national assessment were provided by Battelle Pacific Northwest Laboratories. Data on water-use characteristics was derived from Solley *et al.* (1993).

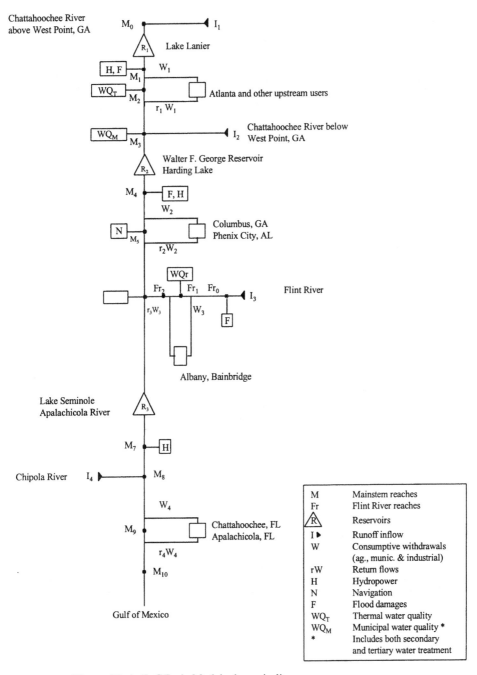

Figure 6.7 A–F–C Basin Model schematic diagram.

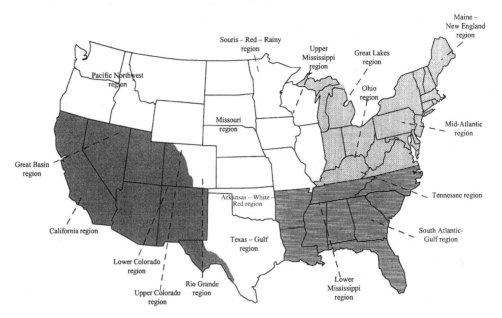

Figure 6.8 United States water resources regions established by the US water resources council in 1970 (from Solley *et al.*, 1993).

i and region *j* (based on estimates from Solley *et al.*, 1993)[6]. $\Delta \overline{W}_{ij}$ is the change in water use by sector *i* in region *j*, and is a function of baseline-water-use patterns in region j_0, the simulated changes in sector water use in the modeled region j_0, and relative runoff changes between target region *j* and modeled region j_0, given as

$$\Delta \overline{W}_{ij} = \overline{W}_{ij}[(1 + \%\Delta W_{ij_0}) \times \frac{(1 + \%\Delta Q_j)}{(1 + \%\Delta Q_{j_0})} - 1], \tag{6.3}$$

where W_{ij0} is the efficient water withdrawal to sector *i* determined in basin model j_0, and Q_j is a measure of simulated runoff conditions.

This procedure accounts for differences in runoff and scale across regions; however, it assumes that the response of water users to price changes (within each economic sector) is the same between the modeled regions and paired regions. For example, this implies that agricultural water use in the upper Mississippi region has the same demand elasticity as agricultural water use in the Missouri region.

It is important to account for nonconsumptive water use at the national level.

---

[6] We recognize the difference between *withdrawals* and *consumptive use*, and that efficient use depends on equalizing the marginal value across consumptive uses, i.e. after accounting for return flows as described in Appendix A6. However, consistent data on consumptive use were not available. If average return flow rates are approximately the same within a given sector across regions, then no particular bias is introduced.

Table 6.3. *Pairing of river basins to US water resource regions*

Modeled basin	Water resource regions
Colorado River	Rio Grande, Great Basin, California
Missouri River	Upper Mississippi, Souris–Red–Rainy, Arkansas–White–Red, Texas–Gulf, Pacific Northwest
Delaware River	Maine–New England, Mid-Atlantic, Great Lakes, Ohio
Apalachicola–Flint–Chattahoochee Rivers	South Atlantic–Gulf, Tennessee, Lower Mississippi

However, it is also subject to greater uncertainty given the difficulties in measuring values associated with nonconsumptive use. A slightly different approach than the one presented above is used because of the absence of water-use data. Instead, the extrapolation is based on the estimated change in nonconsumptive welfare from the regional models; this value is then scaled by two factors to account for regional differences in runoff under climate change, and scale (absolute magnitude) differences across river basins. For the first factor, the ratio of percentage changes in runoff between the two regions is used. Scaling by the ratio of runoff changes, as in the consumptive-use procedure above, accounts directly for regional variation in runoff and water availability. Accounting for regional differences in the nature (and scale) of nonconsumptive water use is more uncertain.

To account for the relative scale of nonconsumptive values across regions, the ratio of water used in hydropower production between the two regions is used.[7] The change in national welfare derived from nonconsumptive water uses is defined as

$$\Delta \ National \ nonconsumptive-use \ welfare = \sum_j \Delta R_{nc,j}, \tag{6.4}$$

where $R_{nc,j}$ is the change in the welfare of nonconsumptive users in region j. This change in welfare is given by:

$$\Delta R_{nc,j} = \Delta(\bar{\$}_{nc,j_0}) \times \frac{(1 + \%\Delta Q_j)}{(1 + \%\Delta Q_{j_0})} \times \frac{H_j}{H_{j_0}} \tag{6.5}$$

---

[7] Hydropower was observed in the model results to relate more directly to the estimates of nonconsumptive welfare than to annual water volume. Hydropower accounted for more than 60% of estimated nonconsumptive welfare in three of four basins (the Delaware was the exception with a relatively low share). At the national level, this assumption does not appear to introduce significant bias into the estimates. However, extrapolating to specific regions, particularly the Northeast and mid-Atlantic, is not advised because of the relatively small capacity for hydropower production in the Delaware basin and the potential for bias at the regional level.

153

where $\overline{\$}_{nc,j_0}$ is the value of nonconsumptive water use in modeled region j_0, Q_j is runoff in region j, and H_j is the quantity of water used in hydropower production in region j in 1990 (Solley *et al.*, 1993).

The accuracy of this procedure depends critically on two premises: first, the assumption that the value of water in a modeled region is largely similar to those in the extrapolated regions; second, the assumption that hydropower is representative of scale differences across regions. These assumptions may be valid for estimating national-level impacts, but could be very misleading if applied to extrapolating specific regional estimates.

6.3   Scenario and model assumptions

The scenarios were developed to project water-use conditions in 2060. The models projected demands in 2060 and simulated water use for 39 years (holding parameters constant and equal to the 2060 projections). The hydrologic data used in the 39-year simulations were based upon the historical period from 1949 to 1987. The assumptions needed to generate baseline projections, climate change scenarios, and an institutional scenario are discussed below.

Projections of water demand in 2060 were derived by scaling current demands using estimated growth rates, which accounted for changes in population, income, and recent historical trends in consumptive water use. These historical data suggested that water demand in the energy and municipal sectors has been growing over time, whereas irrigation demand has been relatively constant. The growth of water demand by thermal energy producers has been considerably less than the growth by municipal users. In addition, we hypothesized that the future growth in electricity demand will be increasingly met by technologies that are less water intensive. Based on these historical trends, demand for water in 2060 is estimated to be 23 percent and 10.2 percent greater than current demand by the municipal and thermal energy sectors, respectively.

Irrigation demand has been relatively constant over the last 20 years. Future projections of irrigation demand assume that no new significant federal water supply projects will be built, and that changes in irrigation technology will offset increases in the demand for irrigation water. That is, the overall demand for irrigation water will remain constant at current levels under baseline climate conditions.

Ten climate change scenarios are analyzed in this study. These scenarios span a wide range of changes in annual average temperature and precipitation. These changes in climate are likely to have an effect on the demand for water by consumptive users. Warmer temperatures and changes in precipitation will affect evapo-transpiration rates in crops as well as gardens and golf courses. A study of municipal water demand in the Great Lakes region found there could be a small rise in water demand under climate change (Cohen, 1987). In modeling climate change, we assume that agricultural water demand will vary with climate change, but that municipal and thermal energy water demand will remain the same. Because M&I (municipal and industrial users) use very little water, a small change in their demand functions is not likely to bias the results to a great degree. We rely upon the results reported by Peterson and Keller (1990) to assess the effect of climate change on agricultural water demand. They use a soil–crop–water simulation model to analyze changes in net irrigation requirements resulting from changes in climate. They show how net irrigation requirements vary across the United States, and how these requirements might be affected by changes in average temperatures and precipitation levels. In the Missouri basin, for example, Peterson and Keller show that the net irrigation requirement for central Montana increases from 400 mm to 500 mm under a $+3\,^\circ$C temperature change and a -10 percent change in precipitation, a 25 percent irrigation increase. Extrapolating their results to our 10 scenarios in an approximately linear fashion, agricultural demand in the Colorado and Missouri models increases between 2 and 26 percent. In the Delaware and A–F–C models, irrigation demands increase between 0 and 25 percent, depending on whether precipitation increases. Carbon fertilization may reduce the demand for irrigation, however, resulting in much smaller demand increases.

We have emphasized the role of economics in both the assessment of impacts and the adaptive response to climate change. By this emphasis we have largely assumed that water institutions will respond to changes in economic conditions, and, more important, that they will transmit these economic signals (i.e. prices) to water users. History and experience, however, have shown that this assumption is probably optimistic and that institutions are more often slow to adapt to changing conditions. As part of our analysis, we used the Colorado model to examine the effects of different assumptions concerning institutional adaptation. Specifically, we used the model to assess the welfare and water-use implications of a regulated scenario depicting current institutional constraints and an unregulated scenario free of institutional constraints under both baseline and $+2.5\,^\circ$C, $+7$ percent precipitation climate change scenarios.

The *regulated* scenario depicts a set of institutions that resemble existing institutions in the Colorado basin. Specifically, this scenario models the Colorado River Compact and Mexican Treaty provisions and a constraint that simulates prior appropriation doctrine. The prior appropriation constraint requires that water allocations in the upper basin of the Colorado River must satisfy at least 85 percent of histor-

HURD *et al.*

ical agricultural uses. These agricultural uses generally have the most senior rights in
the basin, and this constraint restrains the transfer of water to other sectors or downstream users. By comparison, we also simulate an *unregulated* case in which all institutional constraints are removed. Comparing welfare levels under these two extreme
cases highlights the magnitude and importance of institutions in adapting to the
potential consequences of climate change.

6.4   Results

The basin models were programmed using the GAMS language (Brooke *et
al.*, 1988), and were solved with the MINOS solver (Murtagh and Saunders, 1980).
The solutions generated by the models depict perfect foresight in a competitive
market for water. That means that the models solve for the optimal values of all the
decision variables (e.g. water withdrawals and use, reservoir storage and release)
simultaneously, and in a manner consistent with the intertemporal maximization of
total net economic welfare.

In this section, we compare the sensitivity of various measures to different climate
change assumptions, and therefore focus on relative (percentage) changes. The results
for the Colorado River basin are presented in detail because the model was the most
carefully developed basin and the institutional sensitivity analysis was conducted only
on this basin. Discussion of the remaining basins focuses mainly on the welfare results.

Colorado basin results

Tables 6.4 (a and b) show the physical and economic responses to climate
change in the Colorado basin. We distinguish between results for the upper (Table
6.4a) and lower (Table 6.4b) basins to highlight some important geographic and
institutional features within the basin. The second column shows the change in annual
average basin runoff. The hydrologic model is sensitive to both temperature and precipitation. For example, runoff rises by 23.5 percent under the 1.5 °C, +15 percent
precipitation scenario, decreases by 4.2 percent under the 2.5 °C, +7 percent precipitation scenario, and decreases nearly 35 percent under the 5.0 °C scenario. Annual
average reservoir storage patterns track runoff changes across the climate scenarios.
The variability of reservoir storage in some cases increases under the drier scenarios.

The economic responses to changes in runoff are described by the sector prices and
allocations shown in Tables 6.4. Reduced runoff has three important effects: (1) water
prices are higher and withdrawals fall; (2) prices and allocations respond to a greater
degree in the lower basin; and (3) the Colorado River Compact constraint increases

156

Table 6.4a. *Climate change impacts on consumptive use and implicit prices in the Colorado River basin: upper basin*

Climate change scenario	% Change in basin-wide annual average runoff	% Shift in demand	Agriculture		M&I		Thermoelectric (upper basin only)	
			% Change in average withdrawals	Net marginal price ($/af)	% Change in average withdrawals	Net marginal price ($/af)	% Change in average withdrawals	Net marginal price ($/af)
Baseline[a]	17 058 (kaf/yr)	0	903.5 (kaf/yr)	77.3	473.9 (kaf/yr)	80.2	205 (kaf/yr)	82.7
1.5°C +15%P	23.5	2.2	4.5	74.0	0.5	76.6	1.2	79.1
2.5°C +15%P	14.1	3.7	5.2	75.3	0.4	77.9	0.8	80.4
5.0°C +15%P	−6.9	7.4	5.5	80.2	−0.4	83.2	−0.9	85.7
1.5°C +7%P	4.0	3.3	3.6	76.9	0.1	79.7	0.1	82.2
2.5°C +7%P	−4.2	5.5	4.3	79.1	−0.3	82.1	−0.6	84.6
5.0°C +7%P	−22.4	11.0	−0.3	97.3	−3.1	102.1	−6.1	104.5
5.0°C 0%P	−34.7	16.2	−26.8	213.9	−49.4	1012.2	−52.3	1248.4
1.5°C −10%P	−32.1	7.8	−23.4	179.2	−18.5	248.3	−25.7	248.1
2.5°C −10%P	−37.9	12.9	−31.4	218.3	−59.2	1283.9	−60.6	1640.8
5.0°C −10%P	−50.4	25.8	−33.2	253.4	−79.4	1655.8	−79.8	2085.2

Notes:

[a] The figures shown in the row labeled "Baseline" report the baseline level from which the percentage change or absolute difference is calculated.

Table 6.4b. *Climate change impacts on consumptive use and implicit prices in the Colorado River basin: lower basin*

Climate change scenario	% Change in basin-wide annual average runoff	% Shift in demand	Agriculture		M&I	
			% Change in average withdrawals	Net marginal price ($/af)	% Change in average withdrawals	Net marginal price ($/af)
Baseline[a]	17058 (kaf/yr)	0	10490.8 (kaf/yr)	11.5	1367 (kaf/yr)	19.9
1.5°C +15%P	23.5	2.0	37.6	8.1	0.5	16.5
2.5°C +15%P	14.1	3.0	22.4	9.3	0.3	17.7
5.0°C +15%P	−6.9	6.5	−11.6	14.3	−0.4	22.9
1.5°C +7%P	4.0	3.0	6.1	11.0	0.1	19.5
2.5°C +7%P	−4.2	5.0	−7.1	13.3	−0.2	21.8
5.0°C +7%P	−22.4	10.0	−35.3	21.8	−1.4	30.5
5.0°C 0%P	−34.7	12.3	−49.0	28.3	−2.2	37.3
1.5°C −10%P	−32.1	5.5	−47.3	25.3	−1.8	34.2
2.5°C −10%P	−37.9	9.2	−52.6	30.0	−2.4	39.0
5.0°C −10%P	−50.4	18.4	−66.9	71.0	−7.6	83.1

Notes:
[a] The figures shown in the row labeled "Baseline" report the baseline level from which the percentage change or absolute difference is calculated.

losses if runoff falls severely. In most cases higher prices lead to reductions in withdrawal; but there are exceptions. Agricultural withdrawals (in relation to baseline levels) can rise with increasing prices because of the outward shift in agricultural demand due to increases in net irrigation requirements associated with greater temperatures. We observe this phenomenon, for example, in the $+2.5\,°C$, $+7$ percent precipitation scenario results in the upper basin. Under this scenario, upper basin agricultural prices increase by $1.8/af in response to a runoff reduction of 4.2 percent at the same time that withdrawals rise by 4.3 percent. The change in withdrawals in this case is lower than the amount of the 5.5 percent shift in demand, and therefore shows a slight displacement in demand as a result of reduced runoff.

Note that although the absolute change in the shadow prices in both basins is roughly equal, there are large disparities in the relative price changes and relative allocations between the basins. This significant price difference is the result of hydropower's relatively large marginal value for upper basin water. The valuable hydropower is located between the two basins. With a value of nearly $65/af in producing electricity (based on data provided in Gibbons 1996), water in the upper basin has a very high shadow price (opportunity cost) because every unit of water consumed in the upper basin never goes through the dams. This high price in the upper basin discourages low valued upper basin users. It is also worth noting that when runoff falls dramatically, the Compact requirements are violated. This is modeled as a violation cost which also adds to the price of upper basin water.

The bottom line is the change in welfare for the entire Colorado basin presented in Table 6.5. With low temperature and high precipitation increases, runoff increases and so does welfare. However, with low precipitation increases or with high temperature increases, runoff declines and welfare falls as well. In the central case of $2.5\,°C$, $+7$ percent precipitation, welfare declines by $102 million. Two-thirds of these losses are in hydropower and one-third is in salinity damages. Increasing temperature by $5.0\,°C$ with the same precipitation increases damages to $572 million. Over 50 percent of these damages are hydropower losses, with another 35 percent of the losses coming in increased salinity. The remaining damages are shared by agriculture (7 percent), industrial (4 percent), and recreation (1 percent).

We also analyzed the welfare impact of the central case warming using two alternative institutional settings. The regulated scenario characterizes the current institutional setting with the Colorado River Compact and Mexican Treaty constraints and prior appropriation constraints. The unregulated setting removes all of these institutional constraints. With the regulations in place, the damages from the central case warming are $105 million. In the unregulated setting, the damages are only $65 million. The additional flexibility allowed by removing the regulations can

Table 6.5. *Climate change impacts on welfare in the Colorado River basin (millions of 1994$)*

Climate change scenario	Change in total impact welfare	Change in agri. welfare	Change in M&I welfare	Change in thermo-electric welfare	Change in hydropower value	Change in recreation value	Change in salinity damages
Baseline[a]	7744	307.3	5371.3	385.8	1035.5	644.1	−244.7
1.5°C +15%P	486	38	5	0.5	281	1	160
2.5°C +15%P	294	29	3	0.3	159	1	102
5.0°C +15%P	−175	−1	−4	−0.4	−112	−1	−57
1.5°C +7%P	85	13	1	0.1	40	0.2	31
2.5°C +7%P	−102	−2	−3	−0.2	−66	−0.5	−34
5.0°C +7%P	−572	−40	−20	−3	−300	−7	−202
5.0°C 0%P	−1193	−112	−263	−63	−436	−18	−301
1.5°C −10%P	−899	−99	−76	−16	−401	−16	−291
2.5°C −10%P	−1372	−127	−332	−81	−468	−28	−336
5.0°C −10%P	−2087	−221	−494	−110	−654	−96	−512

Notes:
[a] The figures shown in the row labeled "Baseline" are baseline welfare estimates. Negative values indicate baseline damages.

significantly reduce the damages from global warming. Rather than losing hydropower from runoff reductions, the unregulated model forces a larger reduction in consumptive use in the upper basin, saving $40 million.

Missouri basin results

The change in runoff in the Missouri basin due to climate change is summarized in column two of Table 6.6. The changes in runoff are similar in sign and in some cases slightly greater in magnitude than the runoff changes in the Colorado basin. For example, projected runoff falls by over 42 percent under the +5.0 °C scenario compared to a reduction of 35 percent in the Colorado basin. In the +2.5 °C, +7 percent precipitation and +1.5 °C, +15 percent precipitation scenarios, projected runoff changes by − 9 percent and +20 percent, respectively. These changes in runoff alter withdrawals for the agricultural, municipal, and thermoelectric sectors. Agriculture is the most affected sector in the Missouri basin with withdrawals falling by 54 percent under the +5.0 °C scenario. In the same scenario, municipal and thermoelectric withdrawals fall only by 0.9 and 1.8 percent, respectively. This result reflects the much greater price elasticity for agricultural versus municipal and thermo-electric water use.

In scenarios where runoff increases, welfare also increases. However, in most cases, runoff declines and welfare falls as well. With the central case climate scenario, welfare losses are $519 million throughout the basin. Two-thirds of these losses are due to water quality problems and one-quarter of the losses are lost hydropower. With the 5 °C and zero precipitation case, losses climb to $2.2 billion. Sixty percent of these damages are water quality effects whereas only 17 percent are hydropower losses. Agriculture absorbs most of the remaining loss (22 percent).

Delaware basin results

Runoff changes are summarized in column two of Table 6.7. The hydrology simulations predict increases in runoff with little warming and high precipitation increases but otherwise project runoff will decline. Changes in withdrawals are closely linked to changes in runoff. The agricultural sector has the greatest change in withdrawals because of its highly elastic demand, whereas municipal and industrial withdrawals are hardly affected.

Changes in total welfare vary much less in percentage terms than runoff changes. In the central case, runoff falls by 4 percent and welfare falls by only 0.3 percent ($22 million). Even with the severe scenario of 5 °C with no precipitation increase, runoff falls by 34 percent and total welfare falls by only 3 percent ($207 million). Most of the damages are from water quality costs. Under the central scenario, water quality

Table 6.6. *Climate change impact on the Missouri River basin (millions of 1994$)*

Climate change scenario	Runoff % change	Total welfare	Agriculture		M&I		Thermoelectric		Hydropower welfare	Navigation welfare	Flooding welfare	Water quality welfare
			% Withdraw	Welfare	% Withdraw	Welfare	% Withdraw	Welfare				
Baseline[a]	56651 (kaf/yr)	$10 804.8	13322 (kaf/yr)	$1097.8	2124 (kaf/yr)	$4021	12890 (kaf/yr)	$5658.3	$558.8	$3.8	−$14.5	−$520.4
1.5°C + 15%P	20.5	314	14.3	94	0.04	0	−0.4	0.1	66	3	−39	190
2.5°C + 15%P	9.1	2	3.7	47	−0.01	0	−0.3	0	−16	−0.1	−10	−19
5.0°C + 15%P	−15.5	−1172	−27.5	−213	−0.3	−0.3	−0.4	−1	−245	−3	13	−723
1.5°C + 7%P	1.0	−95	1.2	33	−0.04	0	−0.05	−0.1	−40	−1	1	−88
2.5°C + 7%P	−9.1	−519	−9.1	−39	−0.1	−0.1	−0.1	−0.3	−133	−3	12	−356
5.0°C + 7%P	−30.6	−1945	−47.9	−433	−0.7	−2	−1.4	−18	−348	−4	15	−1155
5.0°C 0%P	−42.4	−2239	−54.0	−498	−0.9	−3	−1.8	−24	−381	−4	15	−1344
1.5°C − 10%P	−35.3	−1437	−32.6	−248	−0.3	−0.4	−0.4	−2	−282	−3	15	−917
2.5°C − 10%P	−42.5	−2041	−48.9	−427	−0.7	−1	−1.0	−7	−361	−4	15	−1256
5.0°C − 10%P	−56.8	−2292	−53.7	−480	−0.8	−1	−0.9	−6	−379	−4	15	−1437

Notes:

[a] Figures in this row are baseline welfare estimates. Negative values indicate baseline damages.

Table 6.7. Climate change impacts in the Delaware River basin (millions of 1994$)

Climate change scenario	Runoff % change	Total impact welfare	Agriculture		M&I		Thermoelectric		Hydropower welfare	Flooding welfare	Water quality welfare
			% Withdraw	Welfare	% Withdraw	Welfare	% Withdraw	Welfare			
Baseline[a]	13 660 (kaf/yr)	$6564.9	14.7 (kaf/yr)	$1.0	3399.5 (kaf/yr)	$5085.7	3416.4 (kaf/yr)	$1560.6	$4.0	−$0.0148	−$86.3
1.5°C +15%P	16.8	48	3.4	0	0.2	1	0.5	2	0	0	45
2.5°C +15%P	9.9	25	1.4	0	0.1	0.2	0.2	1	0	0	24
5.0°C +15%P	−8.7	−49	−5.4	−0.1	−0.3	−2	−1.4	−7	0	0	−40
1.5°C +7%P	2.7	6	0.0	0	0.01	0	−0.01	−0.1	0	0	6
2.5°C +7%P	−4.1	−22	−2.7	−0.1	−0.1	−1	−0.6	−3	0	0	−18
5.0°C +7%P	−22.3	−119	−12.9	−0.1	−0.9	−5	−3.5	−20	0	0	−94
5.0°C 0%P	−33.9	−207	−8.2	−0.1	−2.0	−12	−7.4	−45	−0.1	0	−150
1.5°C −10%P	−26.8	−134	−4.8	0	−0.9	−5	−3.7	−20	0	0	−109
2.5°C −10%P	−33.2	−187	−6.1	−0.1	−1.6	−9	−6.2	−36	−0.1	0	−142
5.0°C −10%P	−49.8	−418	−40.1	−0.4	−5.4	−41	−17.6	−128	−0.2	0	−248

Notes:

[a] The figures shown in the row labeled "Baseline" report the baseline value level from which the percentage change or absolute difference is calculated. Negative values indicate baseline damages.

accounts for over 80 percent of the damages and under the severe climate scenario, water quality accounts for almost three-quarters of the damages. The bulk of the remaining damages are from lost thermoelectric generation.

A–F–C basin results

The A–F–C basin results are given in Table 6.8. There are many similarities between the results of the A–F–C and the Delaware basins. In both cases, the available freshwater easily exceeds the consumptive withdrawals. The bulk of the damages in both systems are consequently due to nonconsumptive uses such as water quality and hydroelectricity.

The changes in runoff in response to climate change in this basin are less severe than in the other regions. Runoff changes are summarized in column two of Table 6.8. Average annual runoff increases under five climate scenarios (compared to three in the other basins). Reductions in runoff, when they occur, are smaller in magnitude. Changes in withdrawals in the A–F–C basin are relatively small. Municipal withdrawals hardly vary at all and thermal withdrawals fall by only a small amount even in severe scenarios. Agricultural withdrawals under many of the climate changes actually increase because of the outward shift in agricultural demand for irrigation water.

Changes in welfare across the climate scenarios show two important results. First, runoff reductions result in negligible changes in the welfare of agricultural, municipal, and thermal energy users. Second, the biggest damages once again are in the nonconsumptive sector. In the central case, total damages are $15 million with damages in navigation, hydroelectricity, water quality and flooding. With the more severe 5 °C case, total damages are $31 million with the bulk of damages in hydroelectricity, navigation and water quality and substantial flooding benefits.

6.5   National results

There are important regional differences in the distribution and magnitude of climate change effects on water resources which must be taken into account in estimating national effects. We begin with the basin studies above and extrapolate to the paired regions for each basin using region-wide estimates of runoff by climate scenarios. We then sum these results across regions to arrive at our national estimates. The change in total national withdrawals is presented in Table 6.9.[8] When national

[8] Estimates in Table 6.9 for the zero percent precipitation change scenario are based on linear interpolation between the comparable + 7 percent and –10 percent precipitation scenarios. While the – 10 percent precipitation scenarios are plausible at the regional level, they are unlikely outcomes for uniform national scenarios, and therefore are not presented in the table.

Table 6.8. *Climate change impacts on the A–F–C basin (millions of 1994$)*

Climate change scenario	Runoff % change (kaf/yr)	Total impact welfare	Agriculture		M&I		Thermoelectric		Hydropower welfare	Navigation welfare	Flooding welfare	Water quality welfare
			Withdraw % (kaf/yr)	Welfare	Withdraw % (kaf/yr)	Welfare	Withdraw % (kaf/yr)	Welfare				
Baseline[a]	24363	$2225	84.9	$4.8	771.3	$1575	1304	$589.07	$93.4	$23.8	$−77.9	−$16.6
1.5°C+15%P	18.7	−37	0.1	0	0.03	0	0.2	0	9	1	−53	6
2.5°C+15%P	13.7	−36	0.1	0	0.01	0	0.1	0	6	1	−47	4
5.0°C+15%P	0.5	−36	−1.6	0	−0.08	0	−0.4	−0.2	−8	−4	−18	−6
1.5°C+7%P	5.1	−15	0.1	0	−0.0	0	0.01	0	0.2	−0.1	−16	1
2.5°C+7%P	0.3	−15	−0.8	0	−0.03	0	−0.2	0	−3	−0.3	−10	−2
5.0°C+7%P	−12.4	−28	−1.6	0	−0.12	0	−0.8	−0.3	−20	−8	14	−14
5.0°C 0%P	−23.5	−31	−16.7	1	−0.3	−0.1	−1.6	−1	−31	−12	39	−27
1.5°C−10%P	−23.1	−4	−8.1	0.4	−0.16	0	−0.9	−0.3	−24	−10	48	−18
2.5°C−10%P	−27.5	−12	−15.7	1	−0.23	0	−1.4	−1	−28	−10	51	−25
5.0°C−10%P	−38.9	−55	−17.2	1	−0.6	−0.3	−3.8	−4	−47	−17	65	−53

Notes:

[a] The figures shown in the row labeled "Baseline" report the baseline value level from which the percentage change or absolute difference is calculated. Negative values indicate baseline damages.

Table 6.9. *Welfare impact of climate change on US water users (billions of 1994$)*

Climate change scenario	Total withdrawals (maf/yr)	Agriculture withdrawal (maf/yr)	Total welfare	Agriculture welfare	M&I welfare	Thermo-electric welfare	Hydropower welfare	Other nonconsumptive welfare
Baseline[a]	377	157	$131.87	$13.70	$44.94	$29.88	$14.70	$28.65
1.5°C +15%P	27.1	25.8	9.76	0.07	0.0	0.02	0.69	8.98
2.5°C +15%P	9.8	9.6	2.59	-0.32	-0.01	0.02	-0.78	3.68
5.0°C +15%P	-35.6	-32.5	-17.91	-1.73	-0.01	-0.03	-4.65	-11.49
1.5°C +7%P	0.9	1.2	-1.53	-0.47	0.0	0.01	-1.15	0.08
2.5°C +7%P	-16.4	-14.1	-9.41	-0.94	-0.03	-0.01	-2.75	-5.68
5.0°C +7%P	-66.7	-60.0	-31.76	-2.90	-0.02	-0.12	-6.50	-22.22
1.5°C 0%P	-17.0	-15.0	-9.53	-1.0	-0.02	-0.01	-2.8	-5.7
2.5°C 0%P	-34.0	-31.0	-18.06	-1.8	-0.03	-0.03	-4.7	-11.5
5.0°C 0%P	-83.0	-69.8	-43.14	-3.67	-0.05	-0.57	-7.43	-31.42

Notes:

[a] Figures in this row represent baseline welfare estimates. Results for the three zero percent precipitation scenarios were derived from linear interpolation between results for the comparable +7 percent and -10 percent scenarios.

runoff increases (decreases), total withdrawals increase (decrease). The bulk of the change in withdrawals is limited to the agricultural sector. As water becomes more scarce, farmers cannot afford to pay more for the same amount of water and so are forced to reduce use. Because changes in withdrawal are largely limited to agriculture, agriculture bears the brunt of the welfare losses amongst consumptive users. However, it is important to note that welfare losses to consumptive users of water are actually relatively small in all but the most severe climate scenarios.

The largest source of damages in the model are to nonconsumptive users, specifically water quality. For example, in the central climate scenario, national welfare losses are estimated to be $9.4 billion. Of this amount, over $5 billion was associated with water quality damages, with a remaining $2.8 billion for hydroelectric losses. Only $1 billion of this loss was associated with consumptive users and most of this was agriculture. Even with the 5.0 °C, zero percent precipitation severe scenario, water quality accounts for $31 billion of the total $43 billion damage. Hydroelectricity accounts for another $7 billion and consumptive uses for another $4 billion.

It should be understood that there are many uncertainties inherent in these national estimates. The extrapolation from individual basins to regions is imperfect. The basins represent large complex river systems which have been extensively studied before. There is much less information on the remaining rivers in each region. Further, some regions, such as the Pacific Northwest, are really quite different from the four basins in this study. The estimates in this study presume that water will flow towards the highest value users. However, in cases where existing laws protect low value users, this assumption may be violated, adding to the damages.

6.6   Conclusions

This study uses four carefully planned basin studies in order to estimate the national damages from climate change on water systems. The four basin studies indicate that climate change is likely to have very different regional impacts. The Western states are semi-arid so that water can be a limiting factor for development. If climate change reduces runoff, agriculture in these regions will be affected and could well shrink. The eastern basins, in contrast, withdraw only a fraction of the available water. Reductions in runoff will have only a minimal effect on consumptive uses in the East.

The results also imply that it is not consumptive users but rather nonconsumptive users who will bear the bulk of the damages. Total damage and benefit estimates for virtually all scenarios are most heavily influenced by estimates for the nonconsumptive sectors. The nonconsumptive sector estimates are, in turn, dominated by estimates for changes in water quality (and influenced to a lesser degree by navigation and flooding estimates). Another large nonconsumptive loss is from hydroelectricity. In the central climate scenario, nonconsumptive uses account for 60 percent and hydroelectricity for almost 30 percent of the damages from warming. In the more severe 5 °C scenario, nonconsumptive uses account for over 70 percent of the damages and hydroelectricity another 17 percent.

A third important result is that rising temperature, even with moderate precipitation increases, is likely to lead to average runoff reductions nationwide. As runoff falls, total withdrawals fall proportionately. The severity of these reductions determine the damages. For the central climate scenario, total withdrawals are projected to fall by 4 percent, resulting in damages of $9.4 billion. For the more severe 5 °C scenario, total withdrawals are projected to fall by 22 percent, leading to damages of $43 billion.

Another critical issue in this analysis concerns adaptation. We have modeled changes in water allocations assuming that scarce water goes to the highest bidder. However, our analysis of the Colorado River reveals that low valued water users are protected under current agreements. If these protections are allowed to be sustained even as runoff falls, damages will be higher. The extent of institutional adaptation in the face of long-term water shortages is an area which requires more analysis.

Our estimates of water damages from warming are consistent with previous aggregate estimates. Cline (1992) estimates damages of $7 billion and Fankhauser (1995) estimates damages of $13.7 billion for the central case scenario. In comparison, we estimate only $9 billion from this scenario. For the more severe climate scenario, Titus (1992) estimates damages of between $21 and $60 billion. For this scenario, we estimate damages of $43 billion. However, even though our aggregate estimates are consistent with these previous authors, our estimate of what is causing those damages is different. All of the losses in the Cline and Fankhauser studies were predicted for consumptive users while we predict these users suffer losses of only $1 billion. Only Titus predicted that hydropower and water quality would bear the majority of the damages from warming.

This chapter attempts to improve upon earlier studies to estimate the economic damages from warming on the water sector. However, there are a number of caveats which must be repeated so that readers do not get overconfident in the accuracy of the results. First, the estimate of national runoff reductions are very crude as we have limited information about how many basins will react to climate change. Partly, we are highly uncertain about regional precipitation levels given any global climate forecast. Even if the climate forecast were known, we are also uncertain how runoff across unstudied basins would change in response to vegetation adjustments to the new climate, to carbon dioxide levels, and to the changing hydrology.

Second, we have studied four basins in detail in order to try to understand how the economy surrounding different river systems would adjust to runoff changes. We have discovered that there is substantial regional variation in how different river systems will adjust. We are not confident that the four basins we have studied fully capture the range of responses likely across the country. For example, none of the four river systems studied closely resemble the unstudied Columbia River basin. In addition, smaller river systems in each region may behave quite differently from the larger examples which we studied. The national extrapolation is consequently highly uncertain.

Third, this chapter models water quality effects assuming that rivers will have to maintain current pollution concentrations. As runoff falls, the study assumes that polluters will have to reduce emissions proportionately. These increased abatement costs are assumed to reflect the damages which would occur in each river system. Although reasonable as a first approximation, this methodology is clearly inappropriate in the long run given the large damages in this sector. A more accurate approach to modeling water quality is needed which tries to quantify ecological, recreational, and drinking water damages.

Fourth, the interactions between water and related systems must be carefully modeled. Although this study is careful to be consistent with the other studies in this book, a general equilibrium analysis may be able to provide even more careful interactions. For example, a general equilibrium analysis of agriculture and water might be able to predict a more accurate demand for water under different climate scenarios, taking into account the adjustments by farmers and markets. Although the efforts to remain consistent across the studies in this book have eliminated any first-order effects, theoretical improvements could still be achieved through a general equilibrium approach.

References

Booker, J.F. and Young, R.A. 1991. *Economic Impacts of Alternative Water Allocations in the Colorado River Basin*. Colorado Water Resources Research Institute Completion Report No. 161, Fort Collins: Colorado State University.

Brooke, A., Kendrick, D. and Meeraus, A. 1988. *GAMS: A User's Guide*. San Francisco, CA: Scientific Press.

Cline, W.R. 1992. *The Economics of Global Warming*. Washington DC: Institute for International Economics.

Cohen, S.J. 1987. Sensitivity of Water Resources in the Great Lakes Region to Changes in

Temperature, Precipitation, Humidity, and Wind Speed. In: *The Influence of Climate Change and Climatic Variability on the Hydrologic Regime and Water Resources.* International Association of Hydrological Sciences (IAHS) Publication No. 168. Wallingford, Oxfordshire, UK: IAHS Press.

Fankhauser, S. 1995. *Valuing Climate Change: The Economics of the Greenhouse.* Economic and Social Research Council, Centre for Social and Economic Research on the Global Environment, London: Earthscan Publications Ltd.

Frederick, K.D. 1993. Climate Change Impacts on Water Resources and Possible Responses in the MINK Region. *Climatic Change* 24: 83–115.

Gardner, R.L. and Young, R.A. 1985. Economic Evaluation of the Colorado River Basin Salinity Control Program. *Western Journal of Agricultural Economics* 10: 1–12.

Gibbons, D.C. 1986. *The Economic Value of Water.* Washington, DC: Resources for the Future.

Gleick, P.H. 1987. The Development and Testing of a Water Balance Model for Climate Impacts Assessment: Modeling the Sacramento Basin. *Water Resources Research* 23: 1049–61.

Gleick, P.H. 1990. Vulnerability of Water Systems. In: *Climate Change and U.S. Water Resources,* Waggoner, P.E. (ed.). New York: John Wiley.

Griffin, R.C. and Chang, C. 1990. Pretest Analyses of Water Demand in Thirty Communities. *Water Resources Research* 26(10): 2251–5.

Griffin, R.C. and Chang, C. 1991. Seasonality in Community Water Demand. *Western Journal of Agricultural Economics* 16(2): 207–17.

Gutwein, B.J. and Lang, R.L. 1993. Regional Irrigation Water Demand. *Journal of Irrigation and Drainage Engineering* 119(5): 829–47.

Hartman, L.M. and Seastone, D. 1970. *Water Transfers: Economic Efficiency and Alternative Institutions Resources for the Future,* Baltimore: Johns Hopkins Press.

IPCC. 1996. *Climate Change 1995: Impacts, Adaptations, and Mitigation of Climate Change: Science-Technical Analyses.* Watson, R., Zinyowera, M., Moss, R. and Dokken, D. (eds.). Cambridge: Cambridge University Press.

Lettenmaier, D.P. and Gan, T.Y. 1990. Hydrologic Sensitivities of the Sacramento–San Joaquin River Basin, California to Global Warming. *Water Resources Research* 26(1): 69–86.

Lettenmaier, D.P. and Sheer, D.P. 1991. Climatic Sensitivity of California Water Resources. *Journal of Water Resources Planning and Management* 117: 108–25.

Lettenmaier, D.P. and Wood, E. 1994. *Implementation of the VIC-2L Land and Surface Scheme to Model the Hydrology of Large Continental Rivers.* Palo Alto, CA: Report prepared for Electric Power Research Institute.

Lettenmaier, D.P., Brettmann, K.L., Vail, L.W., Yabusaki, S.B. and Scott, M.J. 1992. Sensitivity of Pacific Northwest Water Resources to Global Warming. *The Northwest Environmental Journal* 8: 265–83.

Liang, X., Lettenmaier, D.P., Wood, E.F. and Burges, S.J. 1994. A simple hydrologically based model of land surface water and energy fluxes for general circulation models, *Journal of Geophysical Research.* 99, D7, 14,415–14,428.

Lohman, L.C., Milliken, J.G., Dorn, W.S. and Tuccy, K.E. 1988. *Estimating Economic*

Impacts of Salinity of the Colorado River. Denver, CO: Colorado River Water Quality Office, US Bureau of Reclamation.

Mendelsohn, R., Nordhaus, W.D. and Shaw, D. 1994. The Impact of Global Warming on Agriculture: A Ricardian Analysis. *American Economic Review* **84**(4): 753,771.

Murtagh, B.A. and Saunders, M.A. 1980. Minos-Augmented-Users Manual. Technical report SOL 80-19, Stanford University: Dept. of Operations Research.

Nash, L.L. and Gleick, P.H. 1991. Sensitivity of Streamflow in the Colorado Basin to Climatic Changes. *Journal of Hydrology* **125**: 221–41.

Nash, L.L. and Gleick, P.H. 1993. *The Colorado River Basin and Climatic Change: The Sensitivity of Streamflow and Water Supply to Variations in Temperature and Precipitation*. Washington: US EPA; Climate Change Division, Office of Policy, Planning, and Evaluation. EPA 230-R-93-009.

Nieswiadomy, M.L. 1992. Estimating Urban Residential Water Demand: Effects of Price Structure, Conservation, and Education. *Water Resources Research* **28**(3): 609–15.

Nijssen, B., Lettenmaier, D.P., Liang, X., Wetzel, S.W. and Wood, E.F. 1997. A Streamflow Simulation for Continental-Scale River Basins. *Water Resources Research.* **33**(4): 711–24.

Němec, J. and Schaake, J. 1982. Sensitivity of Water Resource Systems to Climate Variation. *Hydrological Sciences Journal* **27**: 327–48.

Ogg, C.W. and Gollehon, N.R. 1989. Western Irrigation Response to Pumping Costs: A Water Demand Analysis Using Climatic Regions. *Water Resources Research* **25**(5): 767–73.

Peterson, D.F. and Keller, A.A. 1990. Effects of Climate Change on U.S. Irrigation. *Journal of Irrigation and Drainage Engineering* **116**(2): 194–210.

Revelle, R.R. and Waggoner, P.E. 1983. Effects of a Carbon Dioxide-Induced Climatic Change on Water Supplies in the Western United States. In: *Changing Climate: Report of the Carbon Dioxide Assessment Committee*. Washington, DC: National Academy Press.

Samuelson, P.A. 1952. Spatial Price Equilibrium and Linear Programming. *American Economic Review* **42**: 283–303.

Schneider, M.L. and Whitlatch, E.E. 1991. User-Specific Water Demand Elasticities. *Journal of Water Resources Planning and Management* **117**(1): 52–73.

Solley, W.B., Pierce, R.R. and Perlman, H.A. 1993. Estimated Use of Water in the United States in 1990. USGS Circular 1081, Washington, DC: US Government Printing Office.

Stockton, C.W. and Boggess, W.R. 1979. *Geohydrological Implications of Climate Change on Water Resource Development*. Fort Belvoir, VA: US Army Coastal Engineering Research Center.

Takayama, T. and Judge, G.C. 1964. Spatial Equilibrium and Quadratic Programming. *Journal of Farm Economics* **46**: 67–93.

Titus, J.G. 1992. The Costs of Climate Change to the United States. In: *Global Climate Change: Implications, Challenges and Mitigation Measures*, Majumdar, S.K., Kalkstein, B., Yarnal, L.S., Miller, E.W. and Rosenfeld, L.M. (eds.). Philadelphia, PA: Pennsylvania Academy of Science.

US Army Corps of Engineers. 1981. *National Hydroelectric Power Resource Study, Data Base Inventory*. Hydrologic Engineering Center, Institute for Water Resources, Vol. XII.

171

US Army Corps of Engineers. 1993a. *Annual Operating Plan, Missouri River District, 1993–1994.* Omaha, NE: Missouri River District.

US Army Corps of Engineers. 1993b. *Preliminary Draft Environmental Impact Statement, Missouri River Master Water Control Manual Review and Update.* Omaha, NE: Missouri River Division.

US Army Corps of Engineers. 1994a. *Delaware River Basin Study, Main Report.* Philadelphia, PA: Philadelphia District.

US Army Corps of Engineers. 1994b. *ACF River Basin Water Control Manual.* Draft Revision, Mobile, AL: Mobile District.

US Army Corps of Engineers. 1994c. *1994 Annual Flood Damage Report.* Draft Revision, Mobile, AL: Mobile District.

US Bureau of Reclamation. 1986. *Colorado River Simulation System: System Overview.* Denver, CO: Engineering and Research Center.

US Bureau of Reclamation. 1991. *Colorado River Simulation System: Inflow and Demand Input Data.* Denver, CO: Engineering and Research Center.

US Environmental Protection Agency. 1978a. *Analysis of Operations and Maintenance Costs for Municipal Wastewater Treatment Systems.* EPA 430/9-77-015, Washington, DC: Office of Water Program Operations.

US Environmental Protection Agency. 1978b. Construction Costs for Municipal Wastewater Treatment Plants: 1973–1977. EPA 430/9-77-013, Washington, DC: Office of Water Program Operations.

US Water Resources Council. 1978. *The Nation's Water Resources: 1975–2000.* Volume 1: Summary. Washington DC: US Government Printing Office.

Vaux, H.J. and Howitt, R.E. 1984. Managing Water Scarcity: An Evaluation of Interregional Transfers. *Water Resources Research* **20**: 785–92.

Waggoner, P.E. (ed.) 1990. *Climate Change and U.S. Water Resources.* New York: John Wiley.

Watson, R.T., Zinyowera, M.C. and Moss, R.H. (eds.). 1996. *Climate Change 1995: Impacts, Adaptation, and Mitigation of Climate Change: Scientific–Technical Analyses.* Cambridge: Cambridge University Press.

Appendix A6

A6.1 Spatial effects and valuing return flows

Hartman and Seastone (1970), in their analysis of the consequences of return flows on water-use efficiency, observed that optimal (or efficient) water allocations are a function of return flow rates, and therefore, these return flows affect the marginal value of water in different uses. Specifically, the shadow price for water at the optimum is a function of return flow rates, and therefore, generally differs across users. For example, at the optimum, withdrawals by users with high return flow rates (or conversely, low rates of water consumption) are consistent with low marginal values for water, compared to the withdrawals of users with low return flow rates, which are consistent with high marginal values, as described in the following example.

Consider a river basin with three water users, two upstream consumptive users and a downstream user, and available water that exceeds possible consumptive requirements. The upstream users (e.g. a city and an agricultural user) are assumed to have return flow rates of 80 percent and 50 percent, respectively. Water is freely available to each user (i.e. they divert as much water as they wish until the marginal value for further withdrawals is zero), and water in excess of their demands flows to the downstream user. Further assume that this downstream user is, for example, a hydroelectric producer who has a marginal value (i.e. willingness to pay) for water equal to $40/af.

The welfare of both upstream and downstream users can be improved in this situation. Consider, for example, that the downstream user offers each upstream user a payment of $40 for each additional acre foot of water that is made available for downstream use (i.e. for water that is not consumed upstream). This acre foot is in addition to the flows already received, and importantly, includes return flows. To yield this additional acre foot downstream, the agricultural user could reduce diversions by 2 af (i.e. by $1/(1-\text{return flow rate})$), or the city could reduce its diversions by 5 af. In the first case, the payment of $40 to the agricultural user for reducing diversions by 2 af is in effect a payment of $20/af. Therefore, the agricultural user would be willing to reduce diversions up to the point where the marginal value of using the water for irrigation was equal to $20/af. Similarly, for the city user the payment of $40, for foregoing the use of 5 af, results in an average payment of $8/af. And therefore, the city would be willing to forego diversions up to the point where the net marginal revenue from supplying municipal users equaled $8/af.

In general, total welfare is maximized where the marginal value of water is adjusted for return flows and is equated across all users, such as:

$$MV_1/(1-r_1) = MV_2/(1-r_2) = MV_3, \qquad (A6.1)$$

where MV is the marginal value, r is the return flow parameter (i.e. the share of diverted water that is returned to the river), subscripts 1 and 2 refer to the upstream users, and subscript 3 to the downstream user. At the optimum, users with lower return flow rates have greater implicit marginal values for water than those with higher return flow rates. This is an important result that characterizes optimal allocations in a river basin SE model with return flows.

Analytic derivation

To derive the above result for the two user case, consider an upstream and a downstream user who withdraw W_1 and W_2 from the river, respectively. The economic problem is to maximize the welfare of these two users subject to water availability and flow continuity, for example:

$$\text{Max}_{W_1, W_2} \quad f(W_1) + g(W_2)$$

s.t.

(1) $\quad W_1 \leq K$

(2) $\quad W_2 \leq K - W_1 + r_1 W_1$

(3) $\quad W_i > 0 \quad i = 1, 2,$

where f and g are single-valued functions reflecting the net benefits from water use for an upstream user (subscript 1) and a downstream user (subscript 2), respectively. K is a constant equal to the fixed quantity of water available, and r_1 is the return flow parameter. The first two constraints describe the availability and continuity of water to each of the two users, and the third constraint ensures that water use is positive for both users.

The Lagrangian function for this optimization problem is given as:

$$L = f(W_1) + g(W_2) + \lambda_1(K - W_1) + \lambda_2(K - W_1 + r_1 W_1 - W_2),$$

where λ_1 and λ_2 are the shadow prices for each of the two constraints, respectively. These shadow prices represent the marginal value of additional water to the system at each use. The first-order conditions, characterizing optimal withdrawals are:

$$f_{W1} - \lambda_1 - \lambda_2(1 - r_1) = 0, \text{ and} \tag{A6.2}$$

$$g_{W2} - \lambda_2 = 0, \tag{A6.3}$$

where first derivatives are indicated by subscript notation (i.e. $\delta f / \delta W_1 = f_{W1}$). By substitution, and assuming that W_1 does not deplete the entire flow of the river (i.e. there is slack in the first constraint resulting in ($\lambda_1 = 0$), the following relationship characterizes the optimal allocation of water:

$$f_{W1}/(1 - r_1) = g_{W2}. \tag{A6.4}$$

This is the relationship expressed above in Equation (A6.1).

A6.2   General form of the basin economic models

The general structure and composition of the river basin SE models is presented below, in which we describe some of the technical aspects of the models. The model description is general and contains features of all the models, and therefore, some equations may not be defined for a specific model. The objective function and its components is first defined, and this is followed by descriptions of the constraints. All variables are assumed to be positive except the objective variable (CPS), reservoir releases (R_{nt}), and net reservoir evaporation (E_{rt}) which can all vary freely. Definitions of indices, variables, and parameters follow the model description.

Objective Function: Max CPS by choosing F_{nt}, S_{rt}, X_{it}, H_{rt}

$$CPS = \sum_t DF_t \times \Big[\sum_i (a_{ni} + 0.5\, b_{ni} W_{nit} - 0.5\, c_{ni} W_{nit}^2) W_{nit} \quad \text{consumptive use}$$

$$+ \sum_i W_{ni0}(\bar{V}_{ni} - V_{ni0}) + V_{ni0}\left(\frac{W_{nit}}{W_{ni0}}\right)\beta_{ni} \quad \text{(COBEM only)}$$

$$+ \sum_{r \in N} h_r \times P \times H_{rt} \quad \text{hydropower benefits}$$

$$+ \sum_n (1 + e^{ln + mnFnt}) \quad \text{navigation benefits}$$

$$- \sum_n (f_n + g_n FL_{nt}) FL_{nt} \quad \text{flood damages (above threshold)}$$

$$- \sum_n K_n [1 - (1 - e^{knFnt})] \quad \text{thermal waste heat (opportunity costs)}$$

$$+ \sum_n q_n F_{nt} \quad \text{secondary wastewater treatment benefits}$$

$$- \sum_n \Big[\left(\frac{SL_{nt}}{F_n}\right) \times C \times \sum_i 2r_{ni} W_{nit}\Big] \quad \text{advanced wastewater treatment costs}$$

$$+ \sum_r \$ \times VIS_r \sqrt{\frac{(ar_{r0} + ar_{r1}S_{rt} + ar_{r2}S_{rt}^2)}{S_r^{\max}}} \quad \text{flatwater/reservoir recreation}$$

$$+ \sum_n \$\,(i_1 + i_2 F_{nt} + i_3 F_{nt}^2) \quad \text{instream recreation benefits}$$

$$- \sum_i SD_{ni} W_{nit} \frac{(NA_{n-1,t} + INA_{nt})}{(F_{nt} + W_{nit})}\Big] \quad \text{salinity damages}$$

$$- DC_{n,t} \times D_{n,t} \quad \text{penalty for compact violation (COBEM only).}$$

Subject to:

$$F_{nt} = F_{n-1,t} + I_{nt} + R_{nt} + \sum_i r_{ni} W_{n-1,i,t} - \sum_i W_{nit} \quad \text{flow balance}$$

$$S_r^{\min} \leq S_{rt} - S_{r,t-1} + R_{rt} + E_{rt} \leq S_r^{\max} \quad \text{storage balance}$$

$S_{rT} = S_{r0}$ terminal storage constraint

$FL_{nt} \geq F_{nt} - FT_n$ flood level constraint

$GW_{nit} \leq \overline{GW}_{ni}$ groundwater supply

$H_{nt} - F_{nt} + SP_{nt} \leq \bar{H}_r$ hydropower capacity constraint

$$E_{rt} = 0.5(PET_{rt} + \frac{PET_{rt}}{S_r^{\max}} S_{rt})$$ reservoir evaporation constraint

$$NA_{nt} = NA_{n-1,t} + INA_{nt} - \frac{(NA_{n-1,t} + INA_{nt}) \times \sum\limits_{i \in export} W_{nit}}{(F_{nt} + \sum\limits_{i \in export} W_{ni}t + 1)}$$ salt balance (COBEM)

$F_{n,t} + D_{n,t} \geq 8230$ Colorado River Compact constraint, in force for $n = 8$

where *export* is the set of sectors that include export of salt from the Colorado basin.

Definitions:

Indices

i	consumptive users, i = agriculture, municipal, thermoelectric
n	model nodes (reaches), $n = 0,1,3,\ldots,N$
t	model time step (annual for COBEM, seasonal for all others), $t = 1,2,3,\ldots,T$
r	model reservoirs, $r = 1,2,3,\ldots,R$.

Variables

CPS	consumer plus producer surplus
F_{nt}	river flow leaving node n at time t (includes tributaries)
S_{rt}	reservoir storage volume, reservoir r at time t
R_{nt}	net reservoir release, into node n at time t
H_{rt}	reservoir release for hydropower production, into node n at time t
SP_{nt}	reservoir release spill into node n at time t in excess of hydro capacity
I_{nt}	exogenous inflow (including tributaries) into node n at time t
FL_{nt}	river flow in excess of flood damage threshold at node n at time t
SL_{nt}	slack variable reflecting deviation from minimum flow requirements for water quality
GW_{nit}	groundwater use by user i at node n at time t
W_{nit}	withdrawal of water by user i at node n at time t
E_{rt}	reservoir evaporation at time t
NA_{nt}	salinity quantity (thousands of tons) at node n at time t
$D_{n,t}$	deficit from Colorado River Compact, for $n = 8$.

Constants and parameters

DF_t	discount factor (set to zero for purposes of this analysis)
a_{ni}	intercept of linear demand functions, user i at node n
b_{ni}	slope of linear demand functions, user i at node n
c_{ni}	slope of linear cost (supply) functions, user i at node n
β_{ni}	elasticity coefficient for nonlinear value functions
$\bar{V}_{ni}, V_{ni0}$	value parameters for nonlinear value functions
W_{ni0}	climate adjusted depletion request in COBEM
$ar_{r0}, ar_{r1}, ar_{r2}$	reservoir surface area parameters
r_{ni}	return flow coefficient for user i at node n
f_n	slope of flood damage function, node n
FT_n	flood damage threshold at node n
g_n	quadratic term in flood damage function, node n
l_n, m_n	location and slope coefficients for navigation benefits at node n
K_n	maximum value of OTC power production
k_n	slope term for thermal waste heat opportunity costs
h_n	average reservoir head
P	constant term for power production efficiency, utilization, and valuation
q_n	slope of linear secondary treatment benefits
$\bar{F}_n$	minimum flow requirement to maintain water quality at node n
C	average cost per acre foot of advanced wastewater treatment
$\overline{GW}_{ni}$	groundwater supply capacity
$\bar{H}_r$	hydropower release capacity
$\$$	user day value for recreation benefits
PET_{rt}	exogenous potential evaporation level
VIS_r	historical visitation rates at Colorado basin reservoirs
i_1, i_2, i_3	quadratic parameters for instream recreation benefits
SD_{ni}	salinity damage coefficient for lower Colorado basin users
INA_{nt}	exogenous salt loadings in Colorado basin
$DC_{n,t}$	unit cost of Compact violation, for $n = 8$.

7   The economic damage induced by sea level rise in the United States

GARY YOHE, JAMES NEUMANN,
AND PATRICK MARSHALL

Changes in climate are expected to affect the ocean environment in a variety of ways. The potential effects of a temperature increase include thermal expansion and the melting of polar ice caps, both of which contribute to the causes of sea level rise. Increases in sea level can present problems to people living in coastal and low-lying areas, and can damage structures and beachfront property along the coast. Consequently, a sea level rise may impose economic costs on the United States – the costs of protecting coastal structures and the shoreline, or the lost value associated with abandoning such structures and property.

Early predictions of dramatic greenhouse gas-induced sea level rise have given way over the past decade to more modest expectations. High projections for the year 2100 reached more than 3.5 meters as late as 1983 (Hoffman *et al.*, 1983), but they dropped to 1.5 meters in 1990 (IPCC, 1990), and converged slightly more than 1 meter by 1992 (IPCC, 1992). The mid-range best guess now stands between 38 and 55 cm by 2100 (IPCC, 1996). One recent estimate is presented in Table 7.1 (Wigley, 1995; Wigley and Raper, 1992). The oceans would continue to rise for centuries, even if concentrations were stabilized in the interim. Despite this, the highest best guess reported for the year 2100 is 40 cm. One important contribution of this chapter is to present economic cost estimates for these new lower trajectories (less than 1 meter).

Another important contribution of this chapter is to illustrate the importance of including efficient adaptation. People can adjust to rising seas by constructing barriers and retreating. If these decisions are made rationally, the economic damages from rising seas fall dramatically. Economic damages are calculated as the sum of the value of lost property (valued at the time of loss, net of market-based adaptation that might mitigate against this cost, but including the cost of that adaptation) and the expense involved in protection. We model responses to sea level rise as though they are efficient. Because the efficient response of many protection measures involves cooperation among neighbors, this model implies a degree of public efficiency. Because

Table 7.1. *Sea level consequences of greenhouse gas concentrations*[a, b]

(1) Stabilization level (in ppm CO_2)	(2) Year of stabilization	(3) Sea level rise by the year 2100 (in cm)			(4) Sea level rise by the year 2400 (in cm)		
		Low	Middle	High	Low	Middle	High
350	2050	0	16	39	−18	19	77
450	2100	4	26	56	−10	52	157
550	2150	7	32	65	−3	77	216
650	2200	9	36	72	2	96	261
750	2250	11	40	78	7	112	300

Notes:
[a] Wigley, 1995 (Table 2) See citation for further elaboration of calculation and meaning of low, middle, and high designations.
[b] Sea level rise is measured in cm along scenarios that reach the stabilization levels recorded in column (1) by the date indicated in column (2).

public efficiency is by no means guaranteed, the results may be overly optimistic. Nonetheless, they do represent the lowest cost (smallest aggregate damage) of response to sea level change.

The first section of this chapter provides some context for this analysis by reviewing past estimates of the potential cost to the United States of sea level rise, and Section 7.2 offers a description of the assumptions and methods that frame this work. Section 7.3 provides an example site (Charleston, SC) to illustrate the methods and assumptions that are used in this analysis, as well as to demonstrate how the results were obtained and interpreted. Section 7.4 provides the results of the entire study, which examines the aggregate costs resulting from sea level rise over the entire United States, and Section 7.5 contains concluding remarks that return to the historical context to argue that currently accepted base-case estimates of the cost of sea level rise are much too high, and to discuss the implications of these overstated damage estimates.

Past estimates have been derived from sea level rise trajectories that exceed the upper range of the current scientific consensus. These earlier estimates have also underestimated adaptation. After correcting for both sources of error, estimates of the annual damages in 2060 from rising sea level are over an order of magnitude lower than the earlier estimates in the literature. A sea level rise of 33 cm by 2100 is expected to cause annual national damages of only $57 million in 2060.

179

7.1   Historical estimates of costs

Table 7.2 presents some of the cost estimates that have preceded this work. These studies are based upon estimates of economic vulnerability as opposed to true economic cost. Economic vulnerability measures the gross damages today if sea level rise suddenly occurred. It does not take into account adaptation or the fact that inland property would rise in value as it becomes shoreline property. Economic vulnerability is therefore not an accurate measure of net damages. The first cost estimate of sea level rise assumed a dramatic rise in the seas which in turn led to a tremendous damage of $450 billion (Schneider and Chen, 1980). Using a 1-meter rise by 2100 assumption, Nordhaus (1991) used the 1989 USEPA Report to Congress to predict a new estimate of $2.4 billion in lost land value (adjusted to 1990 for inflation) and $4.9 billion in annual protection costs in the year 2065. Several other authors have made similar estimates to the projections by Nordhaus.

The more recent assessments all roughly agree with the early Nordhaus projection. The consistency of these estimates should not be surprising. All of the estimates use the same USEPA report that was used by Nordhaus, and so they all build on the notion that $73–$111 billion in cumulative protection costs would be incurred up to the year 2100 for a 1-meter rise. They also tend to agree with Cline (1992) that something on the order of 6650 square miles of dry land valued at $4000 per acre would be abandoned, and that approximately 13 000 square miles of wetland valued at $10 000 per acre would be lost. In addition, and perhaps more importantly from an economic perspective, each cost estimate for sea level rise has been constructed from vulnerability measures of real estate losses. Each expresses the potential total cost of abandoning property. In other words, each estimate reflects the total, current value of the real estate that might be lost to a rise in sea level between now and the year 2100. This current value of real estate is computed by summing the constant annual values between now and the year 2100. This method by its very construction is static. It compares a snapshot of coastal property taken for the year 2100 with a snapshot of current development, but it expresses any differences between the two in terms of average annual changes. This procedure averages across a 110-year time span, and so it offers the same picture for the years 2050, 2075, or any other year in its range. Integrated assessments that need to calibrate costs at some point in the future would prefer to employ "transient costs" that are computed for that year, rather than rough averages across a century or so.

The data to which this averaging procedure has been applied were drawn from current conditions. Since total damage estimates were constructed from vulnerability estimates, they ignore future development and land appreciation that can be expected

Table 7.2. *Annualized cost at concentration doubling (billions of 1990$)* [a]

| Source | Year | Assumed changes | | Economic damages (comprehensive annual costs) | |
		Temperature change (°C)	Sea level rise (m)	Sea level	Total
Schneider & Chen (1980)	2100	n/a	4.6	$450.0	n/a
Nordhaus (1991)	2065	3.0	1.0	$7.3	$8.7
Cline (1992)	2065	2.5	1.0	$7.0	$61.1
Titus (1992)	2065	4.0	1.0	$5.7	$139.2
Fankhauser (1994)	2065	2.5	1.0	$9.0	$69.5
Tol (1994)	2065	2.5	1.0	$8.5	$74.2

Notes:

[a] The Nordhaus estimates have been converted to 1990 dollars using the US Consumer Price Index; the Tol estimates include Canada and the United States.

even on vulnerable property in the intervening years. These total damage estimates miss any adaptation that might occur naturally within the market as new information emerges. They also miss any policies that might be enacted to protect or abandon property, and the cost of protection that must be applied to property that merits protection.

Correcting for most of these shortcomings can be expected to reduce the potential cost of sea level rise, since adaptation would not occur unless costs were reduced. There is, nonetheless, some ambiguity that must be explored. Adaptation could reduce the cost of abandoning property, but appreciation over the intervening years might increase the eventual cost of abandonment and thus increase the acceptable cost of protection. In addition, while optimal adaptive decisions that minimize the cost to society as a whole may be made with adequate and timely information, it is possible that society will choose high cost decisions (e.g. social decisions to protect current coastal dwellers regardless of costs). The work reported here presents a range of adaptation from perfect to imperfect foresight. However, social decisions to resist sea level rise could be even worse than the imperfect foresight scenario presented in this chapter. The next section describes the framework in detail.

7.2   Methods

Planning the response to rising seas along a developed coastline can be broken into two distinct decisions that are made in an effort to maximize discounted intertemporal welfare (the net benefits of any protection strategy minus the cost of its implementation). The first is a decision to protect the coastline starting at some time t_0 and the second is a decision to stop protection at some time T. The following subsections will discuss the benefits and costs associated with protection decisions.

Benefits of protection

The benefit side of a decision to protect a shoreline from time t_0 to time T can be modeled as the true opportunity cost of abandoning coastal property. This is calculated here as the economic damage that might be attributed to future sea level rise in the absence of any decision to protect threatened property. True opportunity cost is based on the value of that property at the (future) time of inundation, given any adaptation that might have occurred naturally and efficiently prior to flooding and abandonment. Satisfactory descriptions of how future development might affect coastline real estate values were derived from empirical market analyses of how property values might change as factors such as population and real income change. One estimate of these damages is:

$$d[\ln(P_t)] = \alpha_0 + \beta_L g_L + \beta_y g_y + \beta_{-1} d[\ln(P_{t-1})], \tag{7.1}$$

where g_L and g_y represent the rates of growth of population and per capita income, and P_k represents the real price of property in year k (Abraham and Hendershott, 1993). Constructing scenarios of how these "driving socio-economic variables" might move as the future unfolds, produced historically based portraits of how real property values might change over the same time frame (IPCC, 1992). Applied with care in the absence of any anticipated, fundamental structural change in the real estate market-place, the resulting development trajectories offer reasonable portraits of the evolving context of the sea level rise problem.

Satisfactory descriptions of how real estate markets might respond on a smaller, local level in the face of threatened inundation from rising seas were more difficult to create. On the one hand, the value of the land lost to rising seas should be estimated on the basis of the value of land located inland from the ocean. Any price gradient which placed higher values on parcels of land in direct correlation with their proximity to the ocean would, in a very real sense, simply migrate inland as shoreline property disappeared under rising seas. Ignoring potential significant transfers of wealth, the true economic cost of inundation is the value of the land that will, in an economic sense, actually be lost – the interior land equal in area to the abandoned and inundated coastline property (Yohe, 1989). An exception to this rule would occur with barrier islands which must disappear altogether resulting in a net loss of coastal land.

The value of coastal structures, on the other hand, can be expected to depreciate over time as the threat of impending inundation and abandonment becomes known. Structures will be lost at the moment of inundation, and their true economic value at that point could be zero if markets were equipped with enough advanced warning and with a complete understanding that the property would, indeed, be abandoned. Despite stories of individuals' reluctance to abandon threatened property in, for example, flood plains, investigations into how markets react to low probability – high cost events strongly support the assertion that market-clearing real estate prices do indeed decline over time in response to the pending cost of a growing threat.[1]

True economic depreciation (TED), modeled to start at some fixed time prior to inundation and to finish just when inundation would occur, reflects the efficient market response to the known risk of future sea level rise (see Samuelson, 1964; Stiglitz, 1986).

---

[1] Brookshire, Thayer, Tschirhart, and Schulze (1985) found evidence that real estate values did reflect earthquake risks. MacDonald, Murdoch, and White (1987) found homeowner behavior similarly affected in the face of the threat of flooding. Property prices should, over the long term, reflect the threat of gradually rising seas.

TED is, by definition, a representation of how the value of an asset declines over time as it moves toward its retirement from service. Structures are 30-year assets in the view of the Internal Revenue Service (IRS), so 30 years of (certain) advanced warning was deemed to be sufficient. The application of TED here supports the position that the true economic cost of structures lost to rising seas could be as low as zero.

Uncertain abandonment, caused by the uncertain rate of future sea level rise and/or a disbelief that existing property would actually be abandoned, would affect efficiency. Either a source of imperfect information or an incomplete reaction to the threat of rising seas could, for example, reduce the time period over which markets could react to this threat. The value of lost structures or shorelines under these conditions would not be zero; it would, instead, equal the remaining value of the structure or shoreline at the time of inundation. The worst case of imperfect information and uncertain abandonment would allow absolutely no warning and thus no time for any structural depreciation at all. This case takes the lack of information to an extreme, and is more likely to be caused by a sudden realization that the policy of abandonment would be followed, rather than a sudden realization that the oceans have risen; but it captures the situation in which the cost attributed to rising seas would be maximized, and it allows for the possibility that property that should have been abandoned (given maximum efficiency and perfect information) might actually be protected, instead.

Costs of protection

The cost of protection from time t_0 to time T was easier to frame – it was simply the time trajectory of protection costs along the specified sea level rise scenario. Seven published studies offer specific cost estimates for various protection structures.[2] For protection against a 1-meter rise in sea level, a review of these eight studies suggested that the fixed costs of constructing dikes/levees range from $150 to $800 per linear foot, while seawall and bulkhead construction costs range from $150 to $4000 per linear foot. Costs depend upon engineering and construction specifications, as well as design standards and geological characteristics. The baseline results reported in this analysis were derived from a central estimate of $750 per linear foot for a generic hard structure, but their robustness was also tested in the extreme case where protection costs $4000 per linear foot. Maintenance costs, modeled as the variable cost of protection, were also incorporated in these studies. Since the central fixed cost estimate was drawn from Gleick and Maurer, 1990, their representation of annual maintenance expenditure as a percentage of construction was also adopted. Four percent per year was chosen as the central estimate, but 10 percent was applied to hard

---

[2] Weggel *et al.*, 1989; Sorenson *et al.*, 1984; Gleick and Maurer, 1990; URS Consultants, 1991; San Francisco BCDC, 1988; Leatherman, 1989; Leatherman, 1994.

structures that might be built along coastline open directly to the ocean (Weggel *et al.*, 1989; Sorenson *et al.*, 1984).

Structure and maintenance costs were changed for different scenarios, under the assumption that protection for the full measure of sea level rise expected to the year 2100 would be constructed when it was needed. Weggel *et al.* (1989) and Sorenson *et al.* (1984) both indicate that construction costs increase geometrically with height. Weggel *et al.* suggest a cost factor of 1.5 to reflect the geometric increase in cost with the height of the structure. Nichols, 1984 and Sorenson *et al.* offer more insight into the details of construction. They note that hard structures are typically trapezoidal in shape with 1 : 2 slopes on the sides and with the width of the crown on top matching the height. This information enabled us to compute a relationship between the cost of hard structures and their required height along 33 cm and 67 cm scenarios as fractions of the cost along a 100 cm scenario. Our results suggest a cost factor of nearly two to reflect the exponential increase in cost with the height of the structure. For example, at the Bridgeport, CT, site, the fixed cost of protection for 1 meter of sea level rise is $0.619 million. A protection structure under a 67 cm sea level rise scenario, on the other hand, would cost $0.272 million, and a protection structure under a 33 cm sea level rise scenario would cost $0.068 million. These protection costs illustrate the geometric increase in cost with the height of the structure.

A different methodology from that used for coastal structures was employed to accurately characterize the cost of protecting beaches and beachfront property. The basic idea conveyed by experts in the field was that beach nourishment alone would suffice as a protection strategy, provided that nourishment were an ongoing operation from the very start, and as long as sea level rise did not exceed some threshold; 33 cm was chosen to be that threshold. The cost of nourishment was computed from estimates of the requisite volume and the expected (regional) price of sand.[3] Once the threshold was crossed, however, a hard structure constructed at the back of the beach was required both to preserve the nature of the beach and to protect interior property. The cost assumptions for coastal structures described above were then applied with 10 percent annual maintenance costs. Ten percent, as opposed to 4 percent, is used to account for the increased maintenance necessary at open ocean sites.

7.3   Charleston, SC: an application example

This section describes the results of a careful analysis of the economic costs of future sea level rise at a specific site – Charleston, South Carolina. This section

---

[3] Private communication with Stephen Leatherman, Robert Hallermeier, and Dennis Dare.

should clarify the mechanics of how to apply the general structure of the benefit side of protection, as well as the specifics of the cost side, both of which are presented in the methods section above. Charleston was chosen because it was part of the sample which supported the earlier estimates of national vulnerability to sea level rise. Producing a time series of potential economic costs along a given sea level rise trajectory will allow a direct comparison with the vulnerability estimates derived from the previous Charleston analysis (Yohe, 1990). Moreover, the local geography of the Charleston site allowed the consideration of five distinct and qualitatively diverse "subsites": Downtown Charleston, Mount Pleasant, Avondale, Dorchester, and Sullivan's Island. The versatility of the model and its applicability across a range of sites and options was therefore adequately tested.

The first subsection describes the data and assumptions which frame both the Charleston site and the sea level rise trajectory (i.e. rate of sea level rise over time) to be considered. The second subsection presents results for each of the subsites; protection decisions are identified and supported. The descriptions of these protection decisions include their timing, which could minimize the discounted value of anticipated costs. The third subsection presents the ultimate result – a time profile of the cost of sea level rise along the given trajectory.

Background

We assume a quadratic sea level rise (SLR) scenario:

$$SLR(t) = bt^2. \tag{7.2}$$

For a 100 cm rise by the year 2100, $t = 110$ and SLR $(t) = 100$, yielding a value of b of approximately 0.008. The 100-cm trajectory certainly lies on the high side of the IPCC (1996) best estimates. It serves here, however, to support a diverse set of protection responses, and thus economic cost profiles, across the five Charleston subsites. It is also the middle trajectory in the national sample of vulnerability estimates completed in 1989 by Yohe (1990).

Inundation profiles along the 100-cm scenario over time for each subsite of Charleston are available from the computer-based mapping capability developed by Richard Park and his colleagues at the Holcomb Research Institute for the 1989 EPA Report to Congress (Park *et al.*, 1989). Each site in the Park sample, of which roughly one-third were used in the Yohe sample, represented a 30-minute cell provided by the US Geological Survey. The maps divide each site into 500-meter square partitions; and the mapping technology looks at how each partition changed over time for a specified sea level rise trajectory. If the seas were assumed to rise along, for example, the 100-cm scenario reflected in equation (7.2), then the Park maps would show snap-

Table 7.3. *Time series of inundated partitions for each of the Charleston subsites[a] (number of 500-m by 500-m blocks)*

Year	Charleston	Dorchester	Avondale	Mt Pleasant	Sullivan's Island
2000	0	0.5	0	0.5	0
2010	0	1.0	0	1.0	0
2020	0	1.5	0	0.5	0
2030	0	2	0	0.5	0
2040	0	0	0	0.5	0
2050	0	0	0	1.0	0
2060	0.5	0	0	1.0	0
2070	0.5	0	0	0	0
2080	0	0	0	0	0
2090	1.0	2.0	1.0	1.0	1.0
2100	0.5	1.0	0.5	0.5	2.0
Total	2.5	8.0	1.5	6.5	3.0

Notes:

[a] The number of 500-m by 500-m partitions deemed lost to rising seas along the 100-cm trajectory during the decade ending in the year noted. These values were judged from the Park *et al.*, 1989, mapping technology according to the following convention. Newly inundated partitions were noted for both of the 5-year intervals which comprised the period between one decade and the next; a partition is taken as inundated when more than 50 percent of its area would be under water during mean spring high tide. Any partition seen inundated in the first 5-year interval (say between 2040 and 2045) was assigned to the decade in question (the decade ending in 2050). Any partition disappearing in the second 5-year interval (say between 2045 and 2050) was shared 50 : 50 with the next decade (one-half to 2050 and one-half to 2060).

shots of seawater inundation and other land changes across all of the partitions in 5-year increments along that trajectory. Table 7.3 records, in decadal increments, the resulting time series of inundated partitions for each of the five Charleston subsites. The data in the table reflect the number of partitions deemed to be lost to the rising sea each decade. Applying estimates of how the value of the properties located in these threatened partitions might appreciate over time to the dynamic portraits of the physical impact of sea level rise produces estimates of (1) the potential benefit of protection, (2) the potential cost of abandonment, and/or (3) the cost of protection, all of which are statistics required to calculate the present value of the net benefit of protection. Together, the fixed cost and variable cost components define the cost side of the protection decision.

187

Table 7.4. *Characteristics of the Charleston subsites economic parameters*

	Charleston	Dorchester	Avondale	Mt Pleasant	Sullivan's Island
Initial land value[a]	8.6	0.8	1.9	6.0	10.3
Initial value of structure[a]	25.9	2.5	6.0	18.1	30.8
Fixed cost of protection[b]	11.6	23.2	7.7	27.1	0.3
Variable cost of protection[c]	0.05	0.10	0.03	0.12	0.024

Notes:
[a] Denominated in millions of dollars (1989) per 500-m by 500-m partition.
[b] Denominated in millions of dollars (1989).
[c] Denominated in millions of dollars (1989) per cm of sea level rise.

Table 7.4 reports the fixed and variable costs at each of the five Charleston subsites. They are extrapolated from a detailed estimate for building dikes and nourishing beaches (raising a barrier island), for Long Beach Island, NJ, as well as for a few other sites scattered around the country (Weggel, 1989). The relatively low fixed protection cost for Sullivan's Island corresponds to the small initial cost of preparing to raise the island and nourish its beaches with sand. Diking is simply not an option, so variable costs reflect an ongoing and increasing investment in sand along its entire length. A decision to protect the island would, in fact, really be a decision to begin protection in the year 1990, because delay is not possible. Irreversible, or at least problematical, erosion and inundation of beaches and dunes would begin immediately along a sea level rise trajectory unless some protective strategy were adopted. Nourishing the beach with sand is only viable up to a certain threshold. Once the sea rises beyond this threshold, which in this case was designated as 1 foot, a hard structure, such as a sea wall or dike, is necessary along with the nourishment strategy.

Dikes alone emerge as a potential option in the other four subsites. The lists of protection decisions for each are more complete and more complicated. Dikes can be constructed at any time, so questions of when to start construction must be confronted directly. The fixed cost of the initial construction plays a critical role here, but it should be noted explicitly that a dike would be constructed only along the limited coastline that merits protection. Dikes must be maintained and enlarged over time, though, so variable costs which depend upon the rate of sea level rise create the possibility that even limited protection might not be continued indefinitely. The question is when, if ever, to stop protection and to sacrifice previously protected land as well as new property that is subsequently threatened.

Table 7.4 also reports the initial values of land and structures (per 500-m^2 partition) that were employed to anchor the appreciation of property values in each subsite over time. Structure values were assumed to equal three times the land values (see Poterba, 1984). To preserve comparability, and in the absence of any other reasonable set of estimates, we use the average values for land and structures that supported the earlier Yohe, 1990 vulnerability estimates. Appreciation in the value of threatened property reflects the likely effect of future development – development that will be driven by future changes in real income and population. Abraham and Hendershott (1993) provide a regression result for housing prices, which could be interpolated for land value, assuming only that real construction costs and after-tax interest rates will be roughly stable over the very long term.[4] Given this assumed stability of relative prices, the best fit regression over their full sample is

$$d[\ln(P(t))] = -0.006 + 0.0313g_L + 0.565g_Y + 0.402[\ln(P(t-1))], \qquad (7.3)$$

where g_L and g_Y represent the rates of growth of population and real per capita income, respectively.[5] Equation (7.3) provides a means of proposing the P_t trajectory required to quantify the net benefit to society from protecting property from time t_0 to time T, given the anticipated population and (per capita) income scenarios. Given the income and population forecast for the next century (IPCC, 1992), one can forecast how real estate prices (adjusted for inflation) will increase over that time period.

The literature on property values offers only limited and somewhat contradictory evidence that coastal property values might change at a different rate from non–coastal property values. On the one hand, Frech and Lafferty (1984) and East (1990) have argued that policy factors, such as development moratoria, could constrain the future "supply" of coastal properties relative to other locations, and thereby inflate their relative price; historical data do not support this assertion, however. Parsons (1992) and Beaton (1988) note that the data seem to suggest that historical rates of growth over time for coastal and non–coastal property values are not significantly different. Note that rates of change were most important in drawing moving portraits of future development; differences in the initial (1990) valuation of property are reflected in the site-specific property value data. In the absence of more compelling evidence that rates of growth for coastal property should be different, rates of growth for property

[4] The Poterba (1984) correlation, combined with the IRS convention of a fixed proportional relationship between land and structure values, supports the application of the Abraham and Hendershott (1993) results to land and structures taken separately.

[5] The income elasticity reflected here might appear high, but it is a long-term elasticity and represents a response to changes in real income per working age adult. The corresponding short-term elasticity corresponds well to the lower estimates offered by Peek and Wilcox (1991), Mankiw and Weil (1989), and Hendershott (1991).

values in general were used here; we did not distinguish between coastal and non-coastal property values.

Table 7.5 records the current values of property that would be lost to inundation in the absence of any protection along the trajectories described for each subsite in Table 7.3. This table is divided into two components. The first part (A) reflects the value of interior land equal in area to the coastal property which would be lost over succeeding decades, beginning in the year 2000. The data in the second part (B) reflect the potential cost that could be attributable to sea level rise if there were absolutely no anticipation of impending loss. The estimates combine the value of land with the value of appreciated structures, which have not been depreciated at all as a result of impending loss, but are located on the threatened land at the time of inundation. These estimates therefore reflect the cost of abandoning property to the rising sea level if the market did not adjust to the rising sea level or to a plan to abandon threatened property.

Some representative results

Table 7.6 displays an array of results for the Downtown Charleston subsite – the present values of beginning protection at time t_0 (indicated in the first column) and stopping at time T (indicated in the top row). Positive values appear only in the last column in which the property is never abandoned, at least not before the year 2100; and a present value of slightly more than $900 000 emerges as the highest value in the entire table. Notice that this maximum value corresponds to (1) planning to build the requisite protective dike in the year 2050, just before inundation losses would be felt, and (2) maintaining the dike beyond the year 2100. In terms of the notation described earlier, $t_0^* = 2050$ and $T^* > 2100$ for Downtown Charleston. Note that this is only the net benefit result for the inundation period starting in 2054. There are two other inundation periods that are also part of the Downtown Charleston subsite (see Table 7.3). All of these results are then aggregated to get the overall strategy and associated dollar value for the entire Downtown Charleston subsite.

Mount Pleasant is a subsite for which protection fails the net welfare test when threatened structures efficiently depreciate to worthlessness (or partially depreciate, given that the first inundation starts in the year 2000, which allows for only 10 years of depreciation), just before they are inundated by the rising water. The present values of all of the protection options are negative (the decision array is not presented here). If foresight were not perfect, and the undepreciated structure, as well as land, would be lost to inundation, a different decision could be made. For example, if residents and thus real estate markets simply did not believe that their property would be abandoned, then structures and land might continue to appreciate right up to the very end. The cost of abandonment would then be exaggerated by disregarding the threat of sea

Table 7.5. *Current value loss to sea level rise*[a] *(millions of dollars)*

	A. Perfect foresight[b]					B. No foresight[c]				
Year	Charleston	Dorchester	Avondale	Mt Pleasant	Sullivan's Island	Charleston	Dorchester	Avondale	Mt. Pleasant	Sullivan's Island
2000	0	1.6	0	17.5	0	0	1.7	0	24.5	0
2010	0	1.5	0	14.7	0	0	1.9	0	28.9	0
2020	0	1.3	0	3.3	0	0	5.1	0	12.5	0
2030	0	1.5	0	2.8	0	0	6.2	0	11.3	0
2040	0	0	0	7.4	0	0	0	0	28.9	0
2050	0	0	0	7.6	0	0	0	0	29.8	0
2060	7.6	0	0	6.3	0	30.3	0	0	30.3	0
2070	7.6	0	0	0	0	30.2	0	0	0	0
2080	0	0	0	0	0	0	0	0	0	0
2090	10.7	4.1	2.6	7.7	12.8	42.8	16.2	10.2	29.9	51.0
2100	11.7	1.6	2.8	8.4	31.7	46.9	6.5	11.2	32.7	126.9
Total	37.6	11.6	5.4	75.7	44.5	150.2	37.6	21.4	228.8	177.9

Notes:

[a] The value of land (A) or land plus structure (B) that would be lost in the decade indicated.

[b] The values of the lost land, appreciated up to 30 years short of the point of inundation. The 2000 and 2010 values included 20 and 10 years of undepreciated structure; true economic depreciation with a 3 percent discount rate was applied.

[c] No foresight implied no market reaction until the date of inundation.

Table 7.6. *Decision array for Downtown Charleston: the present value of the net benefits of protection alternatives with perfect foresight*[a]

T:	1990	2000	2010	2020	2030	2040	2050	2060	2070	2080	2090	2100	>2100
t_0													
1990	−11.61	−11.62	−11.69	−11.79	−11.89	−11.99	−12.08	−12.08	−11.70	−11.39	−11.01	−10.64	−9.26
2000		−8.60	−8.68	−8.78	−8.88	−8.97	−9.06	−9.07	−8.68	−8.38	−7.99	−7.63	−6.24
2010			−6.37	−6.47	−6.57	−6.67	−6.75	−6.76	−6.37	−6.07	−5.68	−5.32	−3.93
2020				−4.72	−4.82	−4.92	−5.00	−5.01	−4.63	−4.32	−3.94	−3.57	−2.18
2030					−3.50	−3.59	−3.68	−3.69	−3.30	−3.00	−2.61	−2.25	−0.86
2040						−2.59	−2.68	−2.68	−2.30	−2.00	−1.61	−1.24	0.14
2050							−1.92	−1.93	−1.54	−1.24	−0.85	−0.49	0.90
2060								−1.42	−1.28	−1.15	−0.90	−0.63	0.48
2070									−1.05	−1.11	−0.99	−0.82	0.01
2080										−0.78	−0.66	−0.49	0.34
2090											−0.58	−0.55	−0.12
2100												−0.43	−0.43

Notes:

[a] The present value of the benefit of beginning protection at the time t_0 (indicated in the first column) and stopping at the time T (indicated in the top row) net of the present value of the cost of that protection and the loss involved in abandoning previously protected property at time T. These values assume a quadratic trajectory for sea level rise ending at 100 cm in the year 2100. A discount rate of 3 percent was employed both in the present value calculations and in the definition of the time trajectory of structure depreciation.

level rise and any planned retreat in its wake. The decision array (not presented here) obtained using the resulting inflated property values derived under the assumption of absolutely no market foresight, shows that a decision to begin protection in the year 2000 and to continue past 2100 would be best.

The decision arrays for Dorchester and Sullivan's Island both reveal that threatened property should never be protected. The reasons for this differ for each site. Building protective dikes would have been the correct option for Dorchester, but at present their cost exceeds their value, primarily because prospective losses would be felt so far into the future that only interior land values support the benefit side of the calculations. Even discounting the cost of building a dike equally far into the future is not enough to support a positive difference between discounted benefits and costs. By way of contrast, beach nourishment (in effect, raising the island) would have been the correct option for Sullivan's Island, but any nourishment strategy must begin in 1990, even though the potential losses to sea level rise occur far in the future. Unfortunately for those whose relatively valuable properties are located on this barrier island, protection never results in a positive present value up to the year 2100, given a real discount rate of 3 percent. The trajectory of net benefits climbs towards zero as prospective stopping dates rise to 2100, though; so perhaps a longer time horizon would bring better news.

Avondale is a subsite for which a "partial" protection strategy results, meaning that certain blocks of land are not protected while others are. In the case of Avondale, the first inundation block started in the year 2080. The result was to forego protection. However, the next inundated block (actually half a block) occurred in 2090. At this point in time it was efficient to protect.

An intertemporal cost profile for Charleston

The middle five columns of Table 7.7 record the undiscounted incremental costs that are attributable in successive decades to sea level rise. These estimates incorporate perfect foresight and market adaptation (i.e. structure depreciation) along the 100-cm trajectory for each of the five subsites in the Charleston area. The last column shows the subsite costs across the whole area for each decade. The statistics displayed in the last column are expressed in current dollars (not discounted). These estimates include the cost of protection, when protection is deemed to be appropriate, and the cost of abandoned property when retreat from the rising seas is the better response. The present value of all of these costs, discounted at 3 percent, is nearly 37 million dollars – a sizable sum, to be sure, but certainly a small fraction of the total value of the metropolitan Charleston area.

Notice that over time, the current value cost statistics start high, fall quickly, and

Table 7.7. *Decadal economic cost estimates: the Charleston site[a] (in millions of 1989 dollars)*

Year	Dorchester	Avondale	Mount Pleasant	Sullivan's Island	Downtown Charleston	Total
2000	1.6	0	27.1	0	0	28.7
2010	1.5	0	0.1	0	0	1.6
2020	1.3	0	0.1	0	0	1.4
2030	1.5	0	0.2	0	0	1.7
2040	0	0	0.4	0	0	0.4
2050	0	0	0.6	0	14.4	15.0
2060	0	0	1.0	0	16.1	17.1
2070	0	0	1.5	0	18.3	19.8
2080	0	0	2.2	0	21.5	23.7
2090	4.1	2.6	3.2	12.8	25.8	48.5
2100	1.6	2.8	4.5	31.7	31.8	72.4
Total	11.6	5.4	40.9	44.5	127.9	230.3

Notes:
[a] These costs include the cost of protection for Mount Pleasant (beginning in 1990) and Charleston (beginning in 2050) as well as the value of lost property where abandoned (taken from Table 7.6). Optimal protection decisions and efficient adaptation with perfect information are both assumed given a 3 percent discount rate.

then gradually climb again. There are several reasons why this shape makes sense. Economic costs can start high because the cost of deciding not to protect property in the near term must include a significant proportion of the value of the structure since there is simply insufficient time to depreciate standing structures. Perfect foresight would allow all threatened structures to depreciate to zero after the year 2020, though, so this initial cost inflating effect eventually disappears. The cost of protection will eventually rise in the long run because the sea level is rising at a quadratic rate and the value of properties continues to rise over time.

7.4   National aggregate estimates

The results presented in this section were derived by applying the procedures outlined above for Charleston to the same national sample that supported the original vulnerability estimates (Yohe, 1990). The full sample is described in Table 7.8.

Table 7.8. *Subsample sites by region*

Region	Identification[a]	Major municipality	Northern latitude	Western longitude	Natural subsidence[b]
Northeast (NE)	MEROCKLA	Rockland	44 07 30	69 07 30	1.0
	MAWESTPO	Westport	41 37 30	71 07 30	1.5
	RIWATCHH	Watch Hill	41 22 30	71 52 30	0.6
	CTBRIDGE	Bridgeport	41 15 00	73 15 00	0.9
	NJLONGBE	Long Beach	39 45 00	74 15 00	2.7
	MDEASTON	Easton	38 52 30	76 07 30	2.4
	VABLOXOM	Bloxom	37 52 30	75 37 30	1.9
	VANEWPOR	Newport News	37 07 30	76 30 00	3.1
Southeast (SE)	NCLONGBA	Long Bay	35 00 00	76 30 00	0.6
	SCCHARLE	Charleston	30 00 00	80 00 00	2.2
	GASEAISL	Sea Island	31 22 30	81 22 30	1.8
	FLSTAUGU	St Augustine	30 07 30	81 30 00	1.8
	FLMIAMI	Miami	25 52 30	80 15 00	1.1
	FLKEYWES	Key West	24 37 30	81 52 30	1.0
	FLPORTRI	Port Richey	28 30 00	83 45 00	0.7
Gulf Coast (Gulf)					
	FLAPALAC	Apalachicola	29 45 00	85 07 30	1.2
	FLSTJOSE	St Joseph	29 52 30	85 30 00	0.7
	MSPASSCH	Pass Christian	30 22 30	89 15 00	1.2
	TXPALACI	Palacios	28 45 00	96 15 00	2.8
	TXPORTLA	Portland	27 52 30	97 22 00	2.8
	TXGREENI	Green Island	26 30 00	97 22 00	3.9
	LAMAINPA	Main Pass	29 22 30	89 15 00	9.3
	LABARATA	Barataria	29 45 00	90 22 30	9.3
	LAGRANDC	Grand Chenier	29 52 30	93 00 00	8.5
West Coast (West)					
	CAALBION	Albion	39 15 00	123 52 30	0.0
	CAPTSAL	Point Sal	35 00 00	120 45 00	0.0
	CASANQUE	San Quentin	38 00 00	122 30 00	0.1
	ORYAQUIN	Yaquina	44 45 00	124 07 30	−1.0
	WAANACOR	Anacortes	48 45 00	122 45 00	0.2
	WATACOMA	Tacoma	47 30 00	122 30 00	0.8

Notes:

[a] Site identification codes reflect the state abbreviation in their first two letters and the major municipality in their last six letters.

[b] Rate of shoreline subsidence in mm per year.

195

For each site, the same computer-based mapping technique was applied as in Charleston to interpolate inundation effects for each sea level rise scenario.[6] In those maps, each site was partitioned into square cells usually measuring 500 meters on each side. A computer run for each cell provided specific effects in 5-year increments for designated sea level scenarios. These scenarios were defined by an assumed contribution from greenhouse warming, as well as by a site-specific rate of natural subsidence. Sea level rise in year t upon the shoreline of any site J along sea level rise trajectory K was, more specifically, expressed by:

$$SLR_{JK}(t) = S_J(t - t_J) + GH_K(t - t_J)^2, \qquad (7.4)$$

where t_J represents the year of initialization for site J, S_J represents the rate of local subsidence for site J, and GH_K represents a greenhouse warming coefficient intended to produce the chosen cumulative rise to the year 2100.

Time series of the economic cost of future sea level rise at each site were constructed as the sum of protection costs and abandonment losses, under the assumption that decisions to protect were made on a cell-by-cell basis within each sample site. The size of these cells may or may not fit protection strategies for every site, as these must conform to the contours of the land. However, the units are sufficiently small to judge the economic strategy which fits this problem most closely. Further research may indicate slight improvements in decision making with alternative units but is unlikely to uncover a large bias. Abandonment losses, given property appreciation and market adaptation, were derived by applying the procedures described above in the Charleston example to the same property value data that supported the original vulnerability estimates (Yohe, 1990). The resulting series therefore include the expense of protection or the cost (net of adaptation) of abandonment, applicable not only to each specific sample site, but also to specific regions and areas within that site. There was, for example, no reason to require protection for all of the cells in any site that might eventually be threatened by rising seas, as soon as rising seas reached the first one or two cells. For each cell that might be threatened at some time t, in fact, a decision to protect or not was made on the basis of maximizing the present value of the total (net) benefit of protection with respect to t_0, the time when protection might start, and with respect to $T > t_0$, the time when protection might end. The cost trajectories reported here, therefore, include the cost of protection only during times when protection is warranted on a cell-by-cell basis; and the cost trajectories include the (net) cost of abandonment only at the time of that abandonment.

[6] The mapping technology allows 50-, 100-, 150-, 200-, 250- and 300-cm scenarios to be applied to each of 98 sites chosen systematically from the 980 USGS half degree sites that have some coastline around the United States (Park *et al.*, 1989). The work reported here interpolates from these scenarios to get data for the 33-cm and 67-cm trajectories.

Table 7.9 reports the results for each site under the baseline cost assumptions, with and without foresight, given a 3 percent discount rate along three alternative sea level scenarios.[7] Table 7.10 and Figure 7.1 reflect summary estimates that emerge from these data for the United States, which are based upon the best available estimates of property value appreciation, market adaptation, and protection costs for three sea level trajectories under a variety of circumstances. Estimates of the present value of the true economic cost of the indicated sea level rise trajectories are recorded in column (1). They behave appropriately across the cases, showing larger estimates both for steeper sea level rise trajectories and for circumstances of absolutely no foresight. Perfect foresight is, however, not as valuable as one might think. The small increase in damages associated with imperfect foresight is easy to explain. First, a majority of the property is protected even when the maximum efficiency (minimum abandonment cost) implications of perfect foresight are imposed; improved information has, in these cases, no effect on the ultimate decision of whether or not to protect, and it has no effect on the cost of protection. Second, protection costs limit the value of information, because they cap economic costs for the cells that would not be protected with perfect foresight but that would be protected if the decision were made at the time of inundation, with no advanced adaptation or market response. Finally, most of the protection decisions are made well into the future and so differences in the discounted values of different decisions are small.

Columns (2) and (3) are the most easy to compare with previous estimates. For example, with the 100-cm sea level rise scenario, the annuitized costs run between $100 and $200 million (1990 dollars) – not even 20 percent of the estimated $1.1 billion in annual protection costs projected by earlier studies. The transient cost estimates for the year 2065 of $333 and $384 million are larger; but they, too, fail to cover more than one-third of the previous estimates. Fankhauser (1994) has produced the most comparable evaluation of the cost of sea level rise for the United States. It projects a benchmark protection cost that is in line with the established wisdom. The analysis is based upon smooth inundation patterns that are proportional in area to the assumed rates of sea level rise. Proportional inundation, however, is an oversimplification for any rugged coastline, and could easily produce overestimates of the area of land that is actually threatened. While the accuracy of smooth inundation patterns may be an open question, careful review of Table 7.9 shows no discernible patterns of inundation or protection decisions, and so casts some doubt on the Fankhauser results. His widely applicable systematic analysis, nonetheless, offers an estimate of $104.8 billion for the cumulative cost of protecting the United States from a 100-cm higher sea level, given a 3 percent discount rate.

---

[7] The results, given a 5 percent discount rate, are not reported here. The underlying data and spreadsheet-based methodology are available from the authors of this chapter.

Table 7.9. *National and regional estimates for coastal protection strategy costs (in millions of dollars)*

Region	Site name	100-cm SLR scenario			67-cm SLR scenario			33-cm SLR scenario		
		30 Years foresight	0 Years foresight	Protect?	30 Years foresight	0 Years foresight	Protect?	30 Years foresight	0 Years foresight	Protect?
Northeast	Rockland, ME	1.083	1.083	yes	0.477	0.477	yes	0.119	0.119	yes
	Westport, MA	3.829	3.829	yes	1.386	1.386	yes	0.244	0.244	yes
	Watch Hill, RI	9.259	9.259	yes	3.715	3.715	yes	0.768	0.768	yes
	Bridgeport, CT	6.213	7.598	partial protect (30 yrs) yes (0 yrs)	2.849	2.849	yes	0.477	0.477	yes
	Long Beach Island, NJ[b][c]	27.708	27.708	yes	18.331	18.331	yes	8.945	8.945	yes
	Easton, MD	6.102	6.102	yes	2.263	2.263	yes	0.264	0.264	yes
	Bloxom, VA	0	0	zero inundation	0	0	zero inundation	0	0	zero inundation
	Newport News, VA[a]	32.626	33.817	–	12.687	12.687	–	2.374	2.374	–
	Suffolk	0.254	0.445	no (30 yrs) yes (0 yrs)	0.138	0.138	yes	0.009	0.009	yes
	Hampton	22.322	22.322	yes	8.802	8.802	partial protect	1.693	1.693	yes
	Norfolk	6.477	6.477	yes	2.355	2.355	yes	0.353	0.353	yes
	Portsmouth	3.573	3.573	yes	1.392	1.392	partial protect	0.319	0.319	yes
Southeast	Long Bay, NC	1.282	4.762	no	0.877	2.785	no (30 yrs)[d] partial protect (0 yrs)[d]	0.245	0.246	partial protect (30 yrs) yes (0 yrs)
	Charleston, SC[a]	8.971	18.146	–	3.475	4.215	–	0.952	1.233	–
	Charleston City	1.101	1.214	partial protect (30 yrs)[d] yes (0 yrs)	0.261	0.261	yes	0	0	zero inundation

Region	Location									
	Mt Pleasant	4.057	6.287	no (30 yrs) yes (0 yrs)	2.410	2.550	partial protect (30yrs)[d] yes (0 yrs)	0.498	0.498	yes
	Avondale	0.176	0.185	partial protect (30 yrs)[d] yes (0 yrs)	0	0	zero inundation	0	0	zero inundation
	Dorchester	0.962	1.615	no yes (0 yrs)	0.804	1.404	no	0.454	0.735	no (30 yrs)
	Sullivan's Island, GA[bc]	2.675	8.845	no	4.579	0	zero inundation	0	0	zero inundation
	Sea Island, GA[bc]	7.182	7.182	yes	4.579	4.579	yes	2.171	2.171	yes
	St. Augustine, GA[bc]	2.236	2.801	no	1.700	2.322	no	0.805	1.580	no
	Miami, FL[bc]	15.675	15.675	yes	10.386	10.386	yes	5.068	5.068	yes
	Key West, FL[c]	11.636	11.636	yes	2.906	2.906	yes	0.528	0.528	yes
	Port Richey, FL[c]	5.874	8.422	no	5.023	7.855	no	2.329	2.329	yes
Gulf Coast	Apalachicola, FL	0.081	0.230	no (30 yrs) yes (0 yrs)	0	0	zero inundation	0	0	zero inundation
	St. Joseph, FL	0	0	zero inundation	0	0	zero inundation	0	0	zero inundation
	Pass Christian, MS	1.325	1.900	no[d]	0.955	0.955	yes	0.239	0.239	yes
	Palacios, TX	0.106	0.396	no ([d] 0 yrs)	0.078	0.197	no (30 yrs) yes (0 yrs)	0.024	0.024	yes
	Portland, TX	0	0	zero inundation	0	0	zero inundation	0	0	zero inundation
	Green Island, TX	0	0	zero inundation	0	0	zero inundation	0	0	zero inundation
	Main Pass, LA	0	0	zero inundation	0	0	zero inundation	0	0	zero inundation
	Barataria, LA	11.412	16.623	no ([d] 0 yrs)	9.016	8.174	no (30 yrs) yes (0 yrs)	2.044	2.044	yes
	Grand Chenier, LA	2.662	7.813	no	2.001	5.115	no (30 yrs) yes (0 yrs)	0.700	0.700	yes
West Coast	Albion, CA	0	0	zero inundation	0	0	zero inundation	0	0	zero inundation
	Point Sal, CA	0	0	zero inundation	0	0	zero inundation	0	0	zero inundation

Table 7.9. (cont.)

Region	Site name	100-cm SLR scenario			67-cm SLR scenario			33-cm SLR scenario		
		30 Years foresight	0 Years foresight	Protect?	30 Years foresight	0 Years foresight	Protect?	30 Years foresight	0 Years foresight	Protect?
	San Quentin, CA	4.321	5.335	partial protect (30 yrs)[d] yes (0 yrs)	2.055	2.055	yes	0.324	0.324	yes
	Yaquina, OR	1.903	1.903	yes	0.600	0.600	yes	0.075	0.075	yes
	Anacortes, WA	2.373	4.054	no (30 yrs) yes (0 yrs)	1.412	1.484	partial protect (30 yrs) yes (0 yrs)	0.222	0.222	yes
	Tacoma, WA	5.324	5.343	partial protect (30 yrs)[d] yes (0 yrs)	2.274	2.274	yes	0.389	0.389	yes
Regional estimates	Northeast	$2836	$2920		$1362	$1362		$431	$431	
	Southeast	$1665	$2096		$839	$945		$333	$368	
	Gulf Coast	$509	$881		$394	$472		$98	$98	
	West Coast	$455	$543		$207	$209		$33	$33	
National estimates		$5465	$6440		$2802	$2988		$895	$930	

Notes:

National and Regional estimates are calculated by applying a weight of 32.667 (980/30) to each site.

All values assume a rate of 4 percent was used for variable costs of protection, unless otherwise specified.

[a] Values are taken as a sum of all subsites analyzed at that site.

[b] A site involving a beach nourishment strategy.

[c] A site using 10 percent variable protection cost instead of 4 percent.

[d] Using a 1 percent variable protection cost induced a protect strategy.

Table 7.10. *Economic damage from sea level rise[a] (millions of 1990$; 3 percent discount rate)*

Scenario	(1) Present value	(2) Annuitized annual cost	(3) Transient cost (2065)	(4) Percent protected
100 cm (perfect)	5465	164	333	40
100 cm (none)	6440	193	384	70
67 cm (perfect)	2802	84	170	60
67 cm (none)	2988	90	195	78
33 cm (perfect)	895	27	57	88
33 cm (none)	930	28	57	96

Notes:

[a] Annuitized costs are annual costs that produce the same discounted value as the cumulative calculation. Transient costs are actual costs incurred in the year indicated along the sea level trajectory indicated.

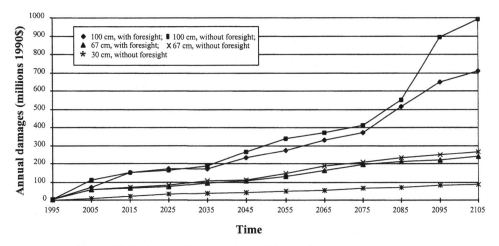

Figure 7.1 Trajectory of estimated annual damages.

Averaging this total over roughly 100 years would put annual costs around $1 billion, and would place him at the upper end of the early Titus and Green (1989) estimates.

The Fankhauser work also suggests that 84 percent of open coastline and 99 percent of US cities would be protected. This level of protection is significantly larger than results indicated in this chapter. Apparently the geographically specific dynamic adaptation modeled in this current analysis reduces the likelihood of protection by deflating the benefit side of any protection decision. It is important to recognize, however, that differences in the frequency of protection do not completely explain the

201

Table 7.11. *Economic damages with high protection costs[a] (millions of 1990$; 3% discount rate)*

Scenario	(1) Present value	(2) Annuitized annual cost	(3) Transient cost (2065)	(4) Percent protected
33-cm trajectory				
Perfect; × 5.33	1922	57	103	54
None; × 5.33	2289	69	110	73
67-cm trajectory				
Perfect; × 5.33	5411	162	324	11
None; × 5.33	7966	238	524	37

Notes:
[a] Damages are calculated based on construction costs equal to $4000 per linear foot; increasing both fixed and variable cost by a multiplicative factor of 5.33.

reported differences in cost. The statistics recorded in Table 7.10 and depicted in Figure 7.1 include not only the expense involved in protection when it is warranted, but also the value of abandoned property when it is lost. Of the $333 million in transient cost recorded in Table 7.10 for the 100-cm sea level rise scenario in the year 2065 under the assumption of perfect foresight, in fact, only about $200 million reflect protection expenditure. The appropriate "apples to apples" comparison shows that the estimate of transient cost reaches only approximately 20 percent of the comparable $1 billion average produced from Fankhauser's cumulative cost calculations.

The results are even more striking along the more likely 33-cm scenario. Annuitized estimates of average cost run from $27 to $28 million per year, and transient costs for the year 2065 round off to $57 million, given a 3 percent discount rate. These are 3 percent and 6 percent of contemporary (100 cm) protection cost estimates, respectively. The 67-cm scenario paints an intermediate case, of course, with transient costs reaching $195 million, given a 3 percent discount rate and zero foresight. With the best guess sea level trajectory somewhere in between these two cases, it would therefore seem that previous estimates were over one order of magnitude too high.

The results presented in this section hinge on protection cost. In Table 7.11 we explore the implications of using the highest protection cost estimate available – $4000 per linear foot estimated for a project in San Francisco Bay (Gleick and Maurer, 1990). New estimates are registered for three sea level trajectories and the summary statistics are reported in Table 7.11 for the 33-cm and 67-cm scenarios. These estimates reflect efficient protection decisions made on a cell-by-cell basis with and without foresight when the fixed and variable costs of protection are 5.33 times higher than the baseline case. This cost factor is derived by dividing $4 000 (the highest per linear foot protection cost estimate available), by $750 (the central per linear foot protection cost estimate used in baseline case results). The percentage of sites that are protected falls dramatically in all cases, and that moving from foresight to no foresight makes a big difference in damages. Annuitized and transient cost estimates also rise, but by less than a factor of 5.33. The value of threatened land and, in the case of no foresight, structures, now cap the cost estimates.

The largest transient cost recorded in Table 7.11 is $524 million, which is associated with no foresight along the 67-cm trajectory. Although it is the largest transient cost in the table, it is still less than 50 percent of the current benchmark protection cost estimate. In addition, it corresponds to a sea level trajectory that is on the high side of current expectations.

7.5   Conclusions

The results reported here are most striking when they are compared with the currently accepted estimates of the potential cost to the United States of greenhouse gas-induced sea level rise, which are recorded in Table 7.2. The results of this current analysis suggest that estimates of the cost of protecting coastal properties against rising seas are about an order of magnitude too high. Earlier estimates miss the cost-reducing potential of natural, market-based adaptation in anticipation of the threat of rising seas and/or the likely decisions to protect or not to protect property on the basis of economic merit. It is difficult to determine what effect these omissions in the modeling of protection decisions for developed property have on the likely total cost of sea level rise. Some thought experiments designed to account for lower sea level trajectories certainly support the qualitative conclusion that current estimates are much too high.

Take, for example, the transient protection and abandonment cost estimate reported here for the year 2065 along the 33-cm linear sea level trajectory; it is $57 million per year with no foresight. This scenario is closest to the most recent IPCC estimates. If we were to proportionately scale current estimates for cumulative dry land and wetland losses expected along a one-meter trajectory, we should add $370 million and $893 million to the estimate, respectively, for a total of about $1.3 billion.[8] This sum clearly falls well short of the accepted $7 to $9 billion range.

----

[8] This scaling process has limitations, because adaptation mechanisms and inundation patterns may differ between wetlands and dry lands. Therefore, using the same scaling factor for both may lead to over- or under-stating protection and abandonment cost estimates.

Because currently accepted cost estimates of protection against greenhouse gas-induced sea level rise appear to be an order of magnitude too high, we should concentrate our efforts on perfecting analytical methods to generate more accurate results. The methodology presented here should be used to conduct more studies on the economic impact of potential sea level rise damages. Ultimately, these results, which must incorporate the cost reducing potential of natural, market-based adaptation in anticipation of the threat of rising seas, as well as the likelihood that decisions to protect or not to protect will be made on the basis of economic merit, should replace earlier cost estimates. In turn, these new results may be used as the basis for specific protection decisions.

There are, finally, many lessons to be drawn from this work – some specific to estimating the cost of greenhouse gas-induced sea level rise, and others that can be applied more generally to impact assessment. It is, first of all, critically important to realize that none of the damage estimates associated with sea level rise, including these, take account of storms and other stochastic events that affect the coastline. The usual response to this criticism is that the jury is still out about whether or not warming will spawn more storms with larger intensities. A second response is that higher seas do not necessarily translate ubiquitously into larger storm surges and thus increased damage. However, a recent and preliminary case study produced by researchers at Carnegie Mellon's Center for Integrated Study of the Human Dimensions of Global Change suggests that storms can be impediments to the orderly market adaptation to rising seas that is embodied in these results (West *et al.*, 1996). Their careful analysis shows that decisions made at the individual level, including decisions to rebuild damaged structures after a storm, can influence the cost of inundation when it finally occurs. Some of the damage simply occurs earlier than envisioned here so that depreciation has not been completed; it thereby increases the present value of cost, but seldom by more than the no foresight case reported above. Costs can, however, also be amplified by individuals' rebuilding structures that should and will be abandoned within the planning horizon. These individual decisions produce a double-cost effect that is not captured in the present analysis but which might, at least according to the mean estimate reported in the case study, nearly triple the cost of abandonment in comparison with the perfect foresight case reported here. Applying this factor to the national estimate would not bring the total cost attributable to sea level rise along the 33- or 67-cm trajectories, even in their transient form, up to the levels of past estimates; but it is certainly enough to warrant further investigation across a wider sample of sites to see if the result can be generalized.

It should also be noted that none of the cost assessments for sea level rise have tracked distributional effects very closely. The method of benefit–cost analysis looks for net effects, assuming implicitly that transfers from "winners" to "losers" are made so that a positive benefit–cost ratio can be pareto improving. These transfers hardly ever happen, though; and severe costs concentrated on a specific group of people can produce pressure to oppose the efficient adaptation envisioned here. Once again, the no foresight case bounds damages estimates on the high side except when that pressure is sufficient to force the protection of property and structure when costs exceed benefits. The Army Corps of Engineers uses benefit–cost calculations to evaluate coastal projects in the United States, and so this might be a small concern for these estimates. Application of the methods described here to coastal zones lying outside the borders of the United States might be more problematic.

The more general lessons to be drawn from these developments can be used to frame the course of impact assessment applied well beyond the coastal zone and beyond the United States. The early impact assessments for sea level rise produced relatively high costs that have been reduced substantially by second round assessments that include adaptation. Adaptation models have usually been constructed within sound and consistent theoretical models of how the world can be expected to work on a tractable micro-level, and so they are the appropriate second steps. Now that they have been completed, third and fourth steps can be expected in two directions. On the one hand, a third step in the evolution of impact assessment should involve application of results to countries where data are more scarce than they are in the United States. One method currently under investigation attempts to produce reduced form estimates of cost functions for the United States that (1) capture a reasonable amount of the variation associated with the adaptation "correction" to vulnerability estimates from a minimal set of data and (2) can support reasonable application to adaptation that will be possible elsewhere around the world as the next century unfolds. Yohe and Schlesinger (1997) have made a first attempt in this direction for US cost estimates tied to specific greenhouse gas-emissions trajectories under a variety of sulfate alternatives. The point here is not to look at conditions today, but rather to try to envision what will exist in a globally integrated world in, say, the year 2050.

On the other hand, a fourth step should look carefully at the modeling assumptions which frame current views of adaptation in the United States to determine if they are sufficiently descriptive of what is possible. In some cases, more potential may exist; in others, informational and institutional impediments may limit adaptation and its ability to reduce costs. In either case, work along these lines will be very data intensive and will certainly not produce modeling candidates for wide application. Moreover, cost estimates depend upon local institutions that can either help or hinder market-based adaptation. In the United States, banks that hold mortgages require insurance coverage; and insurance companies certainly increase premiums as the risk of loss to storms and/or sea level rise climbs. Combining the workings of these two institutions certainly facilitates market depreciation of the value of insured structures even in areas where actual markets are thin. Would this story in support of market-based depreciation apply elsewhere? Nobody will know without careful and expensive analysis on a case-by-case basis. Care should thus be taken to determine when greater detail makes a difference; and efforts should be made to frame significant results in terms of defensible scaling factors (constants or values dependent upon easily perceived parameters) that can be used to increase or decrease aggregate measures of cost.

References

Abraham, J. and Hendershott, P. 1993. Patterns and Determinants of Metropolitan House Prices, 1977 to 1991. In: *Real Estate and the Credit Crunch, Proceedings,* Browne, L.E. and Rosengren, E.S. (eds.). Boston: Federal Reserve Bank of Boston.

Beaton, P. 1988. *The Cost of Government Regulations. A Baseline Study for the Chesapeake Bay Critical Area,* Volume II. New Brunswick, NJ: Rutgers University.

Brookshire, D., Thayer, M., Tschirhart, J. and Schulze, W. 1985. A Test of the Expected Utility Model: Evidence from Earthquake Risks. *Journal of Political Economy* **93**: 369–89.

Cline, W. 1992. *The Economics of Global Warming.* Washington, DC: Institute for International Economics.

East, B. 1990. Paradise Lost: California House Prices are Slumping. *Barron's* July: 15.

Fankhauser, S. 1994. *Valuing Climate Change. The Economics of the Greenhouse Effect.* London: Earthscan.

Frech, H. and Lafferty, R. 1984. The Effect of the California Coastal Commission on Housing Prices. *Journal of Urban Economics* **16**: 105–23.

Gibbs, M. 1984. Economic Analysis of Sea Level Rise: Methods and Results. In: *Greenhouse Effect and Sea Level Rise: A Challenge for This Generation,* Barth, M. and Titus, J. (eds.). pp. 224–51. New York: Van Nostrand-Reinhold.

Gibbs, M. 1986. Planning for Sea Level Rise Under Uncertainty: A Case Study of Charleston, South Carolina. In: *Effects of Changes in Stratospheric Ozone and Global Climates.* Vol. 4. Washington, DC: United Nations Environment Program and the United States Environmental Protection Agency.

Gleick, P.H. and Maurer, E.P. 1990. Assessing the Costs of Adapting to Sea Level Rise: A Case Study of San Francisco Bay. Pacific Institute for Studies in Development and the Environment. Oakland, CA.

Hendershott, P. 1991. Are Real Housing Prices Likely to Decline by 47 Percent? *Regional Science and Urban Economics* **21**: 553–63.

Hoffman, J.S., Keyes, D. and Titus, J.G. 1983. Projecting Future Sea Level Rise. Washington, DC: U.S. Environmental Protection Agency.

IPCC. 1992. *Climate Change: The IPCC Second Scientific Assessment*. Cambridge: Cambridge University Press.

IPCC. 1996b. *Climate Change 1995: Impacts, Adaptations, and Mitigation of Climate Change: Science-Technical Analyses*. Watson, R., Zinyowera, M., Moss, R. and Dokken, D. (eds.). Cambridge: Cambridge University Press.

MacDonald, D., Murdoch, J. and White, H. 1987. Uncertain Hazards, Insurance, and Consumer Choice: Evidence from Housing Markets. *Land Economics* **63**: 361–71.

Mankiw, N. and Weil, D. 1989. The Baby Boom, the Baby Bust, and the Housing Market. *Regional Science and Urban Economics* **19**: 235–58.

Nicholls, R.J. and Leatherman, S. 1994. In: Strzepek, K. and Smith, J. (eds.). *Climate Changes: Potential Impacts and Implications*. Cambridge, UK: Cambridge University Press.

Nordhaus, W. 1991. To Slow or Not To Slow. *Economic Journal* **101**: 920–37.

Park, R., Trehan, J., Mausel, P. and Howe, R. 1989. The Effects of Sea Level Rise on U.S. Coastal Wetlands. Appendix B to *The Potential Effects of Global Climate Change on the United States*. US Environmental Protection Agency Report to Congress.

Parsons, G. 1992. The Effect of Coastal Land Use Restrictions on Housing Prices: A Repeat Sales Analysis. *Journal of Environmental Economics Management* **22**: 25–37.

Peek, J. and Wilcox, J. 1991. The Measurement and Determinants of Single-Family House Prices. *AREUEA Journal* **19**: 353–82.

Poterba, J. 1984. Tax Subsidies to Owner-Occupied Housing: An Asset-Market Approach. *Quarterly Journal of Economics* **99**: 729–52.

Samuelson, P. 1964. Tax Deductibility of Economic Depreciation to Insure Invariant Valuations. *Journal of Political Economy* **72**: 604–6.

Schneider, S. and Chen, R.1980. Carbon Dioxide Warming and Coastal Flooding. *Annual Review of Energy* **5**, 104–40.

Stiglitz, J. 1986. *Economics of the Public Sector*. New York: W.W. Norton.

Sorenson, R.N., Weisman, R., and Lennon, G. 1984. In: Barth, M. and Titus, J. (eds.). *Greenhouse Effect and Sea Level Rise*. New York: Van Nostrand Reinhold Company.

Titus, J. and Green, M.S. 1989. An Overview of the Nationwide Impacts of Sea Level Rise. In Smith, J.B. and Tirpak, D. (eds.). *The Potential Effects of Global Climate Change on the United States. Appendix B: Sea Level Rise*. Washington: Environmental Protection Agency.

Titus, J.G. 1992. The Cost of Climate Change to the United States. In: *Global Climate Change: Implications, Challenges and Mitigation Measures*. Majumdar, S.K., Yarnal, B., Miller, E.W. and Rosenfeld, L.M. (eds.). Pennsylvania: Pennsylvania Academy of Science.

Titus, J., Park, R., Leatherman, S., Weggel, J., Greene, M., Brown, S., Gaunt, C., Trehan, M. and Yohe, G. 1992. Greenhouse Effect and Sea Level Rise: The Cost of Holding Back the Sea. *Coastal Management* **19**: 219–33.

Tol, R.S.J. 1994. *The Damage Costs of Climate Change: Towards More Comprehensive Calculations*. Amsterdam: Free University.

US Environmental Protection Agency. 1989. *The Potential Effects of Global Climate Change on the United States*. Washington.

207

Weggel, J. 1987. The Cost of Defending Developed Shoreline Along Sheltered Waters of the United States From a Two Meter Rise in Mean Sea Level. Appendix B to *The Potential Effects of Global Climate Change on the United States*. US Environmental Protection Agency Report to Congress.

Weggel, J., Brown, S., Escajadillo, J.C., Breen, P., and Doheny, E. 1989. The Cost of Defending Developed Shoreline Along Sheltered Shores. In: Smith, J.B. and Tirpak, D. (eds.). *The Potential Effects of Global Climate Change on the United States*. Appendix B: Sea Level Rise. Washington: Environmental Protection Agency.

West, J., Dowlatabadi, H., Patwardhan, A. and Small, M. 1996. Assessing Economic Impacts of Sea Level Rise. Center for Integrated Study of the Human Dimensions of Global Change. Pittsburgh, PA: Carnegie Mellon University, mimeo.

Wigley, T. 1995. Global Mean Temperature and Sea Level Consequences of Greenhouse Gas Stabilization. *Geophysical Research Letters*. **22**: 45–8.

Wigley, T. and Raper, S. 1992. Implications for Climate and Sea Level of Revised IPCC Emissions Scenarios. *Nature* **347**: 293–300.

Yohe, G. 1989. The Cost of Not Holding Back the Sea – Economic Vulnerability. *Ocean & Shoreline Management* **15**: 233–55.

Yohe, G. 1990. The Cost of Not Holding Back the Sea – Toward a National Sample of Economic Vulnerability. *Coastal Management* **18**: 403–31.

Yohe, G. and Schlesinger, M. 1997. Sea Level Change: The Expected Cost of Protection or Abandonment. Middletown, CT: Wesleyan University, mimeo.

8 The impact of global warming on US energy expenditures

WENDY N. MORRISON AND ROBERT MENDELSOHN[1]

One important impact of global climate change is the effect on energy use. In particular, the residential and commercial energy sectors are expected to be sensitive to climate change due to the impact that climate has on space conditioning. Through changes in space heating and cooling requirements, climate change will play a role in shaping the pattern of energy use in these sectors over the next century. We expect these sectors to adjust energy use as well as space conditioning capital such as heating and cooling equipment, insulation, and conservation features in adapting to climate change. Global warming is expected to yield heating benefits due to the reduced energy and building expenditures necessary to keep interior temperatures at desirable levels in cool locations and seasons. In warm locations and seasons, warming will entail additional cooling costs that include expenditures on energy, building characteristics, and cooling capacity. The net impact of climate change on the energy sector will depend on whether the heating benefits or the cooling damages dominate under a climate change scenario.

To date, all estimates of climate–energy interactions rely on expert opinion, engineering studies, and business–industry studies, focusing heavily on electricity impacts. In addition, theoretical models which lay out the welfare effect from warming have not yet been proposed for the energy sector. This chapter presents the first comprehensive theoretical–empirical model of the impact of climate change on the US energy sector. Section 8.1 provides a brief review of the large body of energy demand research and the small portion of this literature dealing specifically with the climate–energy relationship. Section 8.2 develops a theoretical model to describe how individuals and firms are expected to react to climate change by changing energy use and building characteristics. Section 8.3 describes the empirical model that estimates the climate–energy interactions in these sectors and summarizes the model results, and Section 8.4 presents simulated impacts of climate change on energy based on uniform climate scenarios. Finally, Section 8.5 presents conclusions and suggestions for further research.

[1] We would like to thank the following people who provided comments and suggestions on this analysis: James Neumann of Industrial Economics, Inc., William Nordhaus of Yale University, Kathleen Segerson of the University of Connecticut, and Lara Sheer of Industrial Economics, Inc.

8.1   Literature review

Over the last two decades, researchers using a wide array of econometric techniques have produced a vast body of literature on energy demand (see Taylor, 1975; Griffin, 1992, for reviews). Due to the nature of available data, the majority of these analyses estimate aggregate demand, while only a few use disaggregate data for households and firms. In addition, most energy studies tend to be fuel-specific, often focusing on electricity. Many are sector-specific as well, concentrating on either the commercial, industrial, or residential sector, with the lion's share on the last. Determining price elasticities of demand for the different fuels is the goal of most studies.

There have been several techniques used to estimate energy demand. Methodologically, researchers tend to follow the example set in the seminal article by Fisher and Kaysen (1962), which assumes that electricity demand derives from the household stock and utilization rate of energy using appliances. Balestra and Nerlove (1966), Anderson (1973), Hartman and Werth (1981), and Baker *et al.* (1989) demonstrate the durability of this technique over the last two decades. Baughman and Joskow (1976) develop a variant of this technique to consider both residential and commercial demands for electricity, gas, and oil. They assume individuals and firms engage in a two-stage budgeting process, choosing a level of energy using services in the first stage and then choosing a fuel combination in the second stage. Some researchers, including Houthakker and Taylor (1970), Mount *et al.* (1973), and Halvorsen (1975) estimate household energy demand directly as a function of income and household characteristics. Lyman (1973) and Mount *et al.* (1973) also perform similar analyses for the commercial sector, replacing household characteristics with company information such as type of commercial operation, number of employees, etc. Researchers also use this technique to estimate the demand for space conditioning energy, particularly heating energy (Nelson, 1975; Green *et al.*, 1986; Klein, 1988). While a number of these studies include climate variables in their models, they do so only to control for temperature variations – not to specifically address how changes in climate will affect energy expenditures.

Many estimates of climate-induced impacts on energy demand rely on studies of electricity. Crocker (1976) finds little correlation between electricity use and degree days, holding structural characteristics and substitutable energy sources constant. Conversely, Linder and Inglis (1989) study specific utilities and estimate that climate change will induce increases in electricity peak demands, especially in southern states. On the basis of the Linder study, Smith and Tirpak (1989) argue that electricity demand will increase by 4–6 percent, with larger increases during peak hours.

Nordhaus (1991) and Cline (1992) review the Smith and Tirpak study and conclude that climate change from doubling CO_2 would increase US electricity demand between $2.4 billion and $11.2 billion, and reduce nonelectric heating between $1.7 billion and $1.2 billion, respectively (in 1990$). Degree days are often used in the energy literature to reflect climate. A degree day is defined in terms of a standard temperature. One cooling (warming) degree day is a day in which the temperature exceeds (is less than) the standard by one degree.

While detailed electricity studies are informative, any conclusions of overall energy impacts drawn from these studies may be biased. Studies that incorporate natural gas and fuel oil demand in addition to electricity are necessary to fully understand the responsiveness of energy use to climate. Nelson (1976) finds that degree days are the key explanatory variables in predicting the demand for oil, natural gas, and coal, highlighting the importance of incorporating these fuels into a study of climate change impacts. Rosenthal *et al.* (1995) use an engineering methodology to predict changes in degree days in response to climate change, and then translate these results into impacts on energy expenditures. In contrast to the net damages found in previous studies, their analysis predicts that a global warming of 1 °C will yield net benefits of $5.3 billion in the year 2010 (1990$). Baxter and Calandri (1992) and Scott *et al.* (1993) use an alternative engineering methodology based on detailed building simulation models. In these studies, climate-induced changes in energy are based on projected impacts for a prototypical building. However, due to the tight building specification required, it is difficult to aggregate these impacts to the entire population of buildings. An empirical study of climate-responsive energy demand that is comprehensive across key space conditioning fuels and energy sectors has not been performed.

8.2   Theoretical model

A theoretical model of the impact of climate change on energy demand should reflect changes in energy and building expenditures rather than in stocks and utilization rates of appliances, which is the common approach in the literature. An ideal measure of climate change impacts on energy would predict the willingness to pay to stay at the original climate. In this section, we present a model of energy demand for the household and a similar model for the firm.

Household model

In choosing expenditures on goods, including energy and building characteristics, each household is assumed to maximize utility, subject to a budget

211

constraint. Utility is assumed to be dependent upon interior temperature, T, and upon an index of all other goods, R. Interior temperature, T, is assumed to be a function of climate, C, energy use, Q, and building characteristics, Z, where C, Q and Z could be viewed as vectors. The budget constraint exhausts income, Y, upon purchases of all other goods, energy, and building characteristics, where the price of all other goods is normalized to one, and P_q and P_z are the prices of energy and building attributes, respectively. The household problem is to choose the level of Q, Z, and R for a given climate:

$$\max_{Q,Z,R} U(T,R) \quad \text{s.t. } R + (P_q \times Q) + (P_z \times Z) = Y \tag{8.1}$$
$$T = f(C,Q,Z),$$

where: $C < T^*$ (heating) $\Rightarrow T_Q > 0$, $T_{QQ} < 0$ $C > T^*$ (cooling) $\Rightarrow T_Q < 0$, $T_{QQ} > 0$

$$T_Z > 0, T_{ZZ} < 0 \qquad\qquad\qquad T_Z < 0, T_{ZZ} > 0$$

$$T_{QZ} > 0, T_C > 0 \qquad\qquad\qquad T_{QZ} > 0, T_C > 0$$

where T^* represents the optimal interior temperature, and the subscripts represent first and second partial derivatives. The first-order partial derivatives imply that expenditures on both energy (Q) and thermal enhancing building characteristics (Z) increase (decrease) interior temperatures in buildings that are heating (cooling), at the margin. In addition, an increase in ambient temperature (C) increases interior temperatures. The second-order partial derivatives indicate that the marginal productivities of energy and thermal enhancing capital decline as more interior temperature services are produced. The marginal productivity of energy with respect to changes in building characteristics, however, will depend on whether energy and building characteristics are substitutes or complements in the production of interior temperature services. In the traditional case, one in which building characteristics include thermal enhancing capital such as insulation, shadings, awnings, high intensity lighting, etc., we expect energy and capital to act as substitutes. In this case, the marginal productivity of energy is expected to rise with additions of thermal enhancing capital. On the other hand, climate change may induce increases in cooling capacity, a type of capital that is expected to complement energy use. In this case, the marginal productivity of energy is expected to decline as more interior temperature services are produced. The implications of this dual substitute and complement relationship between energy and capital in the production of interior temperature services are described in more detail later in this chapter.

One of the first-order conditions of Equation (8.1) yields the familiar economic principle that individuals will equate the ratio of marginal productivities of energy and

212

building characteristics with their price ratios. In other words, households are expected to equate the marginal effectiveness of energy per dollar to the marginal effectiveness of capital per dollar in controlling interior temperature:

$$\frac{T_Q}{P_q} = \frac{T_Z}{P_z}. \tag{8.2}$$

The solution to the consumer's utility maximization problem yields the optimal bundle of interior temperature goods, T^*, and all other goods, R^*, demanded, given their budget constraint. With climate change, individuals can respond by changing: (1) expenditures on energy, (2) expenditures on building characteristics, or (3) interior comfort levels. A measure of the welfare impacts of climate change on energy can be described as the change in income necessary to keep utility constant given a change in climate:

$$\left.\frac{\partial Y}{\partial C}\right| U^*(T^*, R^*). \tag{8.3}$$

This represents the change in income necessary to maintain utility as climate changes. If we assume that interior temperatures do not change with climate, then only energy and building expenditures would be altered in the new climate regime.

$$CV = \int_{C_0}^{C_1} \frac{\partial Y}{\partial C} \cong (P_q \times Q_1) + (P_z \times Z_1) - (P_q \times Q_0) - (P_z \times Z_0), \tag{8.4}$$

where subscripts 0 and 1 represent the baseline case and climate change scenario, respectively. Hence, the change in total expenditures is a reasonable prediction of the change in welfare if interior temperatures are held constant. Our data indicate that people choose the same interior temperature during the winter season regardless of climate (Energy Information Administration, 1993). Dewees and Wilson (1990) find a similar effect across four regions of the United States for the winter season. These results support our assumption of constant winter interior temperatures. It does not seem likely that interior temperatures are the same during the summer, however. Not all buildings are equipped for cooling, and cooling is more expensive than heating. If comfort levels are altered during the summer, our model of energy expenditures will underestimate cooling damages since it does not include the value of comfort losses.

It is interesting to note an irony of our theory relative to traditional welfare economics. In general, a shift out in the demand curve represents an increase in welfare. However, in the case of climate change, when the demand for energy and building characteristics increases there is no corresponding increase in interior temperature services, because interior temperature is held constant. This increase in demand is

213

necessary to hold utility constant. In other words, climate change requires an increase in expenditures in order to maintain the same comfort level. Therefore, an increase in energy demand represents a decrease in welfare, while a decrease in energy demand represents an increase in welfare.

Firm model

A parallel model to the household can be constructed for the firm. Instead of maximizing utility, however, the firm is interested in maximizing profit. Rather than being constrained by income, the firm is constrained by their production possibilities set. Given their production technology, the set represents the combination of inputs and outputs that are feasible. Assuming firms take prices as given exogenously in both output and factor markets, the firm chooses the combination of Q, Z and F that maximizes profit (π) subject to their production possibilities:

$$\underset{Q,Z,F}{Max}\ \pi = P_x \times X(T,F) - (P_q \times Q) - (P_z \times Z) - (P_f \times F)\ \text{ s.t. } T = f(C,Q,Z) \tag{8.5}$$

where X is output, F is all other inputs, P_x and P_f are respective prices. Q is energy devoted to heating and cooling as opposed to an input into production. Solving Equation (8.5) for the optimal combinations of Q and Z yields a first-order condition similar to Equation (8.2). In this case the firm must choose the optimal level of production, balancing the value of expenditures on interior temperature and expenditures on all other inputs so that profits are maximized. The change in expenditures necessary to maintain profits given climate change provides an estimate of welfare:

$$\left.\frac{\partial Y}{\partial C}\right| \pi^*(X,T,F), \tag{8.6}$$

where the $*$ indicates optimal values for each variable. Assuming interior temperatures are held constant, changes in energy and building expenditures provide a reasonable estimate of welfare change for both individuals and firms.

8.3   Empirical model

An empirical model that estimates the climate and energy interactions in the residential and commercial sectors should indicate the sensitivity of energy expenditures to climate change. This section describes the empirical model and summarizes the estimation results.

Data

This study relies on data from the Department of Energy's Commercial Buildings Energy Consumption Survey (Energy Information Administration, 1992) and Residential Energy Consumption Survey (Energy Information Administration, 1993). These two surveys provide detailed data on energy expenditures, energy consumption, and building characteristics, in addition to demographic and firmographic information. Baseline climate data from Mendelsohn *et al.* (1994) was matched with each observation by county. The survey collected energy expenditures for several thousand buildings distributed in random clusters across the continental United States and weights are provided for each observation to represent the true population of buildings. A complete list of the variables used in this analysis is presented in Appendix A8.

Estimating welfare

As derived in the previous section, the change in energy and building expenditures indicates the change in welfare impacts on energy. However, only data for energy expenditures and detailed building characteristics are available for this study. Therefore, an alternative welfare measure must be developed to account for the missing building expenditure data. Short and long run models can be used to evaluate the importance of these unobserved building expenditures. In the short run, individuals and firms can only adjust energy expenditures. A model of energy expenditures, holding building characteristics constant, should reflect these short run adjustments. In the long run, individuals and firms can adapt buildings to the warmer climate and hence both energy and building expenditures can be adjusted. Our long run measure of energy expenditures assumes building characteristics are endogenous. The long run measure, therefore, allows for flexibility of building characteristics. The actual estimate, however, does not include building expenditures. The difference between the short run and long run measures provides an indication of how important building adjustments are likely to be. If short run and long run measures are similar, building adjustments are likely to be small and safely ignored. If the disparity between short and long run adjustments using energy alone are large, however, building adjustments are likely to be important. These short and long run adjustments in expenditures on energy are expected to be different for buildings with heating versus cooling dominated energy expenditures, and will depend on whether energy and building characteristics act as complements or substitutes in the climate change adjustment. Identifying these differences illustrates how our short and long run measures of energy expenditures alone are expected to compare to each other and to the ideal measure of energy and building expenditures.

215

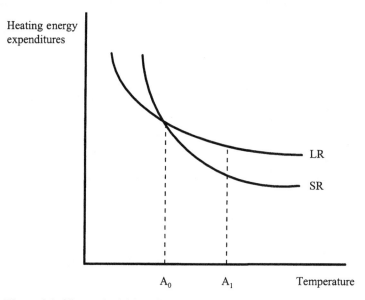

Figure 8.1 Climate elasticities of long and short run heating expenditures energy and capital: substitutes. LR, long run; SR, short run.

Traditionally, energy and capital are viewed as substitutes. In our analysis, the traditional substitute relationship is expected to exist between energy and thermal enhancing capital such as insulation, high intensity lighting, shading, etc. We expect this substitute relationship between energy and capital to dominate the heating impacts. This heating case is illustrated in Figure 8.1, for a building experiencing ambient temperature A_0. If the climate warms to A_1, heating energy expenditures will decline sharply in the short run. In the long run, reductions in thermal enhancing building capital may lead to greater energy expenditures relative to the short run. However, the concomitant savings in building costs are not included in our measures. Hence, total welfare benefits in the form of reduced energy and building expenditures will be greater than even our short run measure. The difference between our long and short run measures will indicate the importance of building adjustments and corresponding building savings.

On the other hand, capital in the form of space conditioning capacity may complement energy use. This is especially true on the cooling side where there is significant potential for greater saturation of cooling equipment as a result of climate change. This cooling case is illustrated in Figure 8.2, for a building experiencing ambient temperature B_0. If the climate warms to B_1, in the short run only expenditures on cooling energy can increase since buildings are capacity constrained. Interior temperatures can only be altered through energy use to the extent that there is cooling capacity to do

216

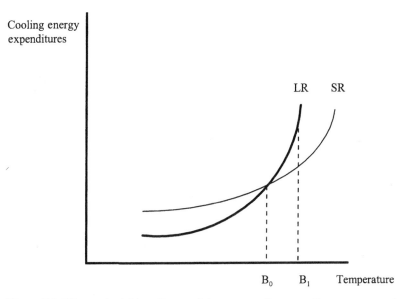

Figure 8.2 Climate elasticities of long and short run cooling expenditures energy and capital: complements. LR, long run; SR, short run.

so. In the long run, there is more flexibility in adjusting energy use due to unconstrained cooling capacity. Hence, the observed long run energy expenditures may be greater than the short run expenditures, reflecting the complementary relationship between energy use and cooling capacity. The total costs of climate change including both energy and building expenditures would be even higher than observed long run expenditures on energy alone due to the unobserved expenditures on cooling capacity. Again, the difference between our long and short run measures indicates the importance of building adjustments. If this type of adjustment to climate change takes place, it is important to recognize the underlying comfort benefits that result. Although long run expenditures (welfare) are greater than those in the short run, the relationship between cooling capacity and energy yields comfort benefits in the long run that could not be achieved in the short run. Hence, individuals and firms are expected to be better off in the long run when both comfort and expenditures are considered. The empirical model should clarify the nature of climate change adjustment for both heating and cooling, and identify whether the hypothesized substitute (complement) relationship between heating (cooling) energy and capital holds.

This model assumes heating expenditures decrease as ambient temperature rises while cooling expenditures increase. Hence, the overall relationship between total energy expenditures should be a quadratic function with the minimum indicating the temperature at which total energy expenditures are lowest.

217

Estimation

We estimate short and long run models to explain how energy expenditures vary with climate for residential and commercial buildings. Ordinary least squares regression analysis is used to estimate the total expenditures on electricity, natural gas, fuel oil, liquid petroleum gas, and kerosene on the residential side, and electricity, natural gas, fuel oil, and district heat on the commercial side (district heat involves centralized furnaces that share heat through steam or hot water with surrounding buildings). Climate variables include average temperature and the summer/winter temperature differential on the residential side, and average temperature and standard deviation on the commercial side. Demographic characteristics on the residential side include average fuel prices, use of alternative fuels, average income range, family size, the age of household head, race, and receipt of government assistance. Company information includes average fuel prices, use of alternative fuels, months open per year, and the types and percentages of various building activities. The building characteristics are divided into *climate-sensitive* and nonsensitive categories, as this allows an important distinction between our short and long run models. Nonclimate-sensitive building characteristics such as square footage, number of floors, and building age are controlled for in all model runs. Climate-sensitive building characteristics such as building material, conservation efforts, the choice of heating and cooling equipment, and high energy-consuming appliances, as well as some aspects of the building structure such as the number of rooms, doors, and windows are held constant in the short run model. The long run model omits these building variables; thus it treats conservation, thermal characteristics, equipment and housing structure as endogenous, allowing them to adjust with climate. By comparing the short run and long run results, we can determine the importance of building adjustments.

A log–linear functional form is used to estimate energy expenditures since it provides the highest predictive power based on F-tests of the overall significance of the regression. The log–linear model is also the most common functional form found in the energy demand literature. The model is joint additive–multiplicative, as only the continuous variables are logged and discrete variables are left in their original form. As mentioned in the previous section, we expect total energy expenditures to exhibit a quadratic relationship with climate since heating expenditures fall as temperature increases while cooling expenditures rise. Therefore, the climate variables are included in linear and quadratic rather than in logged form. The hypothesized expenditure equations for the short and long run, respectively, are represented as follows:

$$\ln \sum_{i=1}^{F} E_i = \alpha + \beta_0 C + \beta_1 C^2 + \beta_2 \ln P + \beta_3 \ln S + \beta_4 \ln Z_{nc} + \beta_5$$
$$+ \beta_6 \ln Z_c + \beta_7, \tag{8.7}$$

218

$$\ln \sum_{i=1}^{F} E_i = \alpha + \beta_0 C + \beta_1 C^2 + \beta_2 \ln P + \beta_3 \ln S + \beta_4 \ln Z_{nc} + \beta_5, \qquad (8.8)$$

where α and β are estimated coefficients, $i = 1 \ldots F$ represents fuels, E_i is energy expenditures, C are climate variables, P are average fuel prices, S are demographic or firm characteristics, and Z are building characteristics. Subscripts c and nc represent portions of Z that are climate-sensitive and nonclimate sensitive, respectively.

A common issue addressed in the energy literature is the identification of the demand equation when a supply equation is not estimated. Historical emphasis on this issue is due mainly to the aggregate nature of most demand models. When estimating aggregate models, one cannot assume that prices are determined exogenously since changes in aggregate demand are expected to exert an endogenous influence on prices. With disaggregate data, on the other hand, one can reasonably assume that prices are exogenous and predict expenditures taking prices as given, since the actions of households and firms are not expected to independently influence the price structure (Berndt and Wood, 1975). Because our equations are estimated using individual data for residences and firms, we can assume that both are price takers; they take prices as given and cannot influence the price structure.

Empirical results

The results suggest that energy expenditures in the residential sector are sensitive to climate. Table 8.1 presents a summary of the regression results for the residential model. Demographic characteristics associated with higher energy expenditures include higher income, larger family size, older age of household head, tenant in home, receipt of heating vouchers, and whether the head of the household is black. On the other hand, expenditures are less if the head is aged over 65 years, Hispanic, receives cash aid for heating, participates in an energy discount program, or burns wood as an alternative fuel.[2] Structural characteristics associated with higher energy expenditures include higher square footage, more rooms, more doors and windows, and the presence of inadequate insulation or leaks. The presence of a basement or more than one unit leads to lower expenditures, as does increasing the number of floors, controlling for square feet. Appliances and electrical equipment that increase energy expenditures include a computer, TV, dishwasher, clothes washer, and clothes dryer. Space conditioning equipment such as central air conditioning, wall or window air conditioning units, central warm air, electric wall or radiator units, and portable

---

[2] Heating vouchers are payments to the individuals in the form of coupons to be mailed to the energy provider, whereas cash aid represents a direct cash outlay for heating expenses. Energy discount programs include peak-load/off-peak pricing as well as interruptible energy service in addition to other utility demand–side management programs.

Table 8.1. *Residential regression model[a]*

Variable	Short run	Long run	Variable	Short run	Long run[b]
Constant	41.48	21.96	Race: Hispanic	-5.51	-7.56
	(8.23)	(4.62)		$\times 10^{-2}$	$\times 10^{-2}$
				(-3.00)	(-3.84)
Average temperature	-3.31	3.42	Age: over 65 years	-3.21	-4.53
	$\times 10^{-3}$	$\times 10^{-3}$		$\times 10^{-2}$	$\times 10^{-2}$
	(-2.30)	(2.45)		(-2.00)	(-2.66)
Average temperature	4.82	3.04	Receives gov't heat cash aid	-0.17	-0.18
	$\times 10^{-4}$	$\times 10^{-4}$		(-3.37)	(-3.32)
	(6.10)	(3.63)			
Winter/summer temp differential	-1.32	1.21	Receives gov't heat vouchers	6.78	6.18
	$\times 10^{-3}$	$\times 10^{-3}$		$\times 10^{-2}$	$\times 10^{-2}$
	(-1.11)	(0.98)		(2.65)	(2.26)
(Winter summer temp differential)^2	-3.79	-3.24	Burns wood as alternative fuel	-3.23	-4.53
	$\times 10^{-4}$	$\times 10^{-4}$		$\times 10^{-3}$	$\times 10^{-2}$
	(-9.95)	(-8.06)		(-0.27)	(-3.70)
Log electricity price	0.30	0.28	Log no. rooms	9.18	NI
	(16.97)	(14.96)		$\times 10^{-2}$	
				(3.85)	
Log natural gas price	0.19	0.21	Log no. doors & windows	9.21	NI
	(10.59)	(11.27)		$\times 10^{-2}$	
				(6.72)	
Log fuel oil price	1.33	1.60	Basement	-9.65	NI
	$\times 10^{-2}$	$\times 10^{-2}$		$\times 10^{-2}$	
	(0.49)	(0.55)		(-7.32)	
Log liquid petroleum gas price	-0.10	-0.11	Poor insulation or leaks	3.91	NI
	(-5.13)	(-5.10)		$\times 10^{-2}$	
				(3.35)	
Log kerosene price	-0.21	-0.26	Interrupt/discount elec. rate	-7.99	NI
	(-4.96)	(-5.82)		$\times 10^{-2}$	
				(-2.94)	
Log household income	4.93	0.10	Color tv	5.71	NI
	$\times 10^{-2}$	(14.31)		$\times 10^{-2}$	
	(6.95)			(10.50)	
Log size of home: square feet	0.17	0.28	Computer	3.11	NI
	(11.92)	(26.27)		$\times 10^{-2}$	
				(2.38)	

Table 8.1. (*cont.*)

Variable	Short run	Long run	Variable	Short run	Long run[b]
Log no. floors	-9.52 $\times 10^{-2}$ (-7.51)	-0.13 (-9.06)	Dish/clothes washer/dryer	9.87 $\times 10^{-2}$ (6.95)	NI
Log year built	-4.82 (-7.27)	-2.36 (-3.76)	Central ac	0.15 (10.34)	NI
Log family size	0.24 (24.43)	0.26 (25.50)	Wall/window ac	4.38 $\times 10^{-2}$ (3.77)	NI
Log age of household head	8.22 $\times 10^{-2}$ (4.76)	0.11 (6.01)	Stove/fireplace heats	-0.23 (-8.85)	NI
More than 1 unit to home	-0.10 (-6.25)	-0.10 (-6.53)	Central warm air heats	4.25 $\times 10^{-2}$ (3.39)	NI
Tenant in home	0.19 (4.83)	0.23 (5.46)	Electric wall/radiator heats	0.11 (6.83)	NI
Race: black	0.10 (3.47)	0.11 (3.65)	Portable kerosene heats	0.21 (3.32)	NI
			Adjusted R^2	0.966	0.961
			Number of observations	5030	5030

Notes:
[a] t-statistics in parentheses.
[b] NI, not included in long run model.

kerosene heating also increase expenditures. Use of a stove or fireplace for heat causes expenditures to be lower.

Characteristics of the commercial operation that increase expenditures include more months open per year, more recently built, type and percentage of various activities, and the use of major energy consuming appliances such as ice, water or vending machines, and commercial refrigerators and freezers. If there is a large percentage of nonrefrigerated warehouse or vacancy, or the tenant controls the heat, energy expenditures are less. Building characteristics that increase expenditures include higher square footage, greater number of floors, and built-up roof material. On the other hand, glass and metal-surfacing are roof materials that lower expenditures, as do certain wall materials such as masonry, siding and shingles. Space

221

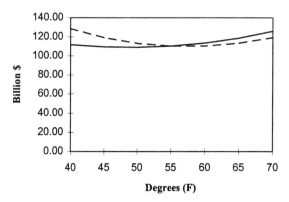

Figure 8.3 Residential model climate–expenditure relationship. Solid line, long run; broken line, short run.

conditioning equipment including computer room air conditioners, boilers, air ducts for heating and cooling, and heat pumps for heating cause expenditures to be higher. If heat pumps are used for cooling, expenditures are decreased.

Climate effect

The climate variables have been de-meaned to allow for easier interpretation. The linear term represents the marginal effect of climate on energy expenditures evaluated at the mean of the sample. The quadratic term shows how the marginal effect changes in movements away from the mean. The climate–expenditure relationships for the residential and commercial models are pictured in Figures 8.3 and 8.4, respectively. The figures show predicted US energy expenditures at various levels of average annual temperature. The results reflect the hypothesized relationship between short run and long run energy expenditures illustrated in Figures 8.1 and 8.2.

The overall relationship between climate and energy expenditures in the residential sector is U-shaped with the minimum expenditure at approximately 49 °F in the long run and 58 °F in the short run. Since average temperature in the sample (as well as the United States) is approximately 55 °F, the marginal effect of warming at the mean is harmful in the long run and beneficial in the short run model. This implies that expenditures are more sensitive to changes in climate in the long run than in the short run. We hypothesize that this is a result of the structure of our long and short run models. While the short run model holds air conditioning capacity fixed, the long run model allows it to vary. If energy and cooling capacity are complements as hypothesized, adding cooling capacity will result in greater long run energy expenditures in warm climates.

For the commercial sector, the results are shown in Table 8.2. The results, as pre-

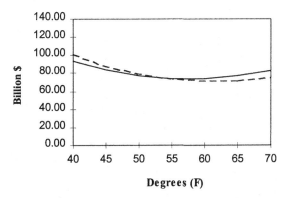

Figure 8.4 Commercial model climate–expenditure relationship. Solid line, long run; broken line, short run.

sented in Figure 8.4, indicate that the minimum expenditure occurs at approximately 58 °F in the long run and 62 °F in the short run. The commercial sector is less sensitive to warming than the residential sector.

Thus, while energy expenditures are negatively affected by changes in temperature at the mean, it does not take a significant increase in mean temperature before expenditures begin to rise. The climate–expenditure relationship in the commercial sector exhibits the same long run sensitivity as in the residential sector, although slightly less pronounced in the commercial case. As we hypothesized earlier, long run expenditures may include greater expenditures on air-conditioning, the capacity for which is held constant in the short run model but allowed to vary in the long run model.

Price elasticities

Table 8.3 compares the price elasticity estimates from our study to those calculated in the literature for both sectors. Since the dependent variable is expenditures and not quantity demanded, the price coefficients represent expenditure elasticities. Hence, demand elasticities must be calculated (see Appendix B8 for details). In most cases we compare our elasticity estimates to long run models because of the nature of our data and specifications. Our short run and long run models are defined to capture the climate change adjustment of energy expenditures and building characteristics. For this reason, in addition to the fact that we rely on cross-sectional data, the elasticity estimates are expected to most closely approximate long run estimates.

On the residential side, our elasticity estimates generally fall within the range of estimates reported in the literature. In addition to the major fuels studied in the literature, we considered liquid petroleum gas and kerosene. The results indicate that the

Table 8.2. *Commercial regression model results*[a]

Variable	Short run	Long run	Variable	Short run	Long run[b]
Constant	−65.01 (−8.87)	−72.01 (−9.16)	% industrial activities	1.17 $\times 10^{-2}$ (3.39)	1.77 $\times 10^{-2}$ (4.66)
Average temperature	−8.99 $\times 10^{-3}$ (−4.30)	−2.94 $\times 10^{-3}$ (−1.34)	% office space	4.62 $\times 10^{-3}$ (10.35)	5.97 $\times 10^{-3}$ (12.75)
Average temperature^2	6.57 $\times 10^{-4}$ (4.44)	7.20 $\times 10^{-4}$ (4.47)	% retail/service activities	3.17 $\times 10^{-3}$ (8.36)	3.90 $\times 10^{-3}$ (9.64)
Standard deviation temperature	2.01 $\times 10^{-3}$ (0.29)	1.99 $\times 10^{-2}$ (2.65)	% educational activities	3.20 $\times 10^{-3}$ (5.62)	4.66 $\times 10^{-3}$ (7.55)
Standard deviation temperature^2	3.07 $\times 10^{-3}$ (2.80)	4.03 $\times 10^{-3}$ (3.37)	Ice/vending/water machines	0.49 (17.44)	NI
Log electricity price	−0.43 (−13.72)	−0.54 (−15.87)	Commercial refrig./ freezer	0.45 (12.44)	NI
Log natural gas price	−5.55 $\times 10^{-2}$ (−1.81)	−5.08 $\times 10^{-2}$ (−1.52)	Air duct – cool	0.11 (3.24)	NI
Log fuel oil price	0.20 (0.99)	−2.97 $\times 10^{-2}$ (−0.13)	Air duct – heat	0.10 (3.21)	NI
Log district heat price	6.27 $\times 10^{-3}$ (0.05)	−3.02 $\times 10^{-2}$ (−0.22)	Boilers	0.28 (7.50)	NI
Log square feet	0.54 (39.93)	0.73 (54.33)	Computer room ac	0.60 (11.11)	NI
Log no. floors	0.16 (5.01)	0.26 (7.35)	Heat pump-cool	−0.17 (−2.31)	NI
Log year built	8.79 (9.05)	9.51 (9.14)	Heat pump-heat	0.17 (2.39)	NI
Months open per year	2.73 $\times 10^{-2}$ (4.87)	2.94 $\times 10^{-2}$ (4.78)	Tenant controls heat	−0.15 (−5.48)	NI

224

Table 8.2. (*cont.*)

Variable	Short run	Long run	Variable	Short run	Long run[b]
Alternative fuel used	−0.61	−0.71	Roof – built up material	0.14	NI
	(−13.25)	(−14.24)		(4.88)	
% food sale/service activities	8.97 $\times 10^{-3}$	1.58 $\times 10^{-2}$	Roof – glass	−1.80	NI
	(14.54)	(27.16)		(−5.59)	
% warehouse/vacant	−2.99 $\times 10^{-3}$	−5.68 $\times 10^{-3}$	Roof -metal surface material	−0.18	NI
	(−6.51)	(11.75)		(−5.12)	
% in-patient/skilled healthcare	6.95 $\times 10^{-3}$	1.09 $\times 10^{-2}$	Wall-masonry/siding	−0.85	NI
	(5.62)	(8.13)		(−4.40)	
% out-patient health/ public safety	3.27 $\times 10^{-3}$	4.44 $\times 10^{-3}$	Wall-shingles/siding	−0.15	NI
	(3.91)	(4.87)		(−4.54)	
% lab. or refrigerated warehouse	1.07 $\times 10^{-2}$	1.25 $\times 10^{-2}$			
	(7.14)	(7.60)			
			Adjusted R^2	0.958	0.949
			Number of observations	5653	5653

Notes:
[a] t-statistics in parentheses.
[b] NI, not included in long run model.

demand for these fuels is most elastic with predicted elasticities of −1.25 and −2.38, respectively. We also note in Table 8.3 whether the elasticity estimate is based on average or marginal prices. Average and marginal rates for fuels often differ due to block rate and other types of quantity-based pricing methods which lead to declining marginal prices. Although marginal prices are the desirable measure, they are not available on a building-specific basis for this study. This problem has received a great deal of attention in the literature. Predicted elasticities based on average rates are expected to be biased (Taylor, 1975). However, Halvorsen (1975) demonstrates that when using a double-log form, the elasticity estimates for average rates are quite comparable to those observed using marginal rates.

For the commercial model, the electricity and natural gas price coefficients are significant. In general, these are the fuels most often studied in the literature. There

Table 8.3. *Comparison of long run fuel price elasticities*

Study	Electricity	Natural gas	Fuel oil	Type of price
Residential models:				
This study	−0.61	−0.59	−0.97	average
Anderson, 1973	−1.12	−2.70		average
Baker, Blundell, & Micklewright, 1989	−0.76	−0.31		average
Balestra & Nerlove, 1966		−0.63		relative
Barnes, Gillingham, & Hagemann, 1981	−0.55			marginal
Baugman & Joskow, 1976	−1.00	−1.01	−1.12	average
Halvorsen, 1975	−1.52			marginal
Hartman & Werth, 1981	−0.19 to −0.40 (SR)			marginal
Houthakker & Taylor, 1970	−1.89			average
Houthakker, Verleger, & Sheehan, 1973	−1.02			marginal
Mount, Chapman, and Tyrrell, 1973	−1.20			average
Wilson, 1971	−1.33			average
Commercial models:				
This study	−2.41	−0.97	−1.0	average
Baugman & Joskow, 1976	−1.00	−1.01		marginal
Lyman, 1973	−2.10			average
Mount, Chapman, & Tyrrell, 1972	−1.36			average

Note:
SR, short run.

are far fewer commercial/industrial studies than residential studies. However, as in the residential case, our elasticity estimates are similar to those predicted in other studies. While relatively few historical studies consider fuel oil and district heat, we perform a comprehensive fuel study including both of these alternatives. The coefficients for fuel oil and district heat are not significantly different from zero at the 95 percent level in both the short and long run regressions. This suggests that the elasticities for these fuels are expected to be close to unity.

8.4 Climate change simulation

This section presents the impacts of climate change on energy. Uniform increases in temperature of 1.5, 2.5, and 5 °C are explored using the predicted energy expenditure equations.[3] This simulation involves inserting the new predicted temperature into the estimated regression equation for each scenario.

In addition to analyzing three climate scenarios, the results are summarized for both a 1990 and a 2060 economy. Although 1990 is not a realistic climate change scenario, it allows for comparisons with other studies in the literature. A series of adjustments are made to predict expenditures in a 2060 economy in order to reflect population, GDP, and fuel price changes. Population is projected to grow by approximately 19 percent between 1990 and 2060 (see Chapter 1). We incorporate a uniform population increase of this magnitude into the model by altering the adjusted weights for each observation in the commercial and residential sector proportionately. GDP per capita is projected to grow by 223 percent over the same period (Chapter 1). We assume energy expenditures will increase in proportion to the income elasticity of energy expenditures predicted in the residential model (0.10) implying that energy expenditures will grow 22 percent. An additional adjustment is made to reflect the age of building stock in the year 2060. The variable for *year built* in both sectors is held constant at the year 2040, reflecting an average age of 20 years for the 2060 building stock.

We use a Hotelling-type model to predict changes in fuel prices over the period between 1990 and 2060. There is likely to be price-induced substitution away from energy over the next century as fossil fuel supplies begin to decline and prices rise. The Hotelling Rule states the net prices for depletable resources must rise with the interest rate to compensate resource owners for holding stocks (Hotelling, 1931):

$$\frac{\dot{P}}{P} = r\left(1 - \frac{c}{P}\right),$$
(8.9)

where P is fuel price, c is extraction costs, and r is the discount rate. We assume that extraction costs are approximately half of gross prices. Given a discount rate of 4 percent, used in this analysis, price is consequently assumed to rise at a rate of 2 percent annually.

These assumptions lead to an increase in residential expenditures in 2060 and a decrease in commercial expenditures, at existing climates. The decline in commercial expenditures is mainly a result of price-induced substitution.

Table 8.4 summarizes the predicted net impacts across scenarios for both sectors in the 1990 and 2060 economies. A positive value indicates a damage associated with a reduction in energy expenditures. Comparing the long run versus the short run resi-

---

[3] In predicting expenditures, the dependent variable is multiplied by a scale factor of $\exp(\sigma^2/2)$ to correct for prediction bias. Since we use a log–linear functional form, the dependent variable must be transformed to estimate expenditures.

Table 8.4. *Predicted change in welfare from climate change*

Sector	Year	Climate change scenario		
		1.5 °C	2.5 °C	5 °C
Residential	1990 Economy			
(billion 1990$)	Short run	0.33	0.09	−2.11
	Long run	−1.44	−2.69	−6.94
	2060 Economy			
	Short run	0.51	0.12	−3.49
	Long run	−2.35	−4.39	−11.30
Commercial	1990 Economy			
(billion 1990$)	Short run	1.63	2.28	2.43
	Long run	0.42	0.23	−1.96
	2060 Economy			
	Short run	1.45	2.04	2.25
	Long run	0.42	0.29	−1.49

dential results reveals more damages in the long run than in the short run. This is consistent with our model of warm climates where cooling expenditures dominate. The expanded cooling capacity which can occur in the long run would increase long run energy expenditures over short run expenditures. The results suggest that these cooling damages dominate heating benefits in the residential sector. In the commercial sector, heating benefits are greater in the short run than the long run. This is consistent with our model of cooler climates where heating dominates. In the long run, firms place less resources in insulation and thus increase energy expenditures. For both sectors, the difference between the long and short run models provides a measure of the importance of the cost of building characteristics. The fact that the estimates are close for most scenarios suggest the estimates are reasonably accurate.

A range of costs and benefits are projected for both sectors. Based on the long run model, the residential sector experiences damages in all climate scenarios. These damages range from $2.4 billion to $11.3 billion in the year 2060. According to our model, for climate change of the order of 1.5 °C to 2.5 °C, the commercial sector experiences benefits. In this case, the heating benefits dominate cooling damages. As illustrated in Figure 8.1, the short run estimate is expected to most closely approximate total welfare benefits, which include building savings. The short run results suggest benefits ranging from $1.5 billion to $2 billion in the year 2060. For a climate change of 5 °C in the year 2060, the commercial sector experiences short run benefits of $2.3 billion and long run costs of $1.5 billion. Long run residential costs are sub-

Table 8.5. *Predicted change in welfare from climate change by climate zone*

Sector/ Zone	(SR)/(LR)	Climate change scenario		
		1.5 °C	2.5 °C	5 °C
Residential				
Zone 1 (coldest)	SR	0.32	0.49	0.78
	LR	0.06	0.08	0.04
Zone 2	SR	0.70	1.03	1.43
	LR	−0.03	−0.13	−0.66
Zone 3	SR	0.71	0.97	0.83
	LR	−0.38	−0.78	−2.30
Zone 4	SR	−0.16	−0.43	−1.71
	LR	−0.72	−1.31	−3.22
Zone 5 (warmest)	SR	−1.05	−1.92	−4.81
	LR	−1.27	−2.23	−5.14
Commercial				
Zone 1 (coldest)	SR	0.66	1.02	1.70
	LR	0.52	0.80	1.27
Zone 2	SR	0.79	1.19	1.82
	LR	0.49	0.70	0.80
Zone 3	SR	0.26	0.37	0.47
	LR	0.09	0.10	−0.10
Zone 4	SR	0.04	0.002	−0.29
	LR	−0.17	−0.35	−1.06
Zone 5 (warmest)	SR	−0.29	−0.54	−1.45
	LR	−0.53	−0.96	−2.40

Notes:
SR, short run; LR, long run. Estimate is in billions $ for a 2060 economy.

stantially higher in the 2060 than in the 1990 economy, reflecting the expansion of the residential energy sector over the period. On the other hand, commercial impacts vary by little over the period and are slightly lower in the 2060 economy, reflecting a contraction in this sector due to price-induced substitution.

These impacts are distributed differentially across climate zones. Table 8.5 details the 2060 welfare impacts by climate zone for both sectors to illustrate the distribution of impacts across the United States.[4] The results by climate zone also follow the

---

[4] The climate zones are defined as: Zone 1: cooling degree days (cdd) to base $65 < 22000$, heating degree days (hdd) to base $65 > 7000$, Zone 2: $cdd65 < 22000$, $hdd65 = 5500 − 7000$, Zone 3: $cdd65 < 22000$, $hdd65 = 4000 − 5499$, Zone 4: $cdd65 < 2000$, $hdd65 < 4000$, Zone 5: $cdd65 > 2000$, $hdd65 < 4000$. On the residential side, 20 observations that have $cdd65 > 2000$ and $hdd65$ slightly more than 4000 are included in Zone 5.

hypothesized relationship between short and long run energy expenditures. Both sectors incur benefits in the colder regions and damages in the warmer regions, with greater damages projected on the residential side.

8.5 Conclusions

This study examines the effect of climate on energy expenditures for all fuels in both the residential and commercial energy sectors. As such, it is the most comprehensive empirical analysis of climate energy effects carried out to date. The theory developed suggests that examining the short run and long run sensitivity of energy expenditures to climate provides an approximation of the welfare effects of climate change on space heating and cooling. The empirical results reveal that short run and long run estimates are close, implying that they provide a good approximation of these welfare effects. Based on the empirical results, a uniform increase in temperature of 2.5 °C is projected to yield net benefits in 2060 in the commercial sector and net damages in the residential sector. The range of total effects lie between a $2 billion gain (short run) to a $4 billion damage (long run). As warming rises to 5 °C, the total impacts become increasingly harmful suggesting damages from $1 billion (short run) to $13 billion (long run). It is important to note that these predicted ranges are subject to uncertainty due to the estimation procedure used to derive them and the assumptions about future climate and economic conditions.

The results support previous studies which projected that there would be energy damages from warming (Smith and Tirpak, 1989; Nordhaus, 1991; Cline, 1992; Fankhauser, 1995). The results confirm that the future economy will be more sensitive to the cooling losses from higher temperatures than to the warming benefits. However, the results suggest that Cline was perhaps too pessimistic in his view of electricity damages. Although this study predicts net damages, it is not necessarily inconsistent with the Rosenthal *et al.* (1995) projection for 2010 that predicts net benefits. First, the warming which will have occurred by 2010 is likely to be small, even less than the 1.5 °C projection in this study, so that warming benefits may still exceed cooling losses. Second, the movement to the 2060 economy with its greater cooling capacity will have only just begun, making the 1990 conditions more applicable. Third, Rosenthal *et al.* assume that the north will warm more than the south and that winter will warm more than summer. All these assumptions would tend to increase net benefits. With the 1990 economy and a 1.5 °C scenario, this study finds that impacts will range from a $2 billion benefit to a $1 billion damage. With an even more optimistic climate scenario, it is indeed plausible that there would be net benefits in the energy sector.

Methodologies that do not incorporate all major fuels and sectors seem to provide a slightly different picture of impacts than studies that do. A number of the earlier studies emphasize electricity impacts. Both our study and Rosenthal *et al.* (1995) are comprehensive across fuels and across the residential and commercial sectors. On the commercial side, our estimates are comparable to Rosenthal *et al.*, particularly since our short run measure is expected to most closely approximate the total welfare impacts. On the other hand, the residential sector impacts diverge, perhaps as a result of the differing treatment of building stocks across studies. It is important to note that in general the distribution of benefits and costs will depend on whether the commercial building or residence has energy expenditures that are dominated by heating or cooling.

Further research should be pursued to expand this analysis to consider more detailed climate data. Although the expected results from a large set of climate models tend to resemble the uniform scenarios examined here, the individual general circulation model (GCM) results suggest a distribution of climate changes across seasons and across the United States. These alternative climate scenarios are likely to alter the distribution of benefits and costs. To investigate this phenomenon, regional climate change predictions from various GCM models should be incorporated into this study. In addition, sensitivity analyses which consider changes in the distribution of buildings across heating and cooling dominated regions are important, particularly if the distribution in building stock is expected to change, since this also would change the magnitude of costs and benefits predicted by our model.

In this study we use a bottom-up approach to estimate the climate sensitivity of the energy sector, based on a reduced-form specification. Further research will compare these results to a structural joint fuel choice–conditional expenditure model. Evaluating the fuel–choice responsiveness to climate change may provide more detail regarding the structure of the climate–energy relationship, and in turn, climate change impacts. Finally, the uncertainty of our estimates is being explored in more detail to consider any additional factors that would alter the magnitude and direction of our predictions. Valuable future research would also consider the energy–climate sensitivity for a developing country to allow for comparison.

References

Anderson, K.P. 1973. *Residential Energy Use: An Econometric Analysis*. Santa Monica, CA: The Rand Corporation (R-719-NF).

Baker, P., Blundell, R. and Micklewright, J. 1989. Modeling Household Energy Expenditures Using Micro-Data. *The Economic Journal* 99: 720–38.

Balestra, P. and Nerlove, M. 1966. Pooling Cross Section and Time Series Data in the Estimation of a Dynamic Model: The Demand for Natural Gas. *Econometrica* **34**(3): 585–612.

Barnes, R., Gillingham, R. and Hagemann, R. 1981. The Short-Run Residential Demand for Electricity. *The Review of Economics and Statistics* **63**: 541–52.

Baughman, M. and Joskow, P. 1976. Energy Consumption and Fuel Choice by Residential and Commercial Consumers in the United States. *Energy Systems and Policy* **1**(4): 305–23.

Baxter, L. and Calandri, K. 1992. Global Warming and Electricity Demand. *Energy Policy* **20**:233–44.

Berndt, E. and Wood, D. 1975. Technology, Prices, and the Derived Demand for Energy. *The Review of Economics and Statistics* **LVII** (3): 259–68.

Cline, W. 1992. *The Economics of Global Warming*. Washington, DC: Institute for International Economics.

Crocker, T. 1976. Electricity Demand in All-Electric Commercial Buildings: The Effect of Climate. In: *The Urban Costs of Climate Modification*, Ferrar, T. (ed.). New York: John Wiley & Sons.

Dewees, D. and Wilson, T. 1990. Cold Houses and Warm Climates Revisited: On Keeping Warm in Chicago, or Paradox Lost. *Journal of Political Economy*. **98**(3): 656–63.

Energy Information Administration. 1992. *1990 Household Energy Consumption and Expenditures*. Data and accompanying documentation and reports. DOE/EIA-0321(90). Washington, DC: Department of Energy.

Energy Information Administration. 1993. *1989 Commercial Buildings Energy Consumption and Expenditures*. Data and accompanying documentation and reports. DOE/EIA-0318(89). Washington, DC: Department of Energy.

Fankhauser, S. 1995. *Valuing Climate Change – The Economics of The Greenhouse*. London: EarthScan.

Fisher, F.M. and Kaysen, C. 1962. *A Study in Econometrics: The Demand for Electricity in the United States*. Amsterdam: North Holland Publishing Co.

Green, R., Sally, A., Grass, R. and Osei, A. 1986. The Demand for Heating Fuels: A Disaggregated Modeling Approach. *Atlantic Economic Journal* **14**(4): 1–14.

Griffin, J. 1992. Methodological Advances in Energy Modeling: 1970–1990. *The Energy Journal* **14**(1): 111–24.

Halvorsen, R. 1975. Residential Demand for Electric Energy. *Review of Economics and Statistics* **57**: 13–18.

Hartman R. and Werth, A. 1981. Short Run Residential Demand for Fuels: A Disaggregated Approach. *Land Economics* **57**(2): 197–212.

Hotelling, H. 1931. The Economics of Exhaustible Resources. *Journal of Political Economy* **39**:137–75.

Houthakker, H.S. and Taylor, L.D. 1970. *Consumer Demand in the United States*, 2nd edition. Cambridge, MA: Harvard University Press.

Houthakker, H.S., Verleger, P.K., and Sheehan, D.P. 1973. *Dynamic Demand Analyses for Gasoline and Residential Electricity*. Lexington, MA: Data Resources, Inc.

Klein, Y. 1988. An Econometric Model of the Joint Production and Consumption of Residential Space Heat. *Southern Economic Journal* **55**(2): 351–9.

Linder, K.P. and Inglis, M.R. 1989. *Potential Impacts of Climate Change on Regional and National Demands for Electricity* in Smith, J. and Tirpak, D. 1989. *The Potential Effects of Global Climate Change on the United States*. Washington, DC: US Environmental Protection Agency.

Lyman, R.A. 1973. *Price Elasticities in the Electric Power Industry*. University of Arizona. Department of Economics.

Mendelsohn, R., Nordhaus, W. and Shaw, D. 1994. The Impact of Global Warming on Agriculture: A Ricardian Analysis. *American Economic Review* **84**: 753–71.

Mount, T.D., Chapman, L.D. and Tyrrell, T.J. 1973. *Electricity Demand in the United States: An Econometric Analysis*. ORNL-NF-49. Oak Ridge: Oak Ridge National Laboratory.

Nelson, J. 1975. The Demand for Space Heating Energy. *Review of Economics and Statistics* **7**(4): 508–12.

Nelson, J. 1976. Climate and Energy Demand: Fossil Fuels. In: *The Urban Costs of Climate Modification*. Ferrar, T. (ed.). New York: John Wiley & Sons.

Nordhaus, W. 1991. To Slow or not to Slow: The Economics of the Greenhouse Effect. *The Economic Journal* **101**: 920–37.

Rosenthal, D., Gruenspecht, H. and Moran. E. 1995. Effects of Global Warming on Energy Use for Space Heating and Cooling in the United States. *Energy Journal* **16**(2): 77–96.

Scott, M., Wrench, L. and Hadley, D. 1993. Effects of Climate Change on Commercial Building Energy Demand. Submitted to *U.S. Department of Energy,* Washington DC.

Smith, J. and Tirpak, D. 1989. *The Potential Effects of Global Climate Change on the United States.* Washington, DC: US Environmental Protection Agency.

Taylor, L.D. 1975. The Demand for Electricity: A Survey. *The Bell Journal of Economics* **6**: 74–110.

Wilson, J.W. 1971. Residential Demand for Electricity. *Quarterly Review of Economics and Business* **11**(1): 7–22.

Appendix A8. Definitions of independent variables used in residential and commercial regressions

A8.1 Residential

Variable	Definition
Average temperature	average annual temperature (de-meaned) – degrees F
Average temperature^2	average annual temperature (de-meaned) squared
Winter/summer temp differential	average winter minus summer temperature (de-meaned)
(Winter/summer temp differential)^2	average winter minus summer temperature (de-meaned) squared

A8.1 Residential (*cont.*)

Variable	Definition
Log electricity price	average electricity price
Log natural gas price	average natural gas price
Log fuel oil price	average fuel oil price
Log liquid petroleum gas price	average liquid petroleum gas price
Log kerosene price	average kerosene price
Log household income	average household income
Log size of home: square feet	home area – square feet
Log no. floors	number of floors in home
Log year built	year home constructed
Log family size	number of household members
Log age of household head	head householder age
More than 1 unit to home	1 if more than 1 unit, 0 otherwise
Tenant in home	1 if a tenant also occupies the residence, 0 otherwise
Race: black	1 if resident is black, 0 otherwise
Race: Hispanic	1 if resident is Hispanic/non-black, 0 otherwise
Age: over 65	1 if age of head householder > 65 years, 0 otherwise
Receives gov't heat cash aid	1 if resident receives cash aid for heat, 0 otherwise
Receives gov't heat vouchers	1 if resident receives heating vouchers, 0 otherwise
Burns wood as alternative fuel	1 if wood is burned as alternative heat source, 0 otherwise
Log no. rooms	number of rooms in home
Log no. doors & windows	number of doors and windows in home
Basement	1 if home has basement, 0 otherwise
Poor insulation or leaks	1 if household has inadequate insulation or window leaks, 0 otherwise
Interrupt/discount elec. rate	1 if household has discounted or interruptible electricity rates, 0 otherwise
Color tv	number of color TVs in household
Computer	1 if household has computer, 0 otherwise
Dish/clothes washer/dryer	1 if household has dishwasher, clothes washer, clothes dryer, 0 otherwise
Central air conditioning	1 if household has central air conditioning, 0 otherwise
Wall/window air conditioning	1 if household has wall or window ac units, 0 otherwise
Stove/Wreplace heats	1 if household uses stove or Wreplace to heat, 0 otherwise
Central warm air heats	1 if household uses central warm air to heat, 0 otherwise
Electric wall/radiator heats	1 if household uses electric wall units or radiators to heat, 0 otherwise
Portable kerosene heats	1 if household uses portable kerosene to heat, 0 otherwise

234

A8.2 Commercial

Variable	Definition
Average temperature	average annual temperature (de-meaned) – degrees F
Average temperature^2	average annual temperature (de-meaned) squared
Standard deviation temperature	standard deviation of temperature (de-meaned)
Standard deviation temperature^2	standard deviation of temperature (de-meaned) squared
Log electricity price	average electricity price
Log natural gas price	average natural gas price
Log fuel oil price	average fuel oil price
Log district heat price	average district heat price
Log square feet	building size – square feet
Log no. floors	number of floors
Log year built	year construction completed
Months open per year	number of months open
Alternative fuel used	1 if alternative fuel used, 0 otherwise
% food sale/service activities	percent food sale and food service
% warehouse/vacant	percent non-refrigerated warehouse or vacant
% in-patient/skilled healthcare	percent in-patient and skilled healthcare
% out-patient health/public safety	percent out-patient healthcare and public safety
% lab. or refrigerated warehouse	percent lab. or refrigerated warehouse
% industrial activities	percent industrial
% office space	percent office
% retail/service activities	percent retail/services
% educational activities	percent education
Ice/vending/water machines	1 if ice, vending or water machines used, 0 otherwise
Commercial refrig./ freezer	1 if commercial freezer or refrigerator used, 0 otherwise
Air duct – cool	1 if air ducts used for cooling, 0 otherwise
Air duct – heat	1 if air ducts used for heating, 0 otherwise
Boilers	1 if boilers used for heating, 0 otherwise
Computer room air conditioning	1 if there is air conditioning in computer room, 0 otherwise
Heat pump – cool	1 if heat pumps used for cooling, 0 otherwise
Heat pump – heat	1 if heat pumps used for heating, 0 otherwise
Tenant controls heat	1 if tenant controls heat, 0 otherwise
Roof – built up material	1 if roof material = built up, 0 otherwise
Roof – glass	1 if roof material = glass, 0 otherwise
Roof – metal surface material	1 if roof material = metal surface, 0 otherwise
Wall – masonry/siding	1 if wall material = masonry/siding, 0 otherwise
Wall – shingles/siding	1 if wall material = siding/shingles, 0 otherwise

Appendix B8. Derivation of demand elasticities

Assume for simplicity that the matrix X contains all independent variables except the price variable for which the elasticity is being calculated (P_1). The subscript j represents variables in the matrix X which are logged, while the subscript k represents variables in X which are left in their original form. The subscript i indexes fuels, for example in the residential case $i = 5$ and in the commercial case $i = 4$. The initial form of the expenditure equation is:

$$\sum_{i=1}^{F} P_i \cdot Q_i = P_1^{\beta_1} \cdot X_j^{\beta_j} \cdot e^{\beta_k X_k}. \tag{B8.1}$$

If we separate expenditures into $P_1 Q_1$ and $P_m Q_m$, the former representing expenditures on the fuel for which the price elasticity is being calculated and the latter representing expenditures on all other fuels, then the above equation can be written as:

$$P_1 \cdot Q_1 + \sum_{m=1}^{F-1} P_m \cdot Q_m = P_1^{\beta_1} \cdot X_j^{\beta_j} \cdot e^{\beta_k X_k}. \tag{B8.2}$$

Simplifying this equation so that only demand for fuel 1 is on the left hand side we have:

$$Q_1 = P_1^{\beta_1 - 1} \cdot X_j^{\beta_j} \cdot e^{\beta_k X_k} - P^{-1} \cdot \sum_{m=1}^{F-1} P_m \cdot Q_m. \tag{B8.3}$$

Solving for the elasticity of demand for fuel 1 we get:

$$\frac{\partial Q_1}{\partial P_1} \cdot \frac{P_1}{Q_1} = \eta_1 = \frac{(\beta_1 - 1) \cdot P_1^{\beta_1 - 1} \cdot X_j^{\beta_j} \cdot e^{\beta_k X_k} + P^{-1} \cdot \sum_{m=1}^{F-1} P_m \cdot Q_m}{P_1^{\beta_1 - 1} \cdot X_j^{\beta_j} \cdot e^{\beta_k X_k} - P^{-1} \cdot \sum_{m=1}^{F-1} P_m \cdot Q_m} \tag{B8.4}$$

Based on Equation (B8.4) and the regression results, the price elasticities are evaluated at the sample means for the range of observations that use the relevant fuel for space conditioning.

9 The economic impact of climate change on the US commercial fishing industry

MARLA MARKOWSKI, ANGELIQUE KNAPP,
JAMES E. NEUMANN, AND JOHN GATES[1]

As scientists become more certain that increases in concentrations of greenhouse gases will change climate, there has been growing interest in the potential economic effect of climate change. While the National Oceanic and Atmospheric Administration (NOAA) has sponsored investigations into the potential consequences of global climate change on fisheries, fisheries have received very limited attention in previous climate change damage assessments. For example, the two most commonly cited surveys of the effect of climate change on the US economy, Nordhaus (1991) and Cline (1992), provide no estimate of potential effects on the fisheries sector, although Nordhaus lists agriculture, forestry, and fisheries as economic sectors that could be severely affected by climate change.

On one level, the lack of attention to the potential effects on fisheries is understandable. First, fisheries and aquaculture are relatively small sectors of the US economy. The US fishery sector is a $3 billion contributor in a $5 trillion economy. Second, while there has been much research observing the changes in climate, there has been much less work done describing how the oceans will respond to increases in concentrations of greenhouse gases. Third, the link between ocean conditions and fisheries is complex and poorly understood. Major uncertainties and gaps in understanding make it particularly difficult to quantify the effect of global climate changes on the stocks of commercially important fish species.

For many reasons, however, natural resource-based industries, including fisheries, continue to strike a resonant chord in the US consciousness. In addition, because of

[1] The authors gratefully acknowledge suggestions and comments by Rick Freeman, Kathy Segerson, and Jon Sutinen. Research support was provided by Bill Lombardi, University of Rhode Island. We owe thanks to many individuals who most generously offered their assistance in helping us assemble the literature, including Keith Criddle and Mark Herrmann of the University of Alaska; Don DeVoretz of Simon Fraser University; Dan Huppert of the University of Washington; Christopher Kellogg of the New England Fisheries Management Council; and several people from the National Marine Fisheries Service, including Amy Buss Gautam, Samuel Pooley, and John Ward. The content and conclusions are the sole responsibility of the authors.

the structure of Congressional Committees and interest groups, fisheries' interests are well positioned to cast oversized shadows on the political landscape. The high level of public resources devoted to fisheries' management and the maintenance of regional fishery-based economies are also motivation for improving our understanding of the potential effect of climate change on fisheries.

This report provides a preliminary economic estimate of the sensitivity of the commercial fisheries sector to changes in the ocean environment that might result from climate change. Using existing studies that estimate demand for the most economically important fish species, along with current landings and price data, we have constructed a crude model of the ex-vessel fisheries market for each major fisheries' region in the United States. We have used this tool to develop an order-of-magnitude approximation of the potential economic welfare effect of climate-induced changes on fish stocks.

Our estimates of economic effects are based on illustrative scenarios of the effects of climate change on fish stocks. Unfortunately, the daunting task of estimating the effects of climate change on the stocks of commercially important fish is not yet complete. The science of estimating physical and biological effects of climate change on the ocean environment is so complex that even preliminary estimates of this relationship are currently unavailable. At this point we can only speculate on the potential magnitude of changes in fish stocks that climate change could cause. Nonetheless, our work provides a starting point for assessing the economic effect of potential changes in the ocean environment. As the science of estimating effects on the ocean environment progresses, we hope tools such as ours can provide insight into the potential market effects of changes in ocean biology, and can be used to guide scientific research towards answering those questions most relevant to economic policy making. While it is reasonable to assume there might be both positive and negative effects occurring simultaneously from climate change, our estimates reflect first-order damage or beneficial effects on fish harvest levels from climate change. Our estimates do not incorporate the possible compensating effects of climate change, such as the benefits of species migration due to climate change. We take this approach largely due to data limitations. Such an endeavor would require a case study approach to consider the biological implications at particular fisheries.

The results of this quantitative analysis show impacts to commercial fisheries could be small compared to the effects on other economic sectors such as agriculture, but they could be substantial to the commercial fishing sector. Harvest scenarios based on historical variation and sensitivity indicate global climate change effects on open access commercial fisheries could range from 2 to 10 percent of total domestic fishing value, or between $0.4 and $1.2 billion annually.

238

The remainder of this chapter provides the details of this analysis. Section 9.1 reviews the literature describing the potential effects of climate change as they relate to commercial fisheries. Section 9.2 presents the methodology we use to develop climate change harvest scenarios and incorporate them with economic demand research to obtain potential welfare change estimates. In Section 9.3 the results of the climate change scenarios on the United States, regional, and species-specific fisheries are presented and Section 9.4 discusses the implications of the analysis. Finally, the technical appendix presents data sources, demand equations, and the units of measure we used to analyze each species in each region.

9.1   Potential climate change effects

A critical element in our approach to estimate potential economic damages to commercial fisheries from climate change involves understanding the effects of climate change on commercial fisheries (see Gates, 1995). Currently, however, the obstacles to developing a clear statement of the climate change effects on commercial fisheries are formidable. In general, the complexity of the physical and biological processes in the oceans, coupled with the uncertainty over the magnitude and timing of the effects of global climate change, have hampered efforts to characterize these effects.

Despite the lack of explicit commercial fish stock effects, current literature provides several broad conclusions regarding the pattern of potential physical and biological effects of climate change. The range of hypothesized physical effects on the ocean environment includes changes in surface temperature of the sea, upwelling, salinity, stratification, and circulation patterns. In particular, the surface temperatures of the sea are expected to increase by an amount ranging between $0.2\,^\circ C$ and $2.5\,^\circ C$, due to an effective doubling of CO_2. North–south gradients of the surface temperatures of the sea will decrease because of greater warming at higher latitudes. In addition, changes in upwelling are expected to vary by location. Paleoecological evidence suggests that the overall intensity of upwelling may decrease, which would probably decrease overall ocean productivity. In general, salinity is expected to increase at temperate latitudes and decrease in high and low latitudes. These general patterns, however, may be reversed by local factors such as hydrogeology and net evaporation. Furthermore, stratification patterns are likely to change as salinity and water temperature change, but the complex relationships that influence stratification make predicting a general trend extremely difficult. Finally, climate change could affect broad circulation patterns in the oceans through changes in the formation of dense, cold water at higher latitudes, and local circulation patterns may also be affected.

The potential physical effects of climate change are likely to result in impacts on ocean biology. Biological effects resulting from climate change could affect fish abundance and location. The predicted increase in ocean temperatures is likely to have the most significant impact on ocean biology. One major effect could be a poleward shift in the distribution of fish populations. Water column stratification changes are likely to affect fish abundance through changes in food supply and related changes in predator/prey balances. Ocean circulation changes are likely to affect transport and food supply mechanisms for fish populations. The transport of larvae from spawning to nursery grounds is the mechanism most sensitive to climate change. Upwelling current changes are expected to affect associated food supply mechanisms. The direction of the changes will vary by region. Sea level changes may result in damage to estuarine habitats and, in turn, affect estuarine-dependent species. These effects could be more extreme if, for example, sea walls are erected to protect property landward of the marshes and estuaries, leading to flooding of estuaries. Sea temperature changes may also affect aquacultural operations. Most areas are likely to benefit from an enhanced potential for aquacultural production, especially the northern coastal regions.

Climate change is likely to have the greatest effect on recruitment rates of fish populations and the survival of young fish: (1) reproduction and juvenile survival are likely to be sensitive to the changes expected in the ocean environment from climate change, and (2) long run adult populations are very sensitive to even small changes in recruitment. In general, recruitment variability is the dominant source of variability in most commercially important marine fisheries.

Selected North American empirical studies

In addition to the general conclusions stated in the fisheries and climate change literature, there have been a number of efforts that attempt to link environmental factors to changes in the abundance of commercial fish stocks in North America. For example, temperature and other environmental conditions that could be linked to climate change appear to influence the abundance of northeastern US fishery resources. Studies which have helped to develop this link include the following:

> Sissenwine (1974, 1977) found that the correlations between yellowtail flounder recruitment and temperature were so strong that no relationship between spawning stock size and recruitment could be detected.
>
> King (1977) examined dominant species in the demersal groundfish complex of New England, with a focus on determinants of a fishery pro-

duction function. The results indicated that current landings were driven by current effort and by a transformed, lagged Ekman transport term. All other measured variables were insignificant, particularly lagged effort and catches.

Jeffries (1994) found warm winters had a negative impact on the recruitment of winter flounder in Narragansett Bay, while cold winters positively influenced the recruitment. The temperature effect appears to be a triggering mechanism which sets off a chain of events affecting species mix.

Dow (1964) found correlations of lobster abundance with seawater temperature for the period from 1905 to the 1960s and also found the center of catch shifting southward with falling temperature.

Nelson et al. (1977) correlated deviations from a Ricker recruitment curve with anomalous Ekman transports. It is believed that the mechanism is an enhanced probability of reaching the inshore nursery grounds.

Flowers and Saila (1972) correlated catches of lobsters with temperature between 1940 and 1970 and found correlations of 0.87–0.94 depending on which temperature station was used.

Gunter and Edwards (1967) reported high correlations of white shrimp landings with rainfall in the current year and lagged previous two years.

Sutcliffe et al. (1977) correlated water temperatures in the northern Gulf of Maine and Bay of Fundy with catches of 17 species in the late National Commission for the Northwest Atlantic Fisheries (now North Atlantic Fisheries Organization) Statistical Area 5.

Taylor et al. (1957) related landings statistics for mackerel, lobster, whiting, menhaden, and yellowtail flounder to air and water temperature records.

Templeman et al. (1953) examined mackerel, lobster, squid, billfish, capelin, and cod and correlated with air and water temperatures.

Based on the literature reviewed, there is an abundance of information to suggest that climate change could dramatically affect fisheries. Some effects which seem intuitively plausible include a poleward shift of species due to a poleward shift of the ecosystems of which they are a part. This general conclusion is limited in that it relies on very simplistic assumptions regarding the shape and topology of the coast, gravitational and rotational forces, and tidal impulses. Furthermore, Austin and Ingham (1979) provide criticisms of the statistical approach leading to this result. Given these limitations, we offer the following two hypotheses to describe the potential impact of climate change on commercial fisheries:

241

Climate change or Ekman transport changes may affect the following three
types of fisheries in the United States:

estuarine/crustacean/anadromous species (lobster, Alaskan king
crab, Gulf shrimp, salmon),

demersal/bivalve mollusk species (Atlantic groundfish, Alaskan
pollock, Atlantic sea scallop),

small pelagic species (Atlantic mackerel, menhaden).

The direction of change will not be the same for each of these fisheries.

9.2   Methodology

For each species of our analysis, we develop welfare measures and analyze
the sensitivity of commercial fisheries to climate change by applying several potential
harvest effects to fisheries demand functions. First, we make the assumption that each
angler works in a harvest-regulated fishery. We develop climate change scenarios
under the assumption that climate change affects fish stock and, subsequently, the
allowable harvest level for each fishery. Second, we assume that under harvest restrictions, anglers work in an open access fishery. The welfare value of reductions in fish
populations are consequently limited to changes in consumer surplus.

The remainder of this section describes the various climate change scenarios we
analyzed, the theory behind measuring welfare changes in commercial fisheries, how
we apply scenario information to economic demand relationships to calculate welfare
changes, and how we extrapolate regional welfare losses for specific species to obtain a
national welfare loss for all domestic commercial fisheries species.

We develop three illustrative harvest impact scenarios to reflect a range of potential consequences. Case 1 is a severe scenario, Case 2 is moderate, and Case 3 is
neutral. For each of these cases, we examine both an increase and a decrease in the
population of fish. Each case assumes that the impact of climate change is different
for each of the three major types of fish: estuarine-dependent, demersal/bottom
dwelling, and pelagic. Current research suggests that estuarine-dependent species are
likely to incur the greatest stock losses from proposed climate change scenarios.
Demersal or bottom-dwelling species are likely to be affected to a lesser degree, and
pelagic species are likely to be the least affected. Although bottom-dwelling species
differ from demersal species, fish of these two groups live in similar habitats and will
experience similar climate changes, probably resulting in similar climate change
effects.

Table 9.1 shows the commercial fishing species which account for significant per-

Table 9.1. *Commercial fish species groupings*

Region	Estuarine-dependent species	Demersal and bottom-dwelling species	Pelagic species
Atlantic	Bluefish	Groundfish (e.g. cod)[a]	Mackerel
	Lobster[a]	Ocean quahog	Herring
	Menhaden[a]	Sea scallops	Squid
	Shrimp (e.g. pink, rock)[a]	Surf clam[a]	Summer flounder
	Spiny lobster		
	Stone crab		
Pacific/Alaska	King crab[a]	Abalone	Herring[a]
	Salmon[a]	Coral mollusk	Mackerel
		Groundfish[a]	Tuna (e.g. skipjack)[a]
		Pollock[a]	
		Reef fish[a]	

Note:

[a] Species represented in the current analysis.

centages of landings and landed value in the US domestic market grouped by species type for both the Atlantic and Pacific/Alaska regions.

We analyze the historic landing data as a proxy for climate effects. Interannual harvests are hypothesized to be a function of fluctuations in ocean temperatures and patterns. The historic fluctuations in harvest rates consequently identify how sensitive species may be to climate changes. Table 9.2 shows the 1 standard error statistics for each species. The 1 standard error suggests average landings for estuarine-dependent species vary from 4 to 21 percent, demersal fish from 5 to 41 percent, and pelagic fish from 11 to 33 percent.

In most cases we were able to obtain standard error statistics from our data sources. When we could not obtain standard errors of the historical data, we attempted to use the most conservative estimates of variation available. In the case of Atlantic surf clam, we used the annual standard error, which is likely to be lower than the unavailable quarterly standard error. In the case of Pacific pollock, it was necessary to choose between using the quarterly standard deviation and the annual standard error. We selected the annual standard error estimate, an estimate more conservative in terms of potential climate change impacts, which gave an historical variation percentage of 38.31 percent rather than the standard deviation estimate of 55.77 percent. Furthermore, in cases where standard error statistics were not available, we used standard deviation as a measure of historical variation.

Table 9.2. *Historical variation in commercial fisheries quantities*

Region	Sub-region	Species	Percent deviation[b] (one standard error from mean)
Atlantic	North Atlantic	Cod	6
		Lobster	12
	Mid-Atlantic	Surf clam	5
	South Atlantic	Rock shrimp	18
	Atlantic	Menhaden	4
	Gulf	Pink shrimp	
		small (>51 tails/lb)	11
		medium (31–50 tails/lb)	6
		large (<30 tails/lb)	7
Pacific	Northwest Pacific	Groundfish	6
		Salmon	
		high value (sockeye, coho)	8
		low value (pink, chum)	12
	South Pacific	Reef fish[a]	41
		Skipjack tuna[a]	33
	Alaska	King crab	21
		Herring[a]	11
		Pollock	38
		Salmon	
		high value (sockeye, coho, chinook)	4
		low value (pink, chum)	5

Notes:
[a] Standard deviation used because standard error not available.
[b] Refer to Appendix A9 for the one standard error absolute value.

The second set of allowable harvest scenarios we develop combines hypotheses in the climate change literature regarding the relative effects among species types with potential magnitudes of effect. We chose effect magnitudes that represent a sufficiently large signal to evaluate the sensitivity of the system. We test the welfare sensitivity for three impact cases for each of the species groups.

In each case, estuarine-dependent species are hypothesized to be affected more than demersal, and demersal species affected more than pelagic. Case 1 assumes climate change harms 35 percent of the estuarine-dependent, 20 percent of the demersal/bottom-dwelling, and 10 percent of the pelagic species populations. Case 2 assumes climate change harms 20 percent of the estuarine-dependent and 10 percent

Table 9.3. *Percentage reduction in commercial fish harvest: impact cases*

	Estuarine-dependent effects			Demersal/bottom-dwelling effects			Pelagic effects		
	Case 1	Case 2	Case 3	Case 1	Case 2	Case 3	Case 1	Case 2	Case 3
Impact	−35	−20	−15	−20	−10	−0	−10	+10	+15

of the demersal/bottom-dwelling species populations, and increases pelagic species population by 10 percent. Case 3 assumes climate change harms 15 percent of the estuarine-dependent species population and increases pelagic species population by 15 percent. Table 9.3 presents the three cases. Case 1, the severe case, is more severe than the 1 standard error historic data, especially for estuarine species. Case 2 slightly exceeds the 1 standard deviation data except for pelagic species where it is well below the historic average. Case 3 exceeds the average historic data for only estuarine species but otherwise predicts low impacts. For each case, we examine both a beneficial and a harmful scenario where populations either increase or decrease.

We use these hypothesized harvest reductions to evaluate the change in consumer surplus for each species in each region of the US. We calculate the welfare change in the system for each climate change scenario using baseline price and quantity data for each species and the hypothesized climate change effects. We then determine a total welfare change by aggregating estimates across species and regions.

Our welfare calculations assume regulated harvest levels and unregulated angler effort levels for each fishery. We assume a perfectly elastic long run supply of effort where a cost increase is associated with a decrease in the fish stock and, hence, allowable harvest. Anglers enter and exit the market in response to changes in profitability. As described in Tietenberg (1996), an open access fishery will result in too much fishing effort. Figure 9.1 shows that the effort in an open access fishery increases until profits are zero (i.e. E^0, where average cost = average revenue), whereas the effort of a sole property owner (or perfectly managed fishery) increases until profits are maximized (i.e. E^*, where marginal cost = marginal revenue). If climate change decreases the stock of a fishery, anglers may either increase their effort in the fishery to the point where costs exceed benefits for some individuals, E^C, or the decrease in catch per effort may significantly lower the revenue and profits for some anglers. Either of these situations could drive anglers out of the market, thus shifting supply upward.

Finally, we assume this open access fishery operates with a homogeneous set of anglers. These assumptions allow us to calculate welfare as a change in consumer surplus, i.e. producer surplus is zero. Homogeneity and open access assumptions lead

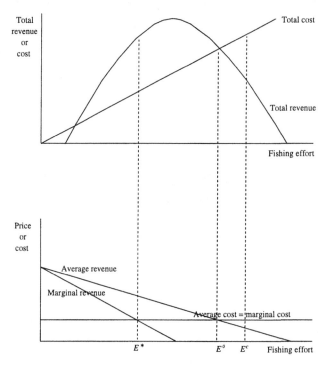

Figure 9.1 Effort in an open access fishery.

to zero producer surplus because inefficient management of a common property resource such as commercial fisheries leads to overcapitalization. That is, anglers will fish until the average cost equals average revenue in the fishery; the excess harvest capacity of the industry dissipates the potential producer rent of the fishery (see Tietenberg, 1996). If fisheries were managed to protect economic rent, the welfare analysis of climate change would also have to include changes in the economic rent of the fishery. Relaxation of the homogeneity assumption would reveal that some anglers may profit substantially more than others due to different inputs and skill levels. Although climate change could potentially affect these income differentials, this effect is likely to be small and would not substantially bias our results. Since these fisheries are regulated to protect long run stocks, the supply of fish in any period is independent of price. That is, the supply function for aggregate catch is perfectly inelastic.

To compute the welfare associated with each shift of supply, we calculate the consumer surplus. Figure 9.2 shows this shift as a price–quantity change from (P_0, Q_0) to (P_1, Q_1). Region P_0AB represents initial consumer surplus. Region P_1CB represents the consumer surplus after climate change. The total value change is the difference in consumer surplus due to the climate change effect, i.e. the shaded region P_1P_0AC. The welfare value, (W), of this change in supply is therefore:

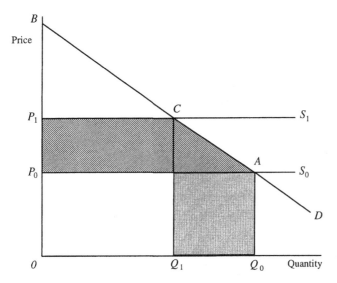

Figure 9.2 Long run welfare effects.

$$W = \tfrac{1}{2}(P_1 - P_0)(Q_0 - Q_1) + Q_1(P_1 - P_0). \tag{9.1}$$

The welfare in Equation (9.1) is measured in the ex-vessel market for fish. In a competitive and closed economy this welfare measure accounts for welfare changes in all other related markets as well (Just *et al.*, 1982).

Although this is a domestic analysis of climate change effects, the US imports many domestically consumed fish products and exports the majority of others, e.g. Alaska pollock. In reality, the domestic industry operates with excess demand curves that have incorporated the effects of international supply. As a result, domestic consumer surplus is overstated for markets of mainly exported species.[2] To adjust welfare values for exports, we assume that the demand for each species has the same price elasticity around the world. The welfare effects will consequently be proportional to consumption. For each principally exported species, we determine the domestic consumption, Y, and exports, M. The domestic welfare effect, W_d is:

$$W_d = W \times [Y / (Y + M)]. \tag{9.2}$$

The economic measures we develop measure the welfare impacts on fisheries as they are managed today. Because future fishery regulations may manage the economic rent of the fishery more optimally, these estimates may underestimate the value of the fishery. If fisheries were regulated efficiently, welfare changes would also have to

---

[2] The authors gratefully acknowledge Jon Sutinen's elucidation of this issue.

include changes in producer surplus. Furthermore, our measures of welfare apply only to domestic impacts. In order to get a more complete analysis, it would be necessary to model both US and foreign demands and supplies. Exploring such a topic in detail is beyond the scope of the current study.

9.3   Empirical results

To use our welfare methodology, we apply each of the climate change scenarios to empirical demand models, considering those species of greatest economic importance in each US region. This section describes the gathering of the appropriate economic demand literature, the method used to evaluate these relationships, and the role substitute species play in our analysis.

The demand relationships we analyze result from an extensive literature search for demand studies of important domestic commercial fish species within the United States. In some cases, this literature was unavailable; thus we could not estimate the demand for all species we consider to be economically important. However, in other cases, several articles were available for a single species. In these cases we selected the most appropriate study according to pre-designated guidelines.

Our guidelines placed greater weight on the more recent studies, on those having price as the dependent variable in the demand equation, and those using ex-vessel prices (as opposed to wholesale prices). Furthermore, we preferred studies including substitute species in the demand specifications, as they give more exact estimates of the coefficients on quantity variables. The studies we found appropriate to use for the analysis are listed in Table 9.4.

To calculate welfare changes in each fishery as described in the previous section, we maintain the functional form of the demand equations from the literature and use the estimated parameters and the various climate change landings scenarios to develop a range of estimates of welfare loss. For each demand relationship, we evaluate a change in quantity at the mean of the sample (i.e. we set all demand shift variables to their means). We then evaluate demand from an initial baseline quantity to a new quantity by integrating the equation. To enable us to use the relationships developed in the literature, we assume that our proposed climate change scenarios represent marginal changes in the demand relationships. That is, the effects on landings do not exceed the range of data for which the relationships are valid. Under this assumption, we are able to construct a measure of lost fisheries value, measured as change in welfare.

To determine baseline price and quantity levels for each fishery, we use historical price and quantity data. We conducted research to obtain historical data that would

Table 9.4. *Regional economic studies of commercial fisheries demand*

Region	Species	Authors (date)
Atlantic	Groundfish	Tsoa, Schrank, and Roy, 1982
	Lobster	Wang and Kellogg, 1988
	Surf clam	Lipton and Strand, 1992
	Rock shrimp	Adams, 1993
	Menhaden	Huppert, 1980
	Pink shrimp	Blomo, Nichols, Griffin, and Grant, 1982
Pacific	Groundfish	Adu-Asamoah, Rettig, and Johnston, 1986
	Salmon	DeVoretz, 1982
	Reef fish	Pooley, 1987
	Skipjack tuna	Hudgins, 1980
Alaska	King crab	Greenberg, Matulich, Mittelhammer, and Herrmann, 1994
	Herring	Mendelsohn, 1993
	Pollock	Herrmann, Criddle, Feller, and Greenberg, 1995
	Salmon (high value)	Herrmann and Greenberg, 1994

represent a long term record of landings. Thus, our primary goal in selecting historical data was to obtain the longest time period available that also used most recent data. In the light of this, we used data from the National Marine Fisheries Service (NMFS) where possible; however, we were constrained in the use of those data by the units of measurement in our demand equations which sometimes did not correspond to those of NMFS. In such cases, we used values provided to us by the studies themselves, or contacted the authors for more recent data.

There were three exceptions to using NMFS data. For Atlantic groundfish, we could not obtain historical data for two of the three species modeled in the literature, and therefore used only price and quantity data for cod. In the case of Pacific king crab, we were unable to obtain data from NMFS, so we used data from the authors' original source. Finally, although we attempted to use the longest time period available to obtain average price and quantity information, it was necessary to shorten the authors' data for Atlantic surf clam.

Once we obtained the averages of historical price and quantity data, we transformed the averages into units compatible with our demand specifications. Usually, this involved simply standardizing the prices into constant 1993 dollars or changing the units of measurement of the raw data. In other cases, the transformations were more involved. The Atlantic cod and lobster NMFS data aggregated annual landings and

249

revenues from regions that did not report monthly values in separate landings and revenues categories. We calculated monthly landings and revenue proportions from disaggregated data and applied them to the aggregate values to obtain a more inclusive measure of historical monthly data. The Atlantic menhaden demand equations use "per unit protein" data which NMFS reported to be 90 percent of the actual fish meal quantities; we adjusted price and quantity values to reflect this "per unit protein" transformation. In the case of Pacific salmon, we had demand equations for each sub-species and calculated historical prices and quantities for each species. To do this we used relative percentages reported for salmon species in the Northwest Pacific and Alaska and applied these percentages to the total salmon catch data reported for the Northwest Pacific region only. Appendix A9 provides the citation list for the baseline data used in this analysis and the values used to calculate welfare changes for all species.

Including substitute species in demand derivation is important since most fish demand equations suggest that substitution exists, and further, most global warming scenarios discuss multiple fish effects. Species that experience stock decline due to climate change are likely to be at least partially replaced by other species, or a compensating shift in production may occur at a different location. Such a biological compensation process will affect the consumer choice set and subsequent demand. If the equations we use to determine welfare changes do not account for these substitution effects, our species-specific approach may not represent the aggregate consequences of global climate change.

Accounting for the substitution effects is difficult, however, given the current literature and data limitations. A study properly accounting for substitution in species demand would involve simultaneous estimation of demand functions for each species that incorporate all relevant substitute species terms. Not all available commercial fishing demand literature estimates such a system of equations. One *ad hoc* method of considering substitutes uses information given in the literature to determine the substitute market, derive estimates of substitute flexibility for each species group where possible, and determine the impact on price from a change in quantity due to climate change for each species type. We pursued this approach and found that the literature provides either elasticity or flexibility measures for substitutes for estuarine and demersal species groups in each region. Table 9.5 shows the substitute market and flexibility/elasticity measures given in the literature for each region and species group.

Because we are considering price effects, we focus on the flexibility measures to calculate an average flexibility measure for substitutes of each species group (i.e. estuarine-dependent and demersal/bottom-dwellers). However, data limitations hinder our ability to determine the effect of substitutes on welfare changes. First, although we could assume the substitute market for a specific species is applicable to a general

Table 9.5. *Substitute elasticity measures from literature[a]*

Region	Estuarine-dependent species		Demersal and bottom-dwelling species	
	Substitute market	Flexibility/ elasticity[b]	Substitute market	Flexibility/ elasticity[b]
Atlantic	Pink shrimp	$g_{p,q} = -0.0306$	Redfish	$g_{q,p} = 0.2237$
	Soybean meal	$g_{q,p} = 0.1203$		
	Pink shrimp (medium)	$g_{p,q} = -0.0005$	Quahog	$g_{p,q} = -0.0029$
	Pink shrimp (small)	$g_{p,q} = -0.0014$		
Pacific	American lobster	$g_{p,q} = -0.0306$	Canadian groundfish	$g_{p,q} = 0.0078$
			Tuna	$g_{p,q} = -35.51$
	Tuna[c]	$g_{q,p} = 4.0700$	Mahi mahi	$g_{p,q} = 0.90$

Notes:
[a] No substitute elasticities were available from pelagic species group literature.
[b] $g_{p,q}$ refers to price flexibility, and $g_{q,p}$ refers to elasticity.
[c] Tuna elasticity is an average of elasticities of substitution from various salmon species.

species group (e.g. that quahog is a substitute for all estuarine-dependent species in the Pacific region), this assumption presents other problems. Namely, some species within a species group have nonfish substitutes (e.g. soybean meal, beef, pork). To use these flexibility measures to create an average measure applied to species having fish-only substitutes could be inappropriate for species of this group. Further, some species might have one substitute whereas others may have many substitutes; thus, an average measure could over- or under-represent the role of substitutes due to these differences. Second, given an aggregate flexibility measure composed of species from several species groups, it is unclear how to define climate change scenarios. For example, for Pacific demersal species, the average flexibility measure is -8.65. We calculate this as an average of the flexibility measures from demersal and pelagic substitute species (i.e. we average quahog, groundfish, tuna, and mahi mahi flexibility measures). Without more biological impact information, it is not clear how to signal a climate change effect for these two species groups simultaneously.

For these reasons, in addition to the lack of substitute information for the pelagic species group, it is currently beyond the scope of this limited screening analysis to account for substitutes in our welfare calculations. We can, however, consider specific cases to provide a cursory assessment of how welfare might change. Not all studies we consider include substitute species in the estimations, and of those that do, only three of those that considered substitute species also analyzed for climate change effects. We looked at two of these studies to provide examples of the potential variation in welfare loss values: rock shrimp (substitute = pink shrimp) and king crab (substitute = American lobster).

For both of these species, we re-evaluated the demand relationship to include substitute species baseline data. We then imposed the climate change effect to determine the substitute species terminal conditions. We re-evaluated the welfare impacts to rock shrimp and king crab by including this substitute species climate effect. Under the Case 1 scenario, rock shrimp welfare loss increased from $228 thousand to $275 thousand; king crab welfare loss increased from $6 million to $17 million.

The above examples show that for a first-order test, the effect of substitutes could be either small or large, depending on the species; but this test is very limited. Consideration of multiple substitute species may increase this welfare effect, while including other species not affected by climate change may mitigate the apparent effect on a particular species. Therefore, the overall welfare effect of excluding substitute species could vary depending on the particular combination of stock effect and elasticity effect. It could also vary depending on whether the substitute species for a commercial fishery is a nonfish species. While we might be able to unravel the effects of substitute species by conducting narrowly defined case studies, we could not extend the results of the case study to all fisheries because the necessary data are incomplete.

To determine a welfare loss value for the entire United States, the results of our species-specific analysis are extrapolated to include the potential effects on fisheries not explicitly included in the analysis. We have excluded species that are not considered the most economically important commercial fish species to each US region. We have also excluded some economically important species because we were unable to obtain appropriate demand studies.

While we would have liked to extrapolate our welfare measures for all the commercially important species (Table 9.1), we were unable to obtain the appropriate baseline data at the species group level necessary for extrapolation. The US Census Bureau provides domestic commercial fisheries revenue calculations by region but does not designate each species' contribution to this value. Further efforts to extend this analysis should include obtaining such data from the National Marine Fisheries Service. As a result, the baseline we use to extrapolate is based on sub-regions. Table 9.6 shows the percentage of fisheries value that the selected fisheries cover, compared with the total domestic commercial fisheries value. This exhibit shows that we were able to obtain good coverage in most regions; however, some regions have poorer coverage (e.g. Gulf Coast)[3].

---

[3] Coverage of total Gulf commercial fisheries value is particularly low because of the lack of published economic studies covering brown and white shrimp, the two highest valued fisheries in the region.

Table 9.6. *Domestic commercial fisheries coverage (millions of 1993$)*

Region	1992 Total revenue	Value represented by species analyzed (percent of total)
New England	$617	$265 (43%)
Mid-Atlantic	$288	$84 (29%)
South Atlantic	$155	$32 (21%)
Gulf Coast	$666	$28 (4%)
Atlantic total	$1726	$409 (24%)
NW Pacific & Alaska	$1936	$1015 (52%)
Hawaii	$72	$5 (7%)
Pacific total	$2008	$1021 (51%)

The species-specific results presented in Table 9.7 indicate which species in each region would be most affected by the damage scenarios. For the Case 1 scenario, in the Atlantic region, menhaden, lobster, and groundfish incur the greatest welfare losses in the region. These three species also account for the greatest impacts to 1992 regional revenue for the analyzed species. In the Pacific region, high value salmon, pollock, and low value salmon incur the greatest welfare losses. They also contribute the greatest impacts to 1992 regional revenue for all analyzed species.

The lower impact Case 3 scenario results in highest impacts in the Atlantic region to lobster and menhaden. In the Pacific region, high and low value salmon incur welfare losses whereas herring incurs a welfare gain. Overall, both scenarios produce results that indicate species demand sensitivity plays a role in determining impacts. For example, a comparison of the groundfish results in the Atlantic and Pacific regions for the Case 1 scenario shows the same impact results in a $13.5 million loss in the Atlantic but a $0.1 million loss in the Pacific region.

The species-specific benefit scenario results presented in Table 9.8 indicate, as in the damage scenarios, certain species to be more drastically affected in each region than others. The key species affected are the same as those of the damage scenarios. For the Case 1 scenario, lobster, groundfish, and menhaden enjoy the greatest benefits in the Atlantic region, and high value salmon, low value salmon, and pollock enjoy the greatest benefits in the Pacific region. For the Case 3 scenario, lobster and menhaden in the Atlantic region and high and low value salmon in the Pacific region incur the greatest welfare impacts. Comparing the impacts to 1992 total regional revenue across species for damage and benefit scenarios shows similar results to the damage scenarios; the species having the greatest welfare impacts contribute the most to the regional revenue.

Table 9.7. *Selected species-specific annual welfare: damage scenario*

Region	Species type	Case 1			Case 3		
		Harvest effect (%)	Annual welfare loss[c]	Percent of 1992 total regional revenue[a]	Harvest effect (%)	Annual welfare loss[c]	Percent of 1992 total regional revenue[a]
Atlantic	Estuarine	−35			−15		
	Lobster		13.6	0.8		6.5	0.4
	Menhaden		17.7	1.0		5.8	0.3
	Pink shrimp		0.1	0.01		<0.1	<0.1
	Rock shrimp		0.2	0.01		0.1	<0.1
	Demersal/bottom-dwelling	−20			−0		
	Groundfish		13.5	0.8		0.0	0.0
	Surf clam		3.6	0.2		0.0	0.0
	Total:		48.7	2.8		12.4	0.7
Pacific	Estuarine	−35			−15		
	King crab		6.3	0.3		3.0	1.5
	High value Salmon		45.5	2.3		20.6	1.0
	Low value Salmon		17.5	0.9		7.7	0.4

Demersal/bottom-dwelling	−20		−0		
Groundfish		0.1	<0.1	0.0	0.0
Reef fish		0.1	<0.1	0.0	0.0
Pollock		24.2	1.2	0.0	0.0
Pelagic	−10		+15		
Herring		12.7	0.6	(21.6)[b]	(1.1)[b]
Tuna		0.1	<0.1	(0.1)[b]	(<0.1)[bb]
Total:		106.5	5.3	80.4	4.0%

Table 9.8. *Selected species-specific annual welfare: benefit scenario*

Region	Species type	Case 1			Case 3		
		Harvest effect (%)	Annual welfare gain[c]	Percent of 1992 total regional revenue[a]	Harvest effect (%)	Annual welfare gain[c]	Percent of 1992 total regional revenue[a]
Atlantic	Estuarine	+35			+15		
	Lobster		19.4	1.1		7.6	0.4
Menhaden			9.5	0.6		4.5	0.3
Pink shrimp			0.1	<0.1		<0.1	<0.1
Rock shrimp			0.3	<0.1		0.1	<0.1
	Demersal/bottom-dwelling	+20			+0		
	Groundfish		16.5	1.0		0.0	0.0
	Surf clam		4.4	0.3		0.0	0.0
	Total:		50.2	2.9		12.2	0.7
Pacific	Estuarine	+35			+15		
	King crab		9.0	0.4		3.5	0.2
	High value salmon		54.1	2.7		22.3	1.1
	Low value salmon		20.4	1.0		8.3	0.4
	Demersal/bottom-dwelling	+20			+0		

Groundfish	0.2	<0.1	0.0	0.0
Reef fish	<0.1	<0.1	0.0	0.0
Pollock	29.5	1.5	0.0	0.0
Pelagic	+10	+15		
Herring	14.0	0.7	(18.5)[b]	(0.9)[b]
Tuna	<0.1	<0.1	<(0.1)[b]	<(0.1)[b..]
Total:	127.3	6.3	15.5	0.8

Notes:

[a] Percentages may not sum to total due to rounding.

[b] Parentheses indicate a loss.

[c] Millions of 1993$.

To achieve a national estimate we extrapolated our estimates for the regional level by dividing our welfare calculations by the percentage coverage for the region. For example, we calculated extrapolated welfare effects for the New England region by dividing the unextrapolated welfare change values by 0.43. For menhaden, the analysis was not specific to either the Mid- or South Atlantic. In this case, we attributed half of the menhaden effects to the Mid-Atlantic and half to the South Atlantic. Because the Gulf region has such poor coverage, we extrapolated from pink shrimp to all shrimp before extrapolating to the 1992 total value of the Gulf fishery. The average revenue for all shrimp species in the Gulf from 1960–1993 as reported by the NMFS is $434 million. We extrapolated the pink shrimp welfare loss to all shrimp species in the Gulf using these data (the pink shrimp welfare loss is 6 percent of the total potential lost value for all shrimp in the Gulf). Since shrimp make up approximately 65 percent of the total value of the Gulf fishery, we then divided the total shrimp calculation by 0.65 to extrapolate to all landings in the Gulf. This extrapolation methodology implicitly assumes that the effects on modeled fisheries are generally representative of the effects on the unmodeled fisheries. In particular we assume the unmodeled fisheries have the same biological sensitivity to climate change as the modeled fisheries. We also assume the demand for the modeled and unmodeled fisheries are the same. These assumptions leave room for further investigation of potential climate change.

Table 9.9 presents the extrapolated upper and lower bound results for the potential damages of climate change on commercial fisheries and the potential benefits. Under these scenarios, US commercial fishery damages range between 2 and 10 percent of total fisheries value, or between $58 and $355 million (1993$) annually. Potential commercial fishery benefits range between 3 and 10 percent of total fisheries value, or between $123 and $386 million annually.

9.4   Conclusion

The analysis presented in this chapter is intended to provide preliminary economic estimates of the sensitivity of the commercial fisheries sector to a broad range of climate change scenarios. Under restrictive methodological assumptions, the conservative climate change scenarios show potential major damages to domestic commercial fisheries. In our results, we find that the effects of climate change on domestic commercial fisheries could be large in terms of the total value of fisheries, although these effects may be considered small in comparison with the effects on other economic sectors (e.g. agriculture). Our screening level results highlight the possibilities for directing future research in the following areas:

Table 9.9. *National annual welfare changes: severe and neutral results, damage and benefit scenarios. (millions of 1993$)*

Region	Case 1 (% of regional revenue)		Case 3 (% of regional revenue)	
	Damage scenario	Benefit scenario	Damage scenario	Benefit scenario
Atlantic	$151	$142	$40	$38
	(9%)	(8%)	(2%)	(2%)
Pacific	$204	$244	$18	$85
	(10%)	(12%)	(<1%)	(4%)
Total	$355	$386	$58	$123
	(10%)	(10%)	(2%)	(3%)

Coverage

More economic demand literature is needed to cover all economically important commercial fish species as laid out in this analysis. Economic demand research is needed to describe estuarine species (bluefish, spiny lobster, stone crab); demersal species (ocean quahog, sea scallops, abalone, coral mollusk); and, especially, pelagic species (mackerel, herring, squid, flounder) which are currently not represented in the analysis for the Atlantic region.

Furthermore, in particular commercial fisheries regions greater coverage of varieties of these economic species would yield more representative results. The Gulf Coast fishery is particularly under-represented in this analysis, because the analysis does not account for brown shrimp demand. The brown shrimp fishery contributes almost 40 percent to the total Atlantic revenue.

Regulatory regime

We have presented an analysis that describes one extreme situation of a fisheries management regime: regulated harvest levels in open access fisheries. In reality the situation is likely to be somewhere between open access and input management; some fisheries may be regulated while others may not. Future consideration of the extent of regulation, and whether or not the regulation is binding for each fishery and region, would provide a richer description of the potential outcome of climate change.

Substitutes

We have excluded the effects of human adaptation to the changing mix of commercial species in this analysis. More economic research is needed to clarify how and in what direction substitutes play a role in demand for specific species. Note that pursuing this type of research itself implies a substantial amount of effort. Applying this research to develop a richer set of climate change scenarios is an even greater undertaking.

Richer scenarios

More biological research is needed to clarify the effects of climate change on various species, including positive effects that may result from species migration. The impact scenarios of the current analysis look at either mainly negative climate change scenarios or mainly positive climate change scenarios, neither of which explicitly considers a combination of these climate change effects across geographic location. Future research would require a more thorough study of particular fisheries to develop richer regional scenarios that incorporate nonuniform biological effects of climate change. A starting point would be to focus on the highest valued fisheries. Previous research (Gates *et al.*, 1994) has identified the Pacific coast and Alaska, the Gulf Coast, and New England as being the most valuable commercial fisheries regions; and salmon, shrimp, and crab as being the most valuable species in the United States.

Future study

The substantial potential loss to commercial fisheries suggests focusing future research on specific species most affected by climate change. Case study analyses would focus scientific and economic research to those fisheries deserving the greatest attention. The species that contribute the most to commercial fisheries losses under our climate change scenarios are Atlantic lobster and Pacific high value salmon, king crab, and pollock. Further examination of climate change effects of these fisheries would provide a richer assessment of the major contributors to economic losses in US commercial fisheries due to climate change.

References

Adams, C. 1993. *A Preliminary Assessment of Ex-vessel Price Movements in the South Atlantic Rock Shrimp Fishery*. Report prepared for the South Atlantic Regional Fisheries Management Council.

Adu-Asamoah, R., Rettig, R.B. and Johnston, R. 1986. *Regional Policy Analysis of the U.S.A. Pacific Groundfish Fishery: Post Extended Fishery Jurisdiction Considerations*. Report prepared for the International Conference on Fisheries, Rimouski, Canada.

Anwar, I.J., Lallemand, P. and Gates, J.M. 1992. Sources of Variance in the Economic

Performance of New England Otter Trawl Vessels. *Proceedings of the Sixth Conference of the International Institute of Fisheries Economics and Trade.* July 6–9, Issy-les-Moulineaux, France. RI Agr. Exp. Sta. Cont. No. 3021.

Austin, H. and Ingham, M.C. 1979. Use of Environmental Data in the Prediction of Marine Fisheries Abundance, pp. 93–106 in *Climate and Fisheries: Proceedings from a Workshop on the Influence of Environmental Factors on Fisheries Production.* The University of Rhode Island, Center for Ocean Management Studies, Kingston.

Blomo, V.J., Nichols, J., Griffin, W. and Grant, W. 1982. Dynamic Modeling of the Eastern Gulf of Mexico Shrimp Fishery. *American Journal of Agricultural Economics* August: 475–82.

Cline, W. 1992. *The Economics of Global Warming.* Washington, DC: Institute for International Economics.

DeVoretz, D. 1982. An Econometric Demand Model for Canadian Salmon. *Canadian Journal of Agricultural Economics* 30: 49–59.

Dow, R.L. 1964. A Comparison Among Selected Species of an Association Between Sea Water Temperature and Relative Abundance. *Conseil Permanent International pour l'Exploration de la Mer* 28: 425–531.

Flowers, J.M. and Saila, S.B. 1972. An Analysis of Temperature Effects on the Inshore Lobster Fishery. *Journal of the Fisheries Research Board of Canada* 29: 1221–5.

Gates, J.M. 1995. Notes on Global Climate Change and Fisheries. RIAES 3175. Eastern Economics Association Meetings, New York.

Gates, J., Amedan, H., Neumann, J. and Unsworth, R. 1994. *Interim Report: Summary of Physical and Biological Effects and Work Plan for Economic Impact Assessment.* Report prepared for the Electric Power Research Institute.

Greenberg, J.A., Matulich, S., Mittelhammer, R. and Herrmann, M. 1994. New Directions for the Alaska King Crab Industry. *Agribusiness* 10(2): 167–78.

Gunter, G. and Edwards, J.C. 1967. The Relationship of Rainfall and Freshwater Drainage to the Production of Penaeid Shrimps (*Penaeus Aviatilis* and *P. aztecus*) in Texas and Louisiana Waters. Proceeding of the World Scientific Conference on Biology and Culture of Shrimps and Prawns. FAO Fish Reports 57(3): 875–92.

Herrmann, M. and Greenberg, J. 1994. A Revenue Analysis of the Alaska Pink Salmon Fishery. *North American Journal of Fisheries Management* 14(3): 537–49.

Herrmann, M., Criddle, K., Feller, E. and Greenberg, J. 1995. Estimated Economic Impacts of Potential Policy Changes Affecting the Total Allowable Catch for Walleye Pollock. Draft.

Hudgins, L.L. 1980. *Economic Model of a Fisheries Market with Endogenous Supply: The Hawaii Skipjack Tuna Case.* Doctoral Dissertation, University of Hawaii.

Huppert, D.D. 1980. An Analysis of the United States Demand for Fish Meal. *Fishery Bulletin* 78(2): 267–76.

Jeffries, P. 1994. The Impacts of Warming Climate on Fish Populations. *Maritimes* 37(1): 12–14. The University of Rhode Island, Graduate School of Oceanography.

Just, R.E., Hueth, D. and Schmitz, A. 1982. *Applied Welfare Economics and Public Policy.* New Jersey: Prentice-Hall, Inc.

King, D.M. 1977. *The Use of Polynomial Distributed Lag Functions and Indices of Surface*

Water Transport in Fishery Production Models with Applications for the Georges Bank Groundfishery. Doctoral Dissertation, The University of Rhode Island, Kingston.

Lipton, D.W. and Strand, I. 1992. Effect of Stock Size and Regulations on Fishing Industry Cost and Structure: The Surf Clam Industry. *American Journal of Agricultural Economics* February: 197–208.

Mendelsohn, R. 1993. *The Effect of the Exxon Valdez Oil Spill on Alaskan Herring Prices.* Report prepared for Private Plaintiffs in conjunction with the *Exxon Valdez* oil spill.

Nelson, W., Ingham, M. and Schaff, W. 1977. Larval Transport and Year Class Strength of Atlantic Menhaden, *Brevoortia tyrannus. Fisheries Bulletin* 75(1): 23–41.

Nordhaus, W.D. 1991. To Slow or Not to Slow: The Economics of the Greenhouse Effect. *Economics Journal* 101: 920–37.

Pooley, S.G. 1987. Demand Considerations in Fisheries Management – Hawaii's Market for Bottom Fish. In: *Tropical Snappers and Groupers: Biology and Fisheries Management*, Polovina, J. and Ralston, S. (eds.). Boulder: Westview Press.

Sissenwine, M.P. 1974. Variability in Recruitment and Equilibrium Catch of the Southern New England Yellowtail Flounder Fishery. *Conseil International pour l'Exploration de la Mer* 36: 15–26.

Sissenwine, M.P. 1977. A Compartmentalized Simulation Model of Southern New England Yellowtail Flounder. *Limanda ferruginea* Fishery. *Fisheries Bulletin* 5(3): 465–82.

Sutcliffe, W.H., Jr., Drinkwater, K. and Muir, B.S. 1977. Correlations of Fish Catch and Environmental Factors in the Gulf of Maine. *Journal of the Fisheries Research Board of Canada* 34: 19–30.

Taylor, C.C., Bigelow, H.B. and Graham, H.W. 1957. Climate Trends and the Distribution of Marine Animals in New England. *Fisheries Bulletin* 57: 293–345.

Templeman, W. and Fleming, A.M. 1953. Long Term Changes in Hydrographic Conditions and Corresponding Changes in the Abundance of Marine Animals. *International Commission for Northwest Atlantic Fisheries.* Annual Proceedings, 3: 78–86.

Tietenberg, T. 1996. *Environmental and Natural Resource Economics.* New York: Harper Collins College Publishers.

Tsoa, E., Schrank, W. and Roy, N. 1982. U.S. Demand for Selected Groundfish Products, 1967–80. *American Journal of Agricultural Economics* August: 483–89.

US Bureau of the Census. 1994. *Statistical Abstract of the United States: 1994.* Washington, DC.

Wang, S.D.H. and Kellogg, C. 1988. An Econometric Model for American Lobster. *Marine Resource Economics* 5: 61–70.

Appendix 9A. *Units of measurement, data sources and demand equations*

Region	Sub-region	Species	Landings units, frequency	Price	Mean price/quantity[a]	Demand equation
Atlantic	North Atlantic	Groundfish (cod)	Million pounds, monthly	Wholesale cents/pound	84.59/6.28	$P = 184.04 - 15.85Q$
		Lobster	Million pounds, monthly	Ex-vessel cents/pound	333.29/5.04	$P = 411.35 - 15.5Q$
	Mid-Atlantic	Surf clam	Bushels, quarterly	Ex-vessel dollars/bushel	$16.22/5.56 \times 10^7$	$P = 21.72 - (6.1 \times 10^{-6})Q$
	South Atlantic	Rock shrimp	Pounds (head-on), monthly	Ex-vessel dollars/pound	3.58×10^5	$P = 1.76 - (5.12 \times 10^{-7})Q$
	Atlantic	Menhaden	Thousands metric tons, annually	Ex-vessel dollars/ metric ton meal	1.80/315.28	$P = \left[\dfrac{(0.11 - Q^{-0.5})}{0.089} \right]$
	Gulf	Pink shrimp (small)	Million pounds (head-off), monthly	Ex-vessel dollars/pound	2.35/0.28	$P = 2.39 - 0.159Q$
		Pink shrimp (medium)	Million pounds (head-off), monthly	Ex-vessel dollars/pound	3.44/0.18	$P = 3.47 - 0.191Q$
		Pink shrimp (large)	Million pounds (head-off), monthly	Ex-vessel dollars/pound	5.06/0.21	$P = 5.07 - 0.046Q$
Pacific	Northwest Pacific	Groundfish	Thousands pounds, monthly	Ex-vessel cents/pound	98.69/384.67	$P = 114.04 - 0.0399Q$
		Salmon (sockeye)	Thousands 48-lb cases, annually	Wholesale dollars/ 48-lb case	137.76/380.85	$P = \exp(6.11 - 0.2 \ln Q)$
		Salmon (coho)	Thousands 48-lb cases annually	Wholesale dollars/ 48-lb case	119.58/77.95	$P = \exp(7.44 - 0.61 \ln Q)$
		Salmon (pink)	Thousands 48-lb cases annually	Wholesale dollars/ 48-lb case	32.27/400.84	$P = \exp(7.85 - 0.73 \ln Q)$

Appendix 9A. (*cont.*)

Region	Sub-region	Species	Landings units, frequency	Price	Mean price/quantity[a]	Demand equation
		Salmon (chum)	Thousands 48-lb cases, annually	Wholesale dollars/48-lb case	58.25/146.32	$P = \exp(7.95 - 0.78 \ln Q)$
	South Pacific	Reef fish	Pound, monthly	Wholesale dollars/pound	$3.02/3.70 \times 10^4$	$P = \exp(3.00 - 0.18 \ln Q)$
		Skipjack tuna	Thousands pounds, monthly	Wholesale dollars/pound	1.72/190.20	$P = 1.95 - 0.0012Q$
	Alaska	King crab	Million pounds, annually	Wholesale dollars/pound	9.30/22.41	$P = 10.28 - 0.04361Q$
		Herring	Thousand pounds, annually	Ex-vessel dollars/pound	$0.38/8.89 \times 10^4$	$P = 1.88 - (1.69 \times 10^{-5})Q$
		Pollock	Kilograms, quarterly	Ex-vessel dollars/kilogram	$0.18/7.86 \times 10^8$	$P = .29 - (3.081 \times 10^{-8})Q$
		Salmon (high value)	Million pounds, annually	Ex-vessel dollars/pound	1.14/281.73	$P = \exp(0.53 - 0.0014Q)$
		Salmon (low value)	Million pounds, annually	Ex-vessel dollars/pound	0.37/335.88	$P = 0.47 - 0.000329Q$

Notes:

[a] Annual means have been used instead of quarterly means for Atlantic surf clam and Alaskan pollock. The quarterly mean values (902 649 bushels for surf clam and 330 080 000 Kg for pollock) are used to evaluate the demand relationship at the mean.

Introduction to recreation

Outdoor recreation activities play a significant economic and social role in the United States. In addition to contributing to total personal consumption and the Gross Domestic Product, outdoor activities support a healthy US lifestyle. The extent of research valuing the effects of environmental changes on outdoor activities reflects these important contributions. A wealth of economics literature exists on valuing the effects of environmental impacts on recreational activities and on developing methods to measure the values of nonmarket goods. Although very little research has been carried out to quantify the potential impacts of climate change on recreation, the effects of climate change on the quality and quantity of natural resources used for outdoor recreation could, in fact, adversely affect the recreation market. Given its limited coverage in past climate literature, we have taken two approaches to studying the effects of climate change on a variety of outdoor recreational activities.

Chapter 10 presents a study by Robert Mendelsohn and Marla Markowski which estimates the direct effects of climate on the demand for seven outdoor recreation activities. Mendelsohn and Markowski consider recreation activities for which climate impacts could be measured on a state-by-state basis. Using a travel cost approach to measure changes in values of all sites within each state, Mendelsohn and Markowski estimate the demand for visits to all sites and explore two econometric models to develop a range of estimates for each recreation category. The authors then apply an average consumer surplus per day to value the estimated change in demand for each recreation activity. The cross-sectional nature of the data provides results that reflect the long-term impacts of climate. This chapter extends earlier climate change research, which focused strictly on skiing, by providing a state-level analysis of skiing as well as several summer activities.

Chapter 11 presents an economic study of climate impacts on selected recreational activities by John Loomis and John Crespi. The authors focus on those recreation activities having both substantial levels of participation and a likelihood of some effect on demand due to climate change. Using regression models and existing literature to measure the direct and indirect effects of warming on recreation demand, Loomis and Crespi analyze eight major categories of recreation, comprising 17 activities. The authors develop a range of demand estimates based on favorable and unfavorable ecological scenarios and determine potential welfare values by employing an average consumer surplus per day. This chapter improves upon previous climate change analysis efforts by providing quantitative measures of climate effects for several economically important outdoor recreation activity groups.

These two analyses provide original, albeit preliminary, assessments of the national impacts of climate on a variety of outdoor recreational activities. The studies take two approaches providing qualitatively similar outcomes. Both studies suggest that climate change will have a positive impact on boating, fishing, and golfing, and a negative impact on camping, hunting, skiing, and wildlife viewing. Despite their differing resource definitions, the range of their quantitative results are similar. Overall, both studies indicate that the positive effects outweigh the negative effects, indicating that the net effect of global warming will be beneficial for the US outdoor recreation sector.

10 The impact of climate change on outdoor recreation

ROBERT MENDELSOHN AND MARLA MARKOWSKI[1]

This study estimates the impact of climate change on the demand for outdoor recreation. The study expands on earlier literature, which only studied skiing, by including analysis of six summer activities. The demand for each activity is estimated by regressing climate and other control variables for each state in the continental United States, using both linear and loglinear functional forms.

The estimated climate relationships are used to forecast the welfare impact of nine alternative climate change scenarios in both 1990 and 2060. With the inclusion of summer activities, the overall net effect of warming is beneficial. An increase in temperature of 2.5 °C and precipitation by 7 percent generates overall net recreation benefits of $2.8 billion (+7 percent) with the linear model and $4.1 billion (+9 percent) with the loglinear demand model. If temperatures increase by 5 °C, net benefits jump to $25.9 billion (+63 percent) with the linear model and $18.9 billion (+40 percent) with the loglinear model. Both models predict large fishing benefits from warming and substantial skiing losses. The linear model also predicts sizable boating benefits and camping and wildlife viewing losses. The higher per capita income in 2060 results in substantially smaller 2060 versus 1990 benefits according to the loglinear model because hunting benefits shrink and skiing damages increase. The results for the linear model are approximately the same in 1990 and 2060.

Climate is expected to affect recreation in three ways. First, warming is expected to lengthen the summer seasons and shorten winter ones. Second, climate could affect the comfort or enjoyment of engaging in any given outdoor activity. For example, while rain could put a damper on the pleasure of being outdoors, fishing may be more successful on cloudy overcast days. Third, climate could alter the ecology of an area and change the quality of the recreation experience. For example, climates conducive to large populations of big game or prize fish are likely to increase hunting or fishing quality.

Existing estimates of the impact of climate on outdoor leisure activities are limited.

[1] We would like to thank the following people who provided comments and suggestions on this analysis: A. Myrick Freeman, III of Bowdoin College; John Loomis of Colorado State University; Shannon Ragland of Hagler Bailley, Inc.; and Kathleen Segerson of the University of Connecticut.

267

Loomis and Crespi, in the next chapter, also explore the impacts of climate change on outdoor recreation. Cline (1992) estimated the potential impacts to the skiing industry using a Canadian study which predicted a loss of 40–70 percent of ski days from climate warming. If skiing in the United States were reduced by this amount, Cline estimates an economic loss of $1.7 billion annually.

This study extends earlier research which focused strictly on skiing by including a number of summer activities as well. It examines the effect of climate on the demand for seven prominent outdoor activities: boating, camping, fishing, golfing, hunting, skiing, and wildlife viewing. Because more recreation occurs in warm weather than in winter, including summer recreation provides a more balanced and comprehensive account of climate impacts on outdoor recreation. Approximately $2.5 billion is spent annually on skiing but $76.3 billion is spent on boating, camping, fishing, golfing, hunting, and wildlife viewing. Since it is likely that warming might have a positive rather than a negative effect on summer versus winter recreation, it is necessary to include summer activities.

In order to measure climate sensitivity, the number of recreation days of each activity are regressed on climate and control variables for each state. We explore both linear and loglinear functional forms for demand. We use the results from the empirical analysis to assess how nine alternative climate change scenarios would affect the consumer surplus for outdoor recreation. We examine warming of $1.5\,^\circ\text{C}$, $2.5\,^\circ\text{C}$, and $5\,^\circ\text{C}$ with current, 7, and 15 percent higher precipitation levels. We then calculate the welfare impacts, the change in consumer surplus, of each climate change scenario for each activity and the net impact across all seven activities.

The cross-sectional nature of the data provides results that reflect long run impacts. Assuming that individuals and ecosystems have adjusted to their local conditions, this "natural experiment" captures long-term behavioral and ecological adjustments. Although this may be a reasonable assumption to make of people (who appear to adjust readily and quickly), it is not clear whether this is an appropriate model for ecosystems. Differences in climates across space may not provide a good prediction of how ecosystems will change over time as climates change. For example, if climate changes to make the Pacific Northwest resemble the current climate of the Southeast, it is not clear whether the species and populations of the Southeast will ever dominate the Pacific Northwest.

The next section describes the theoretical foundation of the analysis. Beginning with a demand function for each activity, we calculate how changing climate would shift this demand function. We then measure welfare using the change in consumer surplus (the area under the curve of the demand function). Section 10.2 presents the results of the empirical models, and Section 10.3 uses the empirical models to calculate the welfare impacts for a set of climate change scenarios. Finally Section 10.4 reviews the overall conclusions of the analysis and recommends directions for further research.

10.1   Theory and data collection

The theoretical demand model we use in this analysis derives from the carefully developed travel cost literature on outdoor recreation (for a review, see Freeman, 1993). This literature measures the value of a recreation site by measuring the demand for visits to the site using distance or travel cost as a measure of price. The value of the site is deduced by observing peoples' actual behavior. Our study mirrors this approach except that we value a collection of available sites to a consumer, rather than a single site. Specifically, we estimate the demand for visits to all sites within a state. We then explore how a change in quality across the state would affect state visits. Thus we shift the focus of travel cost analysis from valuing a specific site to including all sites within a state.

In order to test the sensitivity of the results to various functional forms, we explore the two functional forms most often used in the literature: linear and loglinear. Both models are estimated for each of the seven activities.

We assume that the supply of leisure services is perfectly elastic and insensitive to climate change. Over the range of choices relevant to this problem, we assume the economy can produce more camping gear, hotels, and outdoor clothing without changing price. The cost or price, P, of purchasing recreation is assumed to be the same in every climate state. This implies that growth or shrinkage of overall visits does not change the cost per trip. If people substituted different destinations in response to climate change, however, the cost could either go up or down. For example, if they traveled further (less) to get to the same climate zone they enjoyed before, costs would rise (fall). Although selective sites may be congested, we assume these sites tend to represent a small minority of the available recreation destinations. Over the range of changes we consider in this analysis, these assumptions seem reasonable. If the supply function is price inelastic, this analysis will slightly overestimate the welfare benefits and underestimate the welfare damages of climate change.[2]

Second, in this analysis we are forced to ignore substitution effects across activities due to the absence of suitable price data. It is likely that some of these activities serve as

---

[2] The supply price elasticity is likely to be high whereas the empirical results suggest price inelastic demand. The omitted consumer surplus from assuming perfect price elasticity, with demand shifts of 30 percent or less, is likely to lead to welfare errors of less than 5 percent.

substitutes for each other. If climate change causes across-the-board impacts on many activities, our welfare calculation will fail to include these substitution effects. If the climate impacts are negative (harmful), the damages will be underestimated. If the climate impacts are positive (beneficial), the benefits will be overestimated. The magnitude of this effect increases with the price elasticity of substitution across activities.

Third, we assume a constant cost per activity day across states for boating, camping, golfing, and skiing activities due to lack of suitable price data. While this may be a reasonable assumption for golfing and camping activities, skiing and boating costs per day are likely to vary across states. We do not know how this may bias the results.

Finally, we assume that the real price of recreation remains constant over time. We have little evidence indicating how the price of a leisure-activity day might change relative to the price of all other goods between 1990 and 2060. With slightly higher populations and income, it is possible that prices might rise over time. However, technical change may make outdoor recreation sites more accessible, lowering prices.

Linear model

The linear aggregate demand from each state, s, for participation days in each outdoor activity, i, is:

$$Q_s^i = a_0^i + \sum_1^K a_k^i Z_{k,s} + b^i P_s^i + \sum_1^J c_j^i Y_{j,s} + \sum_1^J d_j^i Y_{j,s}^2, \qquad (10.1)$$

where a_k, b, c_j and d_j are constants, Q_s^i is the quantity of days of activity i in state s; $Z_{k,s}$ is a set of K personal demand shift variables (e.g. per capita income and population) and quality shift variables (e.g. area of public forest and surface water) in state s; P_s^i is the price per day of activity i in state s; and $Y_{j,s}$ is a vector of J climate variables in state s. Because climate variables are expected to have a nonlinear functional form, we include squared terms in the set of climate variables. The coefficients and the specific variables included in each model vary by recreation activity. As mentioned previously, the price per activity day will vary by state only for fishing, hunting, and wildlife viewing activities; for the remaining four activities, the value of this variable will be a constant.

The consumer surplus (CS) associated with each activity, given an initial set of price and demand shift variables, is approximately equal to the area under the demand function above price (see Willig, 1976):

$$CS_{0,s}^i = \int_P^\infty Q_s^i(Z_s, P_s^i, Y_s^0) \, dP. \qquad (10.2a)$$

With the linear demand model, there is a choke price P^*, where demand falls to zero. Net consumer surplus is equal to the area of the triangle between this choke price, the current price, P, and the current quantity:

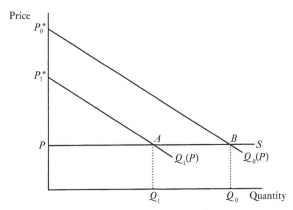

Figure 10.1 Linear model measure of welfare change.

$$CS^i_{0,s} = 0.5 \times (P^{i*}_s - P^i_s) \times Q^i_s. \tag{10.2b}$$

We measure the change in welfare, W, associated with a change from an initial set of environmental conditions, Y^0, to a new set, Y^1, by the difference in consumer surplus:

$$W^i_s = \int_P^\infty Q^i_s(Z_s, P^i_s, Y^1_s)\,dP - \int_P^\infty Q^i_s(Z_s, P^i_s, Y^0_s)\,dP. \tag{10.3}$$

Graphically, the welfare value we measure is the change in consumer surplus. For example, suppose that the demand function begins at an initial level, $Q_0(P_0^*)$, with choke price P_0^* and that climate has a negative impact on demand. After climate change, demand shifts to the left to $Q_1(P_1^*)$ with choke price P_1^*. The loss in welfare is equal to the original area underneath the demand function minus the new area underneath the demand function. The change in welfare is equal to the area $P_0^* P_1^* A B$, shown in Figure 10.1.

Given the statewide aggregate demand function Equation (10.1), the welfare estimate in Equation (10.3) measures the impact in each state. Note that this impact can differ by state depending on the initial and final climate conditions. Some states could experience different impacts because they experience a different climate change. Even with uniform climate change, the impacts could differ depending on initial conditions such as the initial climate, demand shift variables, and, in some cases, state prices. Given that impacts are a function of temperature (precipitation), not temperature change (precipitation change), the starting temperature (precipitation) affects the welfare outcome. Further, the larger the demand in the initial condition, the larger the welfare value of a given shift in demand. For example, as population increases, the welfare value of climate change also increases, *ceteris paribus*.

In order to calculate total welfare, TW^i, for the country for each activity, we sum the welfare effects across all states:

271

$$TW^i = \sum_s W^i_s. \tag{10.4}$$

Because this net impact is measured in common units (billions of dollars), we aggregate these impacts across activities to calculate a recreation sector net estimate.

Loglinear model

Using the same assumptions as in the linear model, the loglinear aggregate demand from each state, s, for participation days in each outdoor activity, i, is of the following form:

$$Q^i_s = \exp\left(a^i_0 + \sum_1^K a^i_k Z_{k,s} + b^i P^i_s + \sum_1^J c^i_j Y_{j,s} + \sum_1^J d^i_j Y^2_{j,s} \right). \tag{10.5}$$

The variables and subscripts are the same as defined for the linear model. We take the integral of Equation (10.5) with respect to price to reveal that consumer surplus is proportional to the quantity purchased:

$$CS^i_{0,s} = \int_P^\infty Q^i_s(Z_s, P^i_s, Y^0_s)\, \mathrm{d}P = \frac{Q^i_s(Z_s, P^i_s, Y^0_s)}{b^i}. \tag{10.6}$$

The more price inelastic is demand (the closer b gets to zero), the larger the consumer surplus. The welfare value of climate change can then be calculated using Equations (10.3) and (10.4).

To estimate the demand for an activity as in Equation (10.5), we take the logarithm of both sides of the equation:

$$\log Q^i_s = a^i_0 + \sum_1^K a^i_k Z_{k,s} + b^i P^i_s + \sum_1^J c^i_j Y_{j,s} + \sum_1^J d^i_j Y^2_{j,s}. \tag{10.7}$$

The linear and loglinear models consequently resemble each other with one exception. The dependent variable has a logarithmic form in the loglinear model.

Because the dependent variable is transformed in the loglinear model, the welfare estimate must be adjusted for the standard error, σ^2:

$$W^i_s = \left[\frac{Q^i_s(Z_s, P^i_s, Y^1_s)}{b^i} - \frac{Q^i_s(Z_s, P^i_s, Y^0_s)}{b^i} \right] \times \mathrm{e}^{\sigma^2/2}. \tag{10.8}$$

For each activity, σ^2 is calculated by dividing the sum of squared errors by the degrees of freedom in the model.

Data

Table 10.1 shows the definitions of all variables used in this study along with the mean of each variable. We obtained statewide participation and expenditure

Table 10.1. *Means and definitions of variables*

Variable	Mean	Definition
Income	17 640	State average per capita income (1991 dollars)
US Forest service area	5 257	Area (square miles)
Population	5 134 265	State population
Total area	64 806	State area (square miles)
Water area	3 149	Water area (square miles)
January temperature	0.0	Demeaned 30-year average, Fahrenheit. US mean=31.6
July temperature	0.0	Demeaned 30-year average, Fahrenheit. US mean=75.8
January precipitation	0.0	Demeaned 30-year average, inches. US mean=2.62
July precipitation	0.0	Demeaned 30-year average, inches. mean=3.63
Boating expenditures	30.33	Per capita, per day (1991 dollars)
Camping expenditures	6.77	Per capita, per day (1991 dollars)
Fishing expenditures	48.43	Per capita, per day (1991 dollars)
Golfing expenditures	21.00	Per capita, per day (1991 dollars)
Hunting expenditures	55.50	Per capita, per day (1991 dollars)
Skiing expenditures	27.66	Per capita, per day (1991 dollars)
Wildlife viewing expenditures	51.59	Per capita, per day (1991 dollars)
Boating consumer surplus	20.00	Per day (1991 dollars)
Camping consumer surplus	11.00	Per day (1991 dollars)
Fishing consumer surplus	23.00	Per day (1991 dollars)
Golfing consumer surplus	21.00	Per day (1991 dollars)
Hunting consumer surplus	22.00	Per day (1991 dollars)
Skiing consumer surplus	18.00	Per day (1991 dollars)
Wildlife viewing consumer surplus	16.00	Per day (1991 dollars)
Boating participation	7 698 146	State average days
Camping participation	10 852 167	State average days
Fishing participation	10 542 542	State average days
Golfing participation	11 359 542	State average days
Hunting participation	4 889 750	State average days
Skiing participation	1 990 896	State average days
Wildlife viewing participation	6 991 292	State average days

273

data for hunting, fishing, and wildlife viewing for 1991 by state from the US Fish and Wildlife Service (1992) national survey. We estimated expenditures per day for each activity by dividing total expenditures of state residents by total days spent by state residents on that activity. We obtained statewide participation data for boating, camping, golfing, and skiing for 1990 from the National Sporting Goods Association (1990). We collected data on average US expenditures per day for boating from the National Marine Manufacturers Association (Gerardi, 1994), for camping from the National Association of RV Parks and Campgrounds (Gorin, 1994), and for skiing from the Ski Industries of America (1994). We obtained an estimate for a golfing day from Loomis and Crespi (Chapter 11, this volume). We compiled additional data on the total area of each state, population, and income per capita from the US Bureau of the Census (1990, 1992). We gathered data on the statewide area of US Forest Service land from the US Bureau of the Census (1993), State and National park land from the National Park Service (1994), and surface water from the US Bureau of the Census (1990). We used climate data from Mendelsohn *et al.* (1994), which extrapolates data from weather stations across the United States to each state. These climate data include average temperature and precipitation over a 30-year period for January (winter), April (spring), July (summer), and October (fall).

Because of the absence of good price data, we are unable to determine the price elasticity of recreation. We consequently rely upon the recreation literature for the appropriate baseline price elasticity estimates. From Bergstrom and Cordell (1991) we obtain estimates of consumer surplus per day for the current climate. From the data set, we have observations about the number of trips and average cost (price) per day. With the linear demand function, consumer surplus is determined by Equation (10.2b). The baseline choke price for each state (under current climate) is consequently equal to:

$$\mathrm{P}_0^i{}^* = \mathrm{P}^i + 2 \times \frac{CS_0^i}{Q^i}, \tag{10.9}$$

where CS_0^i/Q^i is consumer surplus per day for the current climate. We assume that climate change shifts the demand function without changing its slope. Consequently, the choke price for a new climate in each state is equal to:

$$\mathrm{P}_1^i{}^* = \mathrm{P}_0^i{}^* \times \frac{Q_1^i}{Q_0^i}. \tag{10.10}$$

That is, the choke price will shift proportionately with changes in Q. Substituting the new choke price and the new quantity into Equation (10.2b), one can calculate the new consumer surplus according to Equation (10.3). With the loglinear demand function,

274

consumer surplus is Q/b (see Equation 10.8). Dividing by the number of days, Q, the consumer surplus per day is therefore $1/b$ and it is independent of the number of days. We assume that any change in climate does not change this price coefficient but merely results in a shift in the demand function. Note that this procedure results in climate change altering the quantity of trips with both the linear and loglinear models, but that the consumer surplus per trip changes only with the linear model.

In order to facilitate interpretation of the coefficients in the demand models, we have demeaned the climate data. We express the climate data as a deviation from the US mean. Thus, the climate coefficient, c_j is the marginal effect of the climate variable Y_j evaluated at the US mean.[3] The coefficient, d_j shows the curvature of the response function. A positive coefficient, d_j implies a U-shaped relationship for variable Y_j and a negative coefficient implies a hill-shaped relationship.

The unit of observation is recreation days within a state. Some days, however, are spent by nonresidents for fishing, hunting, and wildlife viewing activities. Independent variables such as state population and per capita income will not necessarily explain nonresident behavior. Since resident use dominates total use, however, the presence of nonresident use is not expected to be a serious problem. For example, 88 percent of days fished in each state are by residents of that state, 93 percent of days hunted are by residents, and 80 percent of days viewing wildlife are by residents.

10.2   Empirical results

The full set of demand shift variables was regressed on aggregate days by activity for each state. We tested several alternative ways to measure climate variables. For warm weather activities, we tested average annual, summer seasonal, winter and summer, summer and fall/spring, and just summer temperatures and precipitation. For skiing, we substituted winter measures for summer.

We evaluated regression results on the basis of expected signs, coefficient significance, and goodness-of-fit criteria to determine a "best" model for each activity. Our initial model for each activity included summer and winter temperature and precipitation variables specified as both linear and quadratic terms. Given the limited number of observations in the data set, many of the coefficients were not significantly different from zero. We dropped insignificant coefficients to yield regression results for both the linear and loglinear models (Tables 10.2 and 10.3). All the regressions

---

[3] The partial derivative of days (or log of days) with respect to a climate variable is equal to $\partial Q/\partial Y_j = c_j^i + 2d_j^i Y_j$ for each state s. With de-meaned variables, the mean of Y_j is zero which leaves just c_j^i.

Table 10.2. *Linear regression of climate on activity levels*[a]

Variable	Boating	Camping	Fishing	Golfing	Hunting	Skiing	Wildlife viewing
Constant	-2.21×10^6 (3.49)	-2.80×10^6 (3.91)	3.39×10^5 (0.19)	-6.03×10^5 (0.84)	1.38×10^7 (6.83)	-2.62×10^6 (1.99)	1.76×10^5 (0.27)
Total area		38.23 (3.01)					
Water area	2.06×10^2 (4.22)						
Population	2.08 (13.48)	2.02 (24.54)	2.71 (13.88)	2.62 (14.07)	1.56 (10.68)	0.44 (11.50)	1.16 (16.53)
Population squared	-4.13×10^{-8} (6.96)		-6.45×10^{-8} (8.34)	-2.63×10^{-8} (3.60)	-3.91×10^{-8} (7.11)		
Income					-7.83×10^2 (6.54)	1.34×10^2 (1.73)	
Public forest		1.54×10^2 (2.80)					
January temperature		-2.04×10^5 (4.52)					
January temperature squared			1.20×10^4 (4.35)				

July temperature	3.93×10^4 (2.88)				-1.70×10^5 (2.65)	-2.06×10^5 (4.89)	-2.74×10^5 (3.43)
July temperature squared		9.00×10^5 (2.49)	4.56×10^4 (2.37)				
January precipitation							
January precipitation squared							3.32×10^5 (2.19)
July precipitation squared				-3.91×10^5 (4.46)			
Price			-5.64×10^4 (1.86)				
Adjusted R^2	0.92	0.96	0.89	0.94	0.78	0.81	0.86

Note:

[a] Cross-sectional analysis using lower 48 states of the United States as observations. Absolute value of *t*-statistics are in parenthesis.

Table 10.3. *Loglinear regression of climate on activity levels*[a]

Variable	Boating	Camping	Fishing	Golfing	Hunting	Skiing	Wildlife viewing
Constant	13.69	14.13	14.77	13.99	16.87	9.58	14.6
	(76.03)	(80.82)	(59.22)	(72.90)	(36.56)	(12.57)	(114.40)
Water area	1.91×10^{-5}		2.40×10^{-5}				
	(1.66)		(2.60)				
Population	3.74×10^{-7}	3.74×10^{-7}	2.71×10^{-7}	4.11×10^{-7}	3.20×10^{-7}	3.45×10^{-7}	2.13×10^{-7}
	(8.66)	(9.09)	(7.67)	(10.27)	(9.43)	(6.57)	(6.41)
Population squared	-9.80×10^{-15}	-10.00×10^{-15}	-6.40×10^{-15}	-1.09×10^{-14}	-6.92×10^{-15}	-9.15×10^{-15}	-4.18×10^{-15}
	(5.76)	(5.96)	(4.51)	(6.83)	(5.04)	(4.43)	(3.20)
Income					-1.91×10^{-4}	1.46×10^{-4}	
					(7.75)	(3.38)	
Public forest		4.33×10^{-5}			2.56×10^{-5}	5.34×10^{-5}	
		(4.04)			(2.16)	(3.69)	
January temperature					-4.55×10^{-2}		
					(5.97)		
January temperature squared			-1.04×10^{-3}		-2.12×10^{-3}		
			(2.30)		(4.29)		
July temperature						-0.11	
						(4.45)	

	(1)	(2)	(3)	(4)	(5)	(6)	(7)
July temperature squared	8.52×10^{-2} (2.67)		8.10×10^{-3} (2.83)		9.84×10^{-3} (2.94)		
January precipitation squared			8.63×10^{-2} (3.04)	9.58×10^{-2} (2.83)	6.65×10^{-2} (2.38)		
July precipitation	0.11 (2.03)		0.19 (3.67)		0.42 (5.16)		
Price			-7.91 (1.63)				
Adjusted R^2	0.81	0.75	0.85	0.78	0.84	0.77	0.66

Note:
[a] Cross-sectional analysis using lower 48 states of the United States as observations. Absolute value of t-statistics are in parenthesis.

closely fit the state data, explaining from 75–96 percent of the variation in activity days.

The most consistent, significant explanatory factor was state population. Aggregate days increased dramatically with population. However, with boating, fishing, golfing, and hunting, additional population appears to increase participation days at a decreasing rate. These results imply that days per capita decrease as population increases for these activities.

Other nonclimate, demand-shift variables significantly affect participation in selected activities. With both models, greater per capita income is associated with increased skiing days but decreased hunting days. As we would expect with both models, states with more public forests are associated with higher levels of camping and states with larger amounts of surface water are associated with increased boating days. The linear model also predicted that larger states had more camping. The log-linear model predicted that states with more water also had more fishing and boating, and states with more public forest had more camping, hunting, and skiing.

Climate affects participation differently for each of the seven activities we analyzed. In the final models, we find the specifications of the linear and loglinear model also differ across activities. Many of the effects described below are probably due to the impacts of climate on the ecosystems which, in turn, affect the quality of the experiences.

In the linear model, most summertime activities are associated with July temperatures and precipitation levels, as we would expect. Locations with warmer than average summer temperatures are associated with more boating and fishing activities, e.g. southeastern states. Locations with cooler than average summer temperatures are associated with greater participation in hunting and wildlife viewing activities, e.g. northern states. Greater hunting participation is also associated with locations having drier than average summers, e.g. upper midwest states.

Although it would seem that winter climate conditions should not affect summer participation, the inclusion of winter climate variables for some summer activities helps distinguish between states with similar summer but different winter conditions. This is probably a result of a climate–ecosystem impact. Locations with above average winter (January) precipitation levels are associated with greater camping and wildlife viewing participation, e.g. northeastern states. Camping and fishing are sensitive to winter temperatures. Greater camping participation is associated with states having cooler than average winters, e.g. northern states, while greater fishing participation is associated with states having warmer than average winters, e.g. southeastern states.

Skiing participation is affected by summer temperatures. Although this might seem unrelated given that skiing takes place in the winter, the summer temperature probably limits the length of the skiing season. Areas with longer seasons also have greater investment in resorts and infrastructure, thus increasing participation further.

In the loglinear model, fishing and hunting participation is associated with summer temperatures, and fishing, hunting, and boating participation with summer precipitation, as expected. Locations with warmer than average summers have greater fishing and hunting participation, e.g. southern states. Locations with wetter summer conditions have greater fishing, hunting, and boating participation, e.g. southeastern states. The model type has a significant effect on explaining hunting participation since the July climate variable results differ between the two models. These differences are likely to be attributable to differences in specifications between the two models.

As in the linear model, winter conditions play a role in defining summer activity participation for the loglinear model. Boating, fishing, golfing, and hunting participation is greater in regions with wetter than average winters, e.g. mid-Atlantic states, all else being equal. Especially relevant for boating and fishing activities, this specification could distinguish between coastal and dry inland states having similar temperatures. Similarly, greater fishing and hunting participation is associated with regions having cooler than average winter temperatures, all else being equal, e.g. Great Lake, northeastern, or northwestern states. These fishing results differ from those of the linear model, probably due to the difference in model specification.

Finally, as in the linear model, skiing participation is associated with July temperatures. Results are similar in that greater skiing participation occurs in locations with cooler than average summer temperatures, probably due to longer skiing seasons, e.g. northern and western states.

10.3   Climate change results

To get a sense of the relative importance of climate impacts on recreation, we forecast the climate impacts using the regression results and several different climate scenarios. Following the methodology laid out in Chapter 1, we assess nine uniform temperature and precipitation scenarios for the continental United States. Temperature increases of 1.5, 2.5, and 5.0 °C are examined with precipitation at current and increased levels of 7 and 15 percent for a total of nine climate scenarios.

These scenarios are simulated on a 1990 and a 2060 economy. The difference between the 1990 and 2060 estimates is due to the assumed increase in population and, for both skiing and hunting, income by 2060. For each scenario, we use the estimated relationships (Tables 10.2 and 10.3) to quantify the impact.

Table 10.4 shows the net effect of climate change on each outdoor recreation activ-

Table 10.4. *Impact of climate change on leisure activities (2060)ᵃ: linear demand model (1991$ billion change in welfare)*

Scenario	Boating	Camping	Fishing	Golfing	Hunting	Skiing	Wildlife viewing
+1.5 °C, +0 P	1.1	−0.6	1.8	0.0	0.0	−1.3	−1.8
+2.5 °C, +0 P	3.0	−1.0	5.0	0.0	0.0	−2.0	−2.7
+5.0 °C, +0 P	13.1	−1.8	22.2	0.0	0.0	−3.7	−4.1
+1.5 °C, +7% P	1.1	−0.4	1.8	0.0	0.0	−1.3	−1.4
+2.5 °C, +7% P	3.0	−0.8	5.0	0.0	0.0	−2.0	−2.4
+5.0 °C, +7% P	13.1	−1.7	22.2	0.0	0.0	−3.7	−4.0
+1.5 °C, +15% P	1.1	−0.2	1.8	0.0	0.0	−1.3	−0.9
+2.5 °C, +15% P	3.0	−0.6	5.0	0.0	0.0	−2.0	−1.9
+5.0 °C, +15% P	13.1	−1.5	22.2	0.0	0.0	−3.7	−3.6

Note:

ᵃ US impacts are calculated for uniform climate changes in each scenario using the reduced form models in Table 10.2. Positive numbers imply climate change is beneficial and negative numbers imply damages.

ity in 2060. The sector which is most responsive to warming according to the linear model is fishing.[4] With a 5 °C, 7 percent precipitation increase, fishing dramatically increases in value for a total benefit of $22.2 billion (174 percent). Boating increases 157 percent for a total benefit of $13.1 billion. Camping values fall by approximately $1.7 billion (− 27 percent), skiing declines by $3.7 billion (−51 percent), and wildlife viewing declines by over $4 billion (− 64 percent). Effects are considerably smaller with a 2.5 °C temperature and 7 percent precipitation increase. Fishing increases by only $5.0 billion (39 percent) and boating by $3.0 billion (35 percent). Camping declines $0.8 billion (− 12 percent), skiing shrinks $2.0 billion (− 28 percent), and wildlife viewing loses over $2.4 billion (− 39 percent). Although wildlife viewing and camping positively respond to increased precipitation, these effects are small compared to the temperature sensitivity.

With the loglinear model (Table 10.5), impacts are dominated again by fishing. With a 5 °C, 7 percent precipitation increase, the total value of fishing increases 150 percent for a $22.2 billion benefit. For this same scenario, skiing damages are $4.6 billion (− 62 percent). Boating and golfing increase by less than $1 billion (6 and 4

---

[4] Our results report the climate change effect on salt and freshwater fishing combined. When salt and freshwater fishing are examined independently, participation in both types of fishing responds positively to warming.

Table 10.5. *Impact of climate change on leisure activities (2060):ᵃ loglinear demand model (1991$ billion change in welfare)*

Scenario	Boating	Camping	Fishing	Golfing	Hunting	Skiing	Wildlife viewing
+1.5°C, +0 P	0.0	0.0	1.5	0.0	0.0	−1.9	0.0
+2.5°C, +0 P	0.0	0.0	3.9	0.0	0.0	−2.9	0.0
+5.0°C, +0 P	0.0	0.0	19.0	0.0	0.0	−4.6	0.0
+1.5°C, +7% P	0.7	0.0	3.0	0.6	0.0	−1.9	0.0
+2.5°C, +7% P	0.7	0.0	5.7	0.6	0.0	−2.9	0.0
+5.0°C, +7% P	0.7	0.0	22.2	0.6	0.0	−4.6	0.0
+1.5°C, +15% P	1.7	0.0	5.3	1.7	0.0	−1.9	0.0
+2.5°C, +15% P	1.7	0.0	8.3	1.7	0.0	−2.9	0.0
+5.0°C, +15% P	1.7	0.0	27.1	1.7	0.0	−4.6	0.0

Note:
ᵃ US impacts are calculated for uniform climate changes in each scenario using the reduced form models in Table 10.3. Positive numbers imply climate change is beneficial and negative numbers imply damages.

percent, respectively). With a 2.5 °C, 7 percent precipitation scenario, impacts are smaller with fishing benefits at $5.7 billion (39 percent) and skiing damages at $2.9 billion (−39 percent). Precipitation plays a much larger role in the loglinear model than it plays in the linear model. Boating, fishing, and golfing are all positively affected by increased precipitation, and precipitation contributes all of the benefits associated with boating and golfing.

Table 10.6 presents the net aggregate results for the recreation sector across the nine scenarios and two models for 2060. With low precipitation increases and a 1.5 °C temperature increase, global warming could have a small net negative impact on recreation. With all other scenarios, net impacts are positive. For example, using the IPCC central scenario of a 2.5 °C and 7 percent precipitation increase, net benefits are $2.8 billion (7 percent) with the linear model and $4.1 billion (9 percent) with the loglinear model. With a 5 °C increase, benefits increase substantially to $25.9 billion (63 percent) with the linear model and to $18.9 billion (40 percent) with the loglinear model. Although warming is harmful to winter activities and maybe even to some summer activities, a 2.5 °C or 5 °C warming is beneficial to outdoor recreation as a whole.

Whereas both models have similar temperature sensitivity, the loglinear model is more responsive to precipitation. With a 2.5 °C temperature increase and 15 percent

Table 10.6. *Aggregate impact of climate change on leisure activities (2060) (1991$ billion change in welfare)*

Precipitation change (%)	Temperature change		
	1.5 °C	2.5 °C	5.0 °C
Linear demand model			
0	−0.8	2.3	25.7
7	−0.2	2.8	25.9
15	0.5	3.5	26.5
Loglinear demand model			
0	−0.4	1.0	14.4
7	2.4	4.1	18.9
15	6.8	8.8	25.9

additional precipitation, benefits increase from the central case to only $3.5 billion (8 percent) with the linear model but to $8.8 billion (18 percent) with the loglinear model.

Because the temperature changes explored in this study are likely to happen in the distant future, the impact on the 2060 economy is the most relevant impact to measure, but we also present results for a 1990 economy. Predicting what the economy might look like in 2060 is difficult, consequently early studies concentrated on estimating impacts to a 1990 economy. Tables 10.7 to 10.9 present the results for a 1990 economy in order to compare this study with the earlier literature. With the linear model, 1990 results for most sectors are slightly larger than the 2060 results because the 1990 baseline is smaller but the climate change is the same. This results in a slightly larger welfare impact. With the loglinear model, however, all effects are proportional so that most 2060 estimates are larger than the 1990 values from population growth. Skiing and hunting are also sensitive to income and 2060 incomes are projected to be much larger. Skiing consequently increases and hunting shrinks in 2060. With the linear model, skiing damages are smaller but hunting damages are larger in 1990. The two effects almost offset each other so that 1990 impacts are just slightly smaller than 2060 impacts. These income effects are much stronger with the loglinear model and the climate impact on hunting is beneficial, not harmful.[5] The loglinear model's prediction of a larger hunting benefit and a smaller skiing damage combine to yield a significantly larger predicted benefit in 1990 than in 2060.

In addition to the uniform climate change scenarios, we also simulate the effect of

---

[5] In fact, the predicted income effect on skiing is so large that we constrain skiing participation to increase by only four times the baseline case.

Table 10.7. *Impact of climate change on leisure activities (1990)[a]: linear demand model (1991$ billion change in welfare)*

Scenario	Boating	Camping	Fishing	Golfing	Hunting	Skiing	Wildlife viewing
+1.5°C, +0 P	1.2	−0.6	1.9	0.0	−1.3	−0.8	−1.8
+2.5°C, +0 P	3.1	−1.0	5.1	0.0	−2.0	−1.0	−2.6
+5.0°C, +0 P	14.6	−1.8	22.9	0.0	−3.4	−1.4	−3.9
+1.5°C, +7% P	1.2	−0.4	1.9	0.0	−1.7	−0.8	−1.4
+2.5°C, +7% P	3.1	−0.8	5.1	0.0	−2.4	−1.0	−2.3
+5.0°C, +7% P	14.6	−1.6	22.9	0.0	−3.6	−1.4	−3.7
+1.5°C, +15% P	1.2	−0.2	1.9	0.0	−2.1	−0.8	−0.8
+2.5°C, +15% P	3.1	−0.6	5.1	0.0	−2.7	−1.0	−1.8
+5.0°C, +15% P	14.6	−1.5	22.9	0.0	−3.8	−1.4	−3.4

Note:
[a] US impacts are calculated for uniform climate changes in each scenario using the reduced form models in Table 10.2. Positive numbers imply climate change is beneficial and negative numbers imply damages.

Table 10.8. *Impact of climate change on leisure activities (1990)[a]: loglinear demand model (1991$ billion change in welfare)*

Scenario	Boating	Camping	Fishing	Golfing	Hunting	Skiing	Wildlife viewing
+1.5°C, +0 P	0.0	0.0	1.3	0.0	−0.2	−0.5	0.0
+2.5°C, +0 P	0.0	0.0	3.3	0.0	0.1	−0.8	0.0
+5.0°C, +0 P	0.0	0.0	16.3	0.0	2.9	−1.3	0.0
+1.5°C, +7% P	0.6	0.0	2.6	0.7	0.7	−0.5	0.0
+2.5°C, +7% P	0.6	0.0	4.8	0.7	1.0	−0.8	0.0
+5.0°C, +7% P	0.6	0.0	19.1	0.7	4.3	−1.3	0.0
+1.5°C, +15% P	1.5	0.0	4.6	1.9	1.9	−0.5	0.0
+2.5°C, +15% P	1.5	0.0	7.2	1.9	2.4	−0.8	0.0
+5.0°C, +15% P	1.5	0.0	23.3	1.9	6.3	−1.3	0.0

Note:
[a] US impacts are calculated for uniform climate changes in each scenario using the reduced form models in Table 10.3. Positive numbers imply climate change is beneficial and negative numbers imply damages.

285

Table 10.9. *Aggregate impact of climate change on leisure activities (1990) (1991$ billion change in welfare)*

	Temperature change		
Precipitation change (%)	1.5 °C	2.5 °C	5.0 °C
Linear demand model			
0	−1.4	1.6	27.0
7	−1.2	1.7	27.2
15	−0.8	2.1	27.4
Loglinear demand model			
0	0.6	2.6	17.9
7	4.1	6.3	23.4
15	9.4	12.2	31.7

Table 10.10. *Climate change impacts on outdoor recreation using GCM scenarios: 2060 (1991$ billion change in welfare)*

Scenario	Boating	Camping	Fishing	Golfing	Hunting	Skiing	Wildlife viewing	Total
Linear demand model								
GISS	6.7	−1.8	12.9	0.0	0.0	−2.7	−4.0	11.1
GFDL	30.6	−1.6	40.0	0.0	0.0	−4.8	−5.0	59.2
Loglinear demand model								
GISS	−0.3	0.0	8.7	−1.7	0.0	−3.7	0.0	3.0
GFDL	−1.8	0.0	32.9	−1.6	0.0	−5.7	0.0	23.8

two GCM (General Circulation Model) predicted climates using the GISS and GFDL-R30 models (Table 10.10). The climate models allow regional and seasonal climate change to vary. These two models predict an average temperature and precipitation increase similar to the 5 °C, 7 percent precipitation uniform case. However, the GFDL model predicts a sharp decline of precipitation in the summer and the GISS model predicts a large summer increase. The two models also have different regional forecasts. With the linear model, GFDL predicts larger net benefits ($59.2 billion) compared to the uniform scenario ($25.9 billion) whereas GISS predicts much smaller benefits ($11.1 billion). Similar results occur with the loglinear model. GFDL predicts benefits of $23.8 billion, the uniform model predicts benefits of $18.9 billion, and GISS predicts benefits of $3.0 billion. These results demonstrate that regional and seasonal climate variation can seriously affect the magnitude of impacts.

10.4   Conclusion

This study examines how the demand for outdoor recreation is affected by climate. Some activities such as fishing and boating appear to benefit from warmer temperatures whereas other activities such as skiing are harmed. More precipitation has a net beneficial impact although it is not shared by every sector. The overall net effect from warming appears to be beneficial as the increase in hunting, fishing, and boating consumer surplus outweighs the declines in camping, skiing, and wildlife viewing consumer surplus.

The results of this analysis are qualitatively similar to those of Loomis and Crespi that are presented in the next chapter, who consider the impact of climate on 17 outdoor recreation activities. Loomis and Crespi use a benefits transfer approach to measure welfare effects of climate change for most activities, and for other activities (e.g. golf) they estimate a relationship directly. In their analysis, the authors consider both the direct and indirect effects of the nine climate scenarios, i.e. the change in demand due to climate change and a change in resource quality or quantity. The climate scenarios have a negative impact on skiing, hunting, wildlife viewing, and camping, and a positive impact on boating, fishing, and golfing. Furthermore, the magnitude of the climate effects are similar to this analysis, despite the differing definitions of the resource. The camping values of the current analysis fall within the range of values Loomis and Crespi provide for camping, hiking, and other activities. The Loomis and Crespi results for skiing and golfing are slightly larger than this analysis, attributable to their consideration of both indirect and direct climate effects on participation. Loomis and Crespi consider only coastal hunting, bird viewing, and fishing, so their estimated impacts are smaller than those predicted by the current study which includes all hunting, wildlife viewing, and fishing. Similarly, Loomis and Crespi's combined results for fishing and boating in warm water are lower than those of the current analysis which includes all fishing and boating.

Further empirical studies with more disaggregated data would be desirable because states contain large variations in climate within their borders and the effects of socio-economic variables are hidden in the aggregate data. In particular, better price data across states could allow us to control for differences in costs from region to region and enable us to obtain independent price elasticity estimates. Using aggregate price data was especially problematic for boating, camping, golfing, and skiing, although the problem plagued every activity. Expanding the analysis to other countries would also be important as the results of this study may be specific to the range of climates studied and to US values.

One of the more serious limitations of this cross-sectional analysis is the

287

assumption that ecosystems will freely adjust with climate. It is important to confirm whether this assumption is justified. The similarity of the results in this study with those of Crespi and Loomis is reassuring. However, by integrating a thorough economic analysis with a detailed ecosystem model, as was done with the timber study, one could carefully test the importance of a number of our underlying assumptions and develop a more reliable estimate of net effects. Our confidence in the recreation results would certainly increase if confirmed by an independent detailed ecosystem-based study.

References

Bergstrom, J.H. and Cordell, H.K. 1991. An Analysis of the Demand for and Value of Outdoor Recreation in the United States. *Journal of Leisure Research* **23**(1): 67–86.

Cline, W. 1992. *The Economics of Global Warming*. Washington, DC: Institute for International Economics.

Freeman, A.M. III. 1993. *The Measurement of Environmental and Resource Values: Theory and Methods*. Washington, DC: Resources for the Future.

Gerardi, L. 1994. National Marine Manufacturer's Association, Telephone Conversation with Angelique Knapp, Industrial Economics, Incorporated.

Gorin, D. 1994. National Association of RV Parks and Campgrounds, Telephone Conversation with Angelique Knapp, Industrial Economics, Incorporated.

Mendelsohn, R., Nordhaus, W. and Shaw, D. 1994. The Impact of Global Warming on Agriculture: A Ricardian Analysis. *American Economic Review* **84**(4): 753–71.

National Park Service, Land Resources Division. 1994. *Listing of Acreage, by State and County*. Washington, DC.

National Sporting Goods Association. 1990. *Sports Participation in 1990: State by State*. Mt Prospect, IL.

Ski Industries of America. 1994. *1993–94 Facts and Figures on the On-Snow Industry*. McLean, VA.

US Bureau of the Census. 1990. *Statistical Abstract of the United States: 1990*. Washington, DC: US Government Printing Office.

US Bureau of the Census. 1993. *Statistical Abstract of the United States: 1993*. Washington, DC: US Government Printing Office.

US Bureau of the Census. 1992. *USA Counties on CD-ROM*. Washington, DC: US Government Printing Office.

US Fish and Wildlife Service. 1992. *National Survey of Fishing, Hunting, and Wildlife Associated Recreation: State Overview*. Washington DC: US Department of Interior.

Willig, R.D. 1976. Consumer's Surplus without Apology. *American Economic Review* **66**(4): 589–97.

11 Estimated effects of climate change on selected outdoor recreation activities in the United States

JOHN LOOMIS AND JOHN CRESPI[1]

Most previous published estimates of the economic costs of climate change (i.e. changes in temperature and precipitation) have given little attention to the effects on outdoor recreation. Nordhaus has not included any estimates of effects of climate change on recreation or recreationally used natural resources (Nordhaus, 1991, 1993). The National Academy of Sciences (1992 pp. 607–8) devotes only about two pages out of 900 to the issue. However, they indicate that outdoor recreation is more sensitive to climate change than other sectors of the economy due to recreation's close link to natural resources (National Academy of Sciences, 1992 p. 41). Ewert (1991 p. 366) also states that with respect to the effect of global climate change on recreation ". . . outdoor recreation is an example in which users have a direct interaction with the natural environment . . ." Cline (1992 pp. 122–3) includes only a rough estimate of the effect of climate change on snowskiing in his category of leisure activities. Even Fankhauser's recent book (1995) provides only a one paragraph discussion of recreation before concluding that "Unfortunately, data for monetary valuation are not available for either sector" (1995 p. 43). Fankhauser goes on to say "Nevertheless, the problem of greenhouse damage estimates is perhaps not so much the accuracy of the valuation methods as such, but the fact they have not yet been applied to the problem to a sufficient degree" (1995 p. 21).

[1] We would like to thank Robert Mendelsohn for his guidance and constructive suggestions throughout the period of this analysis. We would like to thank Elizabeth Snell, James Neumann, Marla Markowski, Robert Unsworth, and A. Myrick Freeman, III, for comments. Brian Hurd provided reservoir and stream flow estimates used as inputs for the freshwater recreation analysis. Tim Kittel was most helpful in providing updated VEMAP information for the forestry analysis. Dennis Lettenmaier provided the snow melt statistics used in the skiing analysis. George Parsons provided several valuable suggestions and references early in our research. Shannon Ragland provided several suggestions on the reservoir and golf models that were helpful in performing our sensitivity analysis. Finally Joel Smith provided valuable overall guidance and specific suggestions for improving the skiing analysis.

With nearly nine billion visitor days of recreation in 1987 (Bergstrom and Cordell, 1991), any significant changes in recreation use or the value per day from global warming may result in large changes in economic values that may significantly affect current estimates of US costs from global warming, currently estimated at approximately $50 billion (Fankhauser, 1995). The relative importance of the effects of climate change on outdoor recreation is highlighted in the recent empirical analysis by Mendelsohn and Markowski in Chapter 10 of this book. Their research shows that combining the effects of climate change on both winter and summer recreation accounts for a sizable part of the total economic effects on the US economy resulting from the Intergovernmental Panel on Climate Change (IPCC) central estimate of $+2.5\,^{\circ}\text{C}$ and $+7$ percent precipitation in the year 2060. Therefore, a systematic assessment of how climate change directly affects demand (e.g. desired use levels and benefits) and how climate change alters the availability and quality of natural resources that indirectly affect recreation demand would contribute to a more complete policy analysis.

The remainder of the chapter is organized as follows. In Section 11.1, we provide a brief review of past studies that demonstrate a linkage between climate and recreation use or benefits. Next, we present an overview of our analytical approach relating visitation rates to temperature and precipitation. We then use these response functions to estimate the effects of climate change on the recreational value of natural resources. Section 11.2 provides a discussion of data sources for the values per day and levels of visitor use for each recreational activity. This section also presents the statistical estimates of the visitor-use elasticities for both direct and indirect effects. Section 11.3 presents the results of different climate change scenarios on visitor use and benefits as well as a sensitivity analysis. The chapter concludes with a discussion of the study's limitations and directions for further research.

11.1 Direct and indirect effects of climate change on recreation

Several studies have documented climate effects on recreation. Cato and Gibbs (1973) in a survey of recreational boaters in Florida found that chance of rainfall and expected air temperature had a significant effect on the likelihood of taking a boating trip. In quantifying the willingness to pay (WTP) for beach use, McConnell (1977) as well as Silberman and Klock (1988), found air temperature to have a positive and statistically significant effect on WTP. This partial listing of studies provides a sense that direct climate effects on recreation are likely to be significant. We now turn

290

to developing a more systematic approach in order to evaluate the overall effect of climate change on recreation in the United States.

Increases in temperature and changes in precipitation affect recreation along two major pathways: direct effects on the participants' desired demand and indirect effects on demand through changes in the quantity and quality of natural resources used for outdoor recreation. Temperature has the most direct effect on the desirability or marginal utility of participating in particular recreational activities (we will use boating as an example). For example, as temperature increases, the desirability of water-based recreation such as boating would be expected to increase, shifting the demand curve for this activity in a positive direction, increasing the desired number of trips. In addition, higher temperatures extend the recreation season for many activities, making more trips per year possible.

However, before one can conclude that boating visitor days and the total recreation value of boating would increase by the full amount of the demand shift, one must consider the effect of global climate change on the second pathway, recreation resource availability. If changes in the amount of snowfall result in lower summer river flows and lake levels, then the recreationists' desired increase in demand may not be fully accommodated. While this resource effect could be modeled as a supply change, an alternative approach is to view the change in availability of natural resources due to climate change as an indirect change in demand that would also be translated into shifts in the demand curve. For example, warmer temperatures increase evaporation and evapo-transpiration and may lower stream and reservoir levels. These lower reservoir levels would have several adverse effects on the quality of a boating experience. Lower reservoir levels often make boat ramps and launch facilities unavailable, and the aesthetics of the lake are adversely affected when significant drawdown leaves an exposed steep rocky shoreline. Finally, the smaller surface area of the reservoir results in greater crowding of recreation users such as anglers and waterskiers into the remaining area. Therefore, reservoir level has often been modeled as an indicator of boating quality in studies of boating and reservoir recreation demand (Knetsch *et al.*, 1974; Rosenthal, 1987; Loomis *et al.*, 1995). Thus, the indirect effect of a climate change which reduces reservoir levels would be seen as a partial negative shift in the recreation demand.

To facilitate the discussion of measurement of benefits, it is useful to assume that *demand* will increase due to climate change, and that *availability* does not decrease. A positive shift in the demand for a recreation activity enhanced by temperature increases the number of users and the number of visitor days each user desires. For the moment we assume there is sufficient capacity at existing recreation sites to absorb this additional use. As is required by both economic theory (Just *et al.*, 1982) and Federal benefit–cost procedures (US Water Resources Council, 1983), the economic benefit of recreation is defined as the users' willingness to pay (WTP) over and above their current costs. Using demand curves estimated from visitor data, this net willingness to pay can be measured using consumer surplus. Consumer surplus is the area below the demand function but above the travel costs to a site. An increase in demand causes an increase in consumer surplus. This is the welfare measure associated with the demand increase. A parallel analysis would describe the loss in consumer surplus if demand shifted down because of climate change.

In principle, one can estimate a series of recreation activity demand curves that are a function of the standard demand shifters (i.e. income, tastes, etc.) as well as climate variables such as temperature and rainfall. From these demand curves the change in use and benefits associated with changes in temperature and rainfall could be estimated. However, this would be a major research project itself for the range of recreation activities we evaluate. Therefore, we rely on existing estimates of WTP, derived from past recreation demand studies or recreation surveys, to approximate the area between the demand curves. An estimate of the change in WTP is calculated by the change in visitor days multiplied by the average WTP, net of visitor costs (consumer surplus) per visitor day. This commonly used approach is called "benefit transfers" (Walsh et al., 1992) and is frequently used to perform policy analysis (see Vaughan and Russell, 1982, for their estimation of the national freshwater recreational fishing benefits of water pollution control).

To value the change in visitor days, the literature frequently reports the average consumer surplus per day. Most economists recognize that what is required by theory is the marginal value. Recent research by Morey (1994) suggests that valuing the change in days resulting from a change in site quality by multiplying the predicted number of days in the changed state by a constant average consumer surplus is an upper bound on WTP (Morey, 1994 pp. 268–9). There are just three exceptions to this. First, as Morey notes, the constant value per day assumption is accurate if one adopts a discrete choice type model of recreation site choice such as a multinomial logit model. Second, for those recreation activities in which use is limited by non-priced rationing such as first-come, first-served, lotteries or advance reservation, the average value often equals the marginal (Mumy and Hanke, 1975) since these rationing systems make it equally likely that an additional permit or campsite will go to a high valuing user or a low valuing user. As an empirical matter, the average value per day may be a useful proxy for the marginal value in that many frequently used functional forms for the recreation demand function also imply the average is equal to the marginal for demand shifts. The semi-log model (Adamowicz et al., 1989; Donnelly et al., 1985), the count data model (Creel and Loomis, 1990; Hellerstein and

Mendelsohn 1993), and the double-log model all result in a constant value per day of recreation.

Further countering concerns about overestimation, for our most central base case analysis ($+2.5\,°C$ and $+7$ percent precipitation) the value per day is held constant even though warmer weather shifts the demand curve for recreation to the right implying an increase in value per day. This increase in value is not only for the added days, but for the current level of visitor use as well. Thus there are two sources of benefit for activities enhanced by global warming, just as there are two sources of losses for the adverse effects of climate that shifts the demand curve inward. Walsh *et al.* (1980b) suggest that changes in recreation resource quality will cause the value per day to change at about the same rate as use does. To keep the analysis simple and be conservative, our main scenario analysis values only the change in days and ignores the added benefits or losses associated with a change in value to existing days due to climate change. Thus, for recreation activities enhanced by global warming we will understate increases in benefits and for losing activities we will understate losses. Since our results indicate more gaining than losing activities, overall we underestimate the magnitude of the change in benefits.

Ewert (1991) notes,

What makes outdoor recreation experiences particularly vulnerable to any changes in climate is that certain activities are heavily dependent on site characteristics. For example, "adaptability to climate change and their ability to substitute one site for another naturally arises". Snow is required for skiing, low stream levels can preclude fishing or canoeing, and activities requiring forest lands can be interfered with if forests have given way to grasslands because of increased aridity and temperatures. In another example, rising ocean levels not only would be disruptive for many cities but also would degrade numerous tidal wetlands and beach recreation areas.

He further points out that users "may be forced to" alter the amount of recreation use in response to the changes in natural environments.[2]

Thus, in many respects we believe the recreation sector will be directly and indirectly affected by climate change more so than other economic activities such as manufacturing. Nonetheless, the issue of humans' adaptability to climate change and the substitutability of recreation naturally arises. Sometimes it is argued that if climate

---

[2] The primary exception is beach recreation where the literature indicates beach nourishment mitigation is cheaper than foregoing beach recreation on public beaches when sea level rise is moderate (Bell, 1986 p. 379; National Oceanic and Atmospheric Administration (NOAA), 1994). Yohe *et al.*, in Chapter 7 of this book, suggest few recreationally important beaches would be lost for sea level rises of 0.33 meters or less.

change reduces the natural resources available to support an activity, individuals will not make a net reduction in trips, but rather change to other locations for the same activity or different activities (usually at other locations). This is of course possible, but in fact there are likely to be reductions in benefits from these substitutions. Properly estimated measures of consumer surplus reflect the willingness to pay for participating in one activity compared to: (a) the same activity at a less preferred (i.e. more distant or poorer quality) site and (b) an alternative activity. There is some evidence in the contingent valuation literature that individuals do implicitly consider substitutes when formulating their willingness to pay responses (Boyle *et al.*, 1990) and willingness to pay estimates derived from contingent valuation are consistent with estimates from site substitution values (Thayer, 1981 p. 43). Travel cost models that include the price of substitutes or that are based on a random utility model approach explicitly account for substitutes as well. Thus for the purpose of this analysis, we will use consumer surplus estimates reported in the literature. To the extent these estimates are derived from travel cost and contingent valuation studies that do not fully reflect substitutes, our estimated gains and losses will be overstated. A further qualifier is that our analysis ignores what might be called incremental-dynamic responses to climate change over time. Gradual climate-induced changes in recreational opportunities may, over a long enough period of time, cause people to lose interest in recreation activities that diminish in quality due to climate change (e.g. snowskiing) and seek out ones enhanced by climate change. When these changing tastes are transferred from parents to children, the next generation may not realize they are missing snowskiing opportunities since few people participate in that activity. If tastes change, our estimates would overstate these losses in recreation benefits.

The net effect of ignoring the dynamics of adaptability of recreation users and using average consumer surplus per day to value only the change in days is to overstate losses and understate gains, respectively, from climate change. Overall the results of our main scenario will reflect a conservative estimate of the gains from global climate change. Based on a comparison of the results from Mendelsohn and Markowski (see Chapter 10 of this book), the degree of imprecision appears small, however. They used the same value per day but their cross-sectional approach implicitly assumes recreationists will adapt their activities to the climate in which they live. Since Mendelsohn and Markowski's estimate for $+2.5\,°C$ and $+7$ percent precipitation is nearly identical to ours it does not appear that serious error is introduced by the absence of the dynamic adjustment of recreationists to climate change in our approach.

11.2   Analysis

Detailed analysis of the change in recreation use and benefits caused by climate change is performed for recreation activities that meet three screening criteria. First, the activities must involve substantial recreation visitor days. Second, climate must exercise a direct effect on the demand for the activity, or an indirect effect on recreation demand through the resources used for that recreation activity. In other words, only activities which are likely to see a change in recreation use with changes in temperature, precipitation, or climate-induced natural resource change need be considered for a full analysis. Third, there must be existing data to estimate a model quantifying the direct or indirect links to the recreation activity or empirical estimates in the literature of such linkages.

To facilitate analysis of the indirect effects of climate change on resources used for each recreation activity, the activities are grouped based on shared natural resource inputs that might be affected by global climate change. The categories are land-based, water-based, and snow-based. Of the 41 recreation activities evaluated, there were 17 in which we could document, with quantitative models, the direct effects of temperature and precipitation on participation in the activity or the indirect effects of climate change on natural resources such as wetlands or forests.

Data sources

Several data sources were used to estimate the recreation use by activity and the value per visitor day. The national estimate of total days in each activity comes from Bergstrom and Cordell (1991 p. 79, Table 3) and is based on USDA Forest Service's Public Area Recreation Visitor Survey (PARVS) data which estimate the number and percentage of the US population over the age of 12 years that participates in each activity. The national estimate of total participation days is expressed in millions of days. The data for hunting and fishing come from the 1991 National Survey of Fishing, Hunting, and Wildlife-associated Recreation (US Department of Interior, 1993). For golf, we used estimates of the number of rounds played each year (Balogh and Walker, 1992) and then assumed one round per individual is equal to one golfing day per individual. Beach recreation use by region of the country was developed from NOAA data on use per mile of sampled public beaches throughout the United States and the number of miles of coastal public beaches in each region of the United States.

Projections of future recreation visitation in the absence of climate change were developed using four sources. To be consistent with other climate impact assessments, we used future population and income changes from the IPCC (Houghton *et al.*,

Table 11.1. *Current visitation, value per day and climate elasticities*

Activity	Visitor days (millions)	Value per day (1992$)	Temperature or precipitation or resource elasticity
Snowskiing	156	17.68	Temperature effect on season length
Coastal waterfowl hunting	16	30.45	0.275 Coastal wetlands
Coastal bird viewing	169	29.91	0.173 Coastal wetlands
Beach visitation	192	16.30	1.6 to 2.1 Temperature −0.008 to −0.41 Rainfall +0.09 to 0.43 Shoreline
Reservoir recreation (warmwater fishing, boating, waterskiing, swimming)	1359.5	19.97	1.45 Temperature, −0.02 Precipitation 0.39 Surface area
Golf	488	21.25	1.9 Temperature, −0.237 Precipitation
Camping	520.7	11.22	0.3 for forest acreage
Backpacking and hiking	274.12	19.80	0.3 for forest acreage
Picnicking	480.72	15.00	0.26 for forest acreage
Stream fishing	115.0	22.47	0.60 for stream flow
Rafting	11.6	30.43	0.55 for stream flow
Canoeing and kayaking	64.8	16.00	0.62 for stream flow

1992). These projections show the United States stabilizing at a population of around 294 million in 2050. The projections also include substantial increases in income.

The US Forest Service (USFS) makes long-term recreation participation forecasts for its Resources Planning Act (RPA) program. We derived the USFS estimate of population change from their future recreation forecasts to arrive at a pure change in future participation rates and then applied the IPCC's population changes. We adjusted the USFS estimates of future recreation to take into account IPCC projected future population changes. For hunting, fishing, and wildlife viewing we relied upon the Walsh *et al.* (1987) forecasts. For golf, we estimated our own regression equation that related the number of golfers to population and income.

The baseline values per day are presented in Table 11.1. The majority of the average consumer surplus per day estimates came from the Bergstrom and Cordell (1991) travel cost demand analysis. This analysis used survey data from the PARVS data that was collected from 1985 to 1987. The value of waterfowl hunting was developed from Hay (1988), who used the National Survey of Fishing, Hunting, and Wildlife-associated Recreation. The value of beach recreation was calculated from the Leeworthy (1989) study of coastal beach recreation for NOAA. We were unable to locate a specific value for golf in the literature. However, we were able to find the value of recreation at resorts (USDA 1990). Since many resorts have golf courses and these are often a significant attraction at these resorts, this value will be used as a proxy for golf. All values were converted to 1992 dollars using the Consumer Price Index.

Our screening analysis indicated that we could quantify the direct effects of climate on the desirability of participating in several recreation activities such as golf, beach use, and reservoir recreation (e.g. boating, swimming, waterskiing). This section presents the statistical models used to estimate the visitor-use elasticities for these three categories of activities.

Golf

We estimated an equation that related rounds of golf played in each of the contiguous states and the District of Columbia to income (Inc), number of golf courses ($Courses$), temperature ($Temp$), and precipitation ($Precip$). That equation is:

$$\ln(Rounds) = -25.796 + 1.24\ln(Inc) + 1.038\ln(Courses)$$
$$+ 1.924\ln(Temp) - 0.237\ln(Precip) \tag{11.1}$$

$$(t\text{-statistics}) \quad (-5.14) \quad (4.544) \quad (24.93) \quad (5.00) \quad (-3.14)$$
$$n = 49 \; r^2 = 0.94$$

Temperature was statistically significant at the 0.01 level and positive (elasticity equals 1.924), while precipitation was significantly negative (at the 0.01 level) with an elasticity equaling -0.237. We used this model to predict the change in rounds of golf with each climate change scenario. The coefficient on temperature is generally robust to model specification. For example, treating the number of courses as an endogenous variable and including population as an instrument yields a temperature elasticity of 1.795 ($p < 0.01$), which is fairly close to our original. For purposes of sensitivity analysis, a model that includes both number of courses and population is estimated (in spite of their multicolinearity). This model results in a temperature elasticity of 0.72 ($p = 0.08$) where p is the probability the coefficient is not different from zero. This is our lower bound temperature elasticity for the sensitivity analysis.

Beach recreation

Data on monthly beach visitation to state, county, and city beaches throughout the United States in 1988 were obtained from one of the databases used by

NOAA to support the Natural Resource Damage Assessment Model for Coastal and Marine Environments. The database contains visitor-use statistics collected through NOAA's Public Area Recreation Visitors Survey (NOAA, 1993). The average monthly visitation was derived from coastal state park visitor-use data. From this same database we obtained information on the length in linear meters of each of the beaches as well. This data was combined with data on monthly temperature and precipitation to estimate the following regression equations for each of the three coastal census regions:

Northeastern United States

$$\ln(Mthly\ visits) = 0.302 + 1.903 \ln(Temp) - 0.414 \ln(Rain)$$
$$+ 1.15\ (Summer) + 0.425 \ln(Meter)$$

$$(11.2a)$$

$$(t\text{-statistics})\quad (0.22)\quad (5.43)\quad (-1.65)\quad (2.82)\quad (4.70)$$
$$n = 84,\ r^2 = 0.57$$

Southern United States

$$\ln(Mthly\ visits) = 2.89 + 1.618 \ln(Temp) - 0.307 \ln(Rain)$$
$$+ 0.469\ (Summer) + 0.096 \ln(Meter)$$

$$(11.2b)$$

$$(t\text{-statistics})\quad (1.90)\quad (4.28)\quad (-2.29)\quad (2.09)\quad (2.31)$$
$$n = 168,\ r^2 = 0.21$$

Western United States

$$\ln(Mthly\ visits) = 1.53 + 2.126 \ln(Temp) - 0.0085 \ln(Rain)$$
$$+ 0.1145\ (Summer) + 0.147 \ln(Meter)$$

$$(11.2c)$$

$$(t\text{-statistics})\quad (0.48)\quad (2.72)\quad (-0.09)\quad (0.46)\quad (2.15)$$
$$n = 48,\ r^2 = 0.49$$

where *Mthly visits* is the total number of activity days per month, *Temp* is average daily temperature, *Rain* is inches of rainfall during the month, *Summer* is a dummy variable to reflect summer vacation, and *Meter* is the length of the beach in meters.

Since these are double-log models, the coefficients can be interpreted as elasticities. These regression coefficients are used to predict changes in recreation use under two scenarios. The first scenario relies on the findings that few recreationally important beaches would be lost for moderate levels of sea level rise (Yohe *et al.*, Chapter 7). This is because beach nourishment is technically effective and cost-effective for this change.

Further, Yohe *et al.*, note that for greater sea level rise, there is really no *net* loss of beaches as the beach simply moves inland. That is, sea level rise will not really change the linear amount of coastline in the United States. In addition, Leatherman (1989) indicates that beach nourishment is a likely response for beaches near urban areas and that abandonment is simply not a realistic assumption. Since beaches near urban areas account for the vast majority of beach recreation, our primary scenario evaluates the case of no beach acreage lost. To provide a sensitivity analysis, we use an estimate of beach loss by Fankhauser (1995) which states that about 16 percent of the beaches would be lost using the coefficient on *Meter* in the above equation.

Reservoir recreation

Pooling monthly visitation and reservoir level data from the Sacramento, California District of the US Army Corps of Engineers for nine reservoirs throughout California (three northern reservoirs and six south-central ones), we estimated the effects of changes in temperature, precipitation, and reservoir levels on monthly visits to these reservoirs. Reservoir levels and visitation were represented as monthly data from January to September. The monthly temperature and precipitation data for each reservoir come from the Western Regional Climatology Center in Reno, Nevada. From this data the following model was derived:

$$\ln(Visit) = 2.94 + 1.45 \ln(Htemp) - 0.02 \ln(Arain) + 0.39 \ln(Acre)$$
$$+ 0.01 \,(Summer)$$

(11.3)

$(t\text{-statistics})$ (1.41) (3.24) (-1.34) (3.45) (0.09)
$n = 81, r^2 = 0.784, F\text{-stat.} = 53.6$

where $\ln(Visit)$ is the natural log of the number of monthly visits to each reservoir, $\ln(Htemp)$ is the natural log of the mean monthly high temperature at each reservoir, $\ln(Arain)$ is an augmented precipitation variable, and $\ln(Acre)$ is the natural log of the average surface acreage at each site during the month. The regression shows a significant relationship between monthly visits and monthly high temperature, whereas precipitation is significant only at an 18 percent level. As Walsh *et al.* (1980a) and Loomis *et al.* (1995) found, reservoir levels are positively related to visitation, however, after accounting for temperature the number of visits occurring in July or August seems to be no different than visits in any other month.

Our analysis was for California only, under the assumption that California has such a wide climate range that it may act as a good proxy for the United States in general. For purposes of sensitivity analysis we rely upon the Ward *et al.* (1995) study of the impact of climate on site visitation at 115 Corps of Engineers' sites throughout the

United States. Ward *et al.* ran regressions in double-log format (so that the coefficients may be interpreted as elasticities) on annual reservoir visitation (day-use and camping) with the following independent variables proving significant: *Acre* is the surface acreage of the reservoir, *Dist* is the distance in miles from the site to the nearest metropolitan statistical area, *Pop* is the population of the nearest metropolitan statistical area, *Cdd* is the average annual cooling degree days, and *Jhum* is average July humidity. The final equation is:

$$\ln(Visits) = -1.89 + 0.47 \ln(Acre) - 0.31 \ln(Dist) + 0.35 \ln(Pop)$$
$$+ 0.47 \ln(Cdd) + 1.09 \ln(Jhum) \tag{11.4}$$

with all estimates significant at or below the 0.10 level.

The regression model indicates, for example, that for a 10 percent increase above 65 °F over a year, the number of annual visits at a site increases by 4.7 percent. The Ward *et al.* (1995) nationwide model indicates a smaller temperature elasticity than our model, although this may be due to the less than direct specification of the temperature variable (cooling degree days is an indirect measure). For purposes of estimating a lower bound on increases in reservoir visitation due to climate change, we use their elasticity of 0.47 in our sensitivity analysis.

The other critical element in quantifying the impact on reservoir recreation is the change in reservoir surface areas due to climate change. Hurd *et. al.*, in Chapter 6, adapted a series of hydrologic models for the Colorado Basin, Missouri River Basin, Delaware River Basin, and the Apalachicola–Flint–Chattahoochee River Basin. We used the percentage change in net water storage in reservoirs in each of these river basins as proxies for reservoir storage levels for the West, Northcentral, Northeast, and Southeast census regions, respectively.[3] Hurd, *et al.* provided copies of these predictions of annual average storage levels for each scenario and the baseline. These were translated into percentage change in reservoir surface area and then used in the reservoir recreation demand model presented above.

The 1987 base activity days, the California regression results, and national average consumer surplus values for reservoir activities were used to construct a spreadsheet whereupon we can alter temperature, precipitation, and reservoir surface acreage in order to determine the net result on reservoir recreation activities in the United States. For a second sensitivity analysis, the main spreadsheet model (using the 1.45 temperature coefficient) was augmented with the recreation value per day elasticity to reflect reductions in the value per day as reduced aesthetics and increased congestion

---

[3] After accounting for increased evapotranspiration associated with higher temperatures and increases in priority water demands.

occur with reservoir drawdown (Walsh *et al.*, 1980a). This provides a second sensitivity scenario of the effect of climate change on reservoir recreation.

Our screening analysis suggested that activities such as stream recreation, forest recreation, snowskiing, and wildlife recreation were probably affected both directly and indirectly through climate change. Unfortunately, we were not successful in locating or estimating any elasticities reflecting the direct effect of temperature and precipitation on visitor use. However, climate change does have an indirect effect on visitor use through changes in the natural resources available for the recreation activity. For reservoir recreation we included these indirect effects of reservoir drawdown and shoreline loss into our single model. In the following we develop the indirect visitor-use elasticities in response to changes in streamflow on river recreation, snow availability on skiing, coastal wetlands on waterfowl hunting and bird viewing, and forests on picnicking, camping, and hiking.

Stream recreation

Since no data were available to estimate the effect of higher temperatures on the likely increases in stream fishing, rafting, and whitewater canoeing/kayaking, the primary effect modeled is the effect of climate-change-induced streamflow on recreation use. The hydrology analysis of stream flows was estimated by Rosenberg (1995) using Texas A&M's HUMUS model. This analysis gives the net change in streamflow after deducting infiltration, evapo-transpiration, and evaporation. The detailed watershed sub-areas were aggregated into four US census regions for application to the regional recreation use data. Using the visitor-use elasticities from Walsh *et al.* (1980b) for these activities, the change in recreation use was calculated with different climate-change-induced streamflows.

Downhill and cross-country skiing

The number of snowskiing days is hypothesized to be influenced by: (a) direct effects such as temperature and precipitation (e.g. snowfall) on the desirability of skiing and (b) indirect effects such as the effect of higher temperatures on snowmelt and hence the length of the season.

To model the direct effects we estimated a national demand and supply of skiing model using national data on skier days from 1979 to 1991 as a function of lift ticket costs, gasoline prices, income, temperature, and precipitation. The model was unsatisfactory. The price and income coefficients were poorly behaved, and temperature and precipitation were insignificant. By contrast, a recreation site choice model developed by Morey (1981) to estimate the elasticity of visitation with respect to snowfall at a variety of downhill ski areas in Colorado performed much better. Generally speaking increased snowfall had a strong positive effect on intermediate and advanced skiers but a small negative effect on novices. The overall net positive effect of precipitation on skier visits is so small, relative to the indirect effect of higher temperatures on snowmelt and length of the season, that we have chosen not to incorporate this small positive effect into our modeling effort. The indirect effect of higher temperatures on a later start and an earlier end to the ski season was evaluated for a sample of major ski areas throughout the United States.[4]

The snowmelt analysis provided the number of total days in the season in which snow on the ground exceeded critical depths such as 6 inches and 12 inches to make downhill skiing possible. The percentage change in the length of the ski season was calculated using the mean number of days with the critical amount of snow under baseline conditions and each climate change scenario. Specifically, if Sun Valley currently averages a 100-day season and would have a 70-day season under a particular climate change scenario, then current skier days would be multiplied by 70 percent to arrive at the estimate of skier days under the climate change scenario. This may overstate the effect of climate change on skier days for a number of reasons. Most importantly, it ignores the fact that some skiers may have some ability to shift days and ski more often during the new shortened season. To the extent this is possible, our analysis overstates the loss. However, given that much skiing takes place during major holidays at both ends of the ski season (i.e. Thanksgiving and Spring Break) these days will simply be lost with a shorter season. While increased snowmaking might partially compensate for increased snowmelt, scenarios involving 5 °C temperature increases are likely to preclude snowmaking. Further, snowmaking is not effective for cross-country skiing trails and these trails are often located at lower elevation and in less favorable snow areas than downhill areas. Cross-country skiing represents about 20 percent of all snowskiing days. The analysis that follows simply uses the full reduction of skier days by the length of the season, recognizing there may be a slight overstatement of the losses.

Waterfowl hunting

The connection of sea level rise to the loss of coastal wetlands in the Northeastern and Southern United States and the net gain of coastal wetlands in the West has been quantified by Smith and Tirpak (1989). Coastal wetlands represent between 10 and 13 percent of total wetlands in the Northeast and South, respectively. Miller and Hay (1981) estimate a regression equation that relates the days of waterfowl hunting to several independent variables including income, hunter preferences, and

---

[4] The authors are indebted to Dennis Lettenmaier and Eric Wood for the use of their snowmelt analysis for major ski areas.

waterfowl habitat in the respondent's state of residence. Combining this model with their estimate of the change in probability of being a waterfowl hunter as a function of wetland acres, we estimate that a 1 percent change in wetland acres results in a 0.275 percent change in hunter days. Thus as global warming causes sea levels to rise, we link this to the change in wetland acres as reported in Smith and Tirpak (1989) to calculate the change in waterfowl hunting days.

Bird viewing

Using data from Cooper and Loomis (1991) we estimated a regression equation relating number of bird viewing trips in California to the number of birds seen per trip. This double-log model produces an elasticity of 0.173, meaning that a 10% increase in the number of sighted birds would increase visits by almost 2%. To apply this to loss of coastal wetlands we presumed that a given percentage of change in habitat would be translated into the same percentage change in coastal bird populations and that into an equivalent change in birds seen on viewing trips in coastal states. As global warming causes sea levels to rise, we link this to the change in wetland acres reported in Smith and Tirpak (1989) to calculate the change in bird viewing in coastal states.

Forest recreation

These activities include camping (both in developed and semi-developed sites), hiking, backpacking, and picnicking. These activities generally take place in forests and changes in forested acres in the region would affect the use levels. Current levels of use of these activities were obtained from Bergstrom and Cordell (1991). To estimate how the level of use of these activities would be affected by climate change several steps were necessary. First, to estimate the change in visitation with change in forested acres we developed an elasticity for each activity based on a visitor survey by Walsh and Olienyk (1981). Their study asked visitors engaged in different forest recreation activities how their participation in that activity would change with different amounts of trees. The scenes were represented in photographs which ranged from no trees to a very high density of trees per acre. Using their tabular results, we calculated how visitation would change with a transition from trees to no trees. Given the relatively low elasticities estimated by Walsh and Olienyk (averaging 0.3), our sensitivity analysis evaluates elasticities of twice this amount and then an elasticity that assumes a proportional response of forest recreation to forest cover.

The available climate–forest models predict the change in acres in different vegetation types. Using these model predictions we estimated a net change in forested acres for each of the four census regions in the United States. This net change in forested

303

acres ignored changes in the composition of tree species resulting from climate change and focused on the change from forests (boreal, coniferous, temperate mixed, temperate deciduous, and tropical evergreen) to non-forested land such as grassland (savannas) and shrubland. This net change in forested acres can be related to the current forest acreage to calculate the percentage change. This percentage change in forested acres was multiplied by the elasticity of participation for each activity to obtain a percentage change in use. Finally this percentage change in visitor use was applied to current participation in the census region to estimate the new level of visitation in that activity.

Estimating the change in forested acres involves pairing one of several possible biogeographical models of vegetation change with a particular climate change scenario generated from a general circulation model (GCM). A series of these types of model runs have been performed in a simulation modeling research effort called VEMAP (1995). Our worst case forest loss scenario (-14 percent) arises from pairing the BIOME2 biogeographical model with the Oregon State University (OSU) GCM. Our middle forest loss case and forest gain case involve the use of the MAPPS biogeographical model. This biogeographical model is responsive to CO_2 fertilization effects. These CO_2 effects increase forested acres or reduce the decrease beyond just the temperature and precipitation effects. To be consistent with the Chapter 5 analysis of timber, we chose one of their MAPSS forest scenarios for the $+2.5\,°C$ and $+7$ percent precipitation scenario. They combined MAPSS with UKMO as the GCM. This combination produces a national loss of 7.2 percent of forest cover. This is what we used as input to our middle case forest recreation analysis. A best case alternative scenario combines the MAPSS biogeographical model with the OSU GCM, resulting in a 23 percent gain in forest cover. A sensitivity analysis of least favorable (BIOME2–OSU) and most favorable (MAPSS–OSU) was used to estimate the upper and lower bound effects on recreation.

We also drew upon Chapter 5 to estimate forestry effects associated with other climate change scenarios. In particular we used three of the other MAPSS and GCM model combinations, used by Sohngen and Mendelsohn to estimate the effects on forests, in order to calculate the effects on forest recreation for the $+2.5\,°C/+15$ percent precipitation, $+5\,°C/+7$ percent precipitation, and $+1.5\,°C/+7$ percent precipitation scenarios. The regression coefficients or resource elasticities were used to create a series of spreadsheets for each major recreation activity group. These spreadsheet models were then used to calculate the change in visitor days and economic value for the major recreation activities using current (1990) visitor-use statistics as one base and expected visitor use in the year 2060 as the other.

304

11.3   Results

Results for central climate scenario

The current (baseline) visitation, value per day, and elasticities are shown in Table 11.1; results for the central climate scenario are shown in Table 11.2. As can be seen in Table 11.2, beach recreation and golf have the largest percentage gain in visitor days, increasing by 14 percent. Reservoir recreation has a 9 percent gain in visitation, even when reduction in reservoir surface area effects are accounted for. In contrast, snowskiing shows a 50 percent loss in use due to the delayed start and premature end of the ski season. Overall there is a 3 percent gain in visitor days using the base visitation expected in 1990.

Table 11.2 also shows the change in economic value of the recreation activities associated with our best estimate using the $+2.5\,°C$ and $+7$ percent precipitation scenario. Not surprisingly the largest loss is in snowskiing, with a \$1.4 billion loss given 1990 use levels and a \$4 billion loss given baseline use levels for 2060. Reservoir recreation has the largest gain, increasing by \$2.5 billion at 1990 use levels. The total effect of all eight groups of activities is a gain of \$2.75 billion at 1990 use levels, and \$2.5 billion in 2060. This is quite similar to the gains of \$2.8 to \$4 billion estimated by Mendelsohn and Markowski (see Chapter 10) using an empirical, state-level cross-sectional approach. Their estimates showed little effect of climate change on golf, but substantial gains in boating and fishing (\$8 billion), and larger losses for their more comprehensive nationwide estimates of hunting and wildlife viewing.

Sensitivity analysis for central case scenario

Table 11.3 displays the effect of changing key parameters on significant recreation activities and the overall total recreation effect summed across all activities. These effects can be compared with those in Table 11.2. There are smaller overall net gains in recreation benefits with lower bound estimates of coefficients for reservoir and golf recreation or "worst case" scenarios for forest losses. The largest drop in benefits occurs in reservoir recreation if we use the lower temperature elasticity of Ward et al. (1995). The other large change is in forest-based recreation. Different biogeographical models produce substantially different estimates of changes in forests (ranging from gains of 21 percent to losses of 14 percent). This causes forest-based recreation to either gain \$2.1 billion or lose \$2.6 billion in 2060 using the more extreme scenarios. Narrowing down this range of impacts in the vegetative models is a clear priority for better understanding the magnitude of ecological effects. Clearly, more research in the vegetative modeling is a prerequisite to better estimates of the effects on forest-based recreation.

Table 11.2. *Effect on recreation visitation and value for CO_2 doubling scenario: $+2.5\,^{\circ}C$ and 7% precipitation*

Activity	Year	Visitor days (millions)			Change in economic value (millions 1992$)	% Change in days
		No change	Climate change	Change in days		
Reservoir recreation (Fishing, boating swimming, waterskiing)	1990	1359.45	1484.97	125.52	2514.00	9.2
	2060	1789.18	1953.40	164.22	3267.00	9.2
Forest-based recreation (Camping, hiking, picnicking under mid-level loss estimate)	1990	1238.45	1213.56	−24.89	−357.00	−2.0
	2060	2163.52	2119.55	−43.97	−658.00	−2.0
Beach recreation (Beach nourishment and protection scenario)	1990	191.70	218.65	26.95	337.90	14.1
	2060	256.10	292.15	36.05	451.48	14.1
Golf recreation	1990	487.90	554.40	66.50	1412.70	13.6
	2060	1119.70	1272.20	152.50	3241.80	13.6
Snowskiing (Downhill and cross-country)	1990	155.77	74.35	−81.42	−1439.50	−52.3
	2060	464.07	222.09	−241.98	−4278.22	−52.1
Waterfowl hunting (Coastal wetland – sea level rise)	1990	15.96	15.76	−0.20	−5.80	−1.2
	2060	19.08	18.85	−0.23	−6.94	−1.2
Bird viewing (Coastal wetland – sea level rise)	1990	169.34	169.26	−0.08	−2.26	−0.05
	2060	277.03	276.88	−0.15	−3.77	−0.1
Stream recreation (Stream fishing, kayaking, rafting)	1990	191.18	197.83	6.64	288.27	3.5
	2060	371.11	383.79	12.68	555.47	3.4
Total effect	1990	3809.75	3928.79	119.03	2748.31	3.1
	2060	6459.79	6538.91	79.12	2568.82	1.2

Table 11.3. *Sensitivity analysis of recreation activity to differing assumptions: scenario: +2.5°C and 7% precipitation*

Activity	Year	Visitor days (millions)			Change in economic value ($)	New total effect
		No change	Climate change	Change in days		
Reservoir recreation (smaller temp. elasticity)	1990	1359.45	1376.31	16.85	343.77	578.08
	2060	1789.18	1810.60	21.42	435.97	−262.21
Reservoir recreation (Reservoir drawdown and crowding affecting value per day)	1990	1359.45	1484.97	125.52	703.47	937.78
	2060	1789.18	1953.40	164.22	871.54	173.36
Forest-based recreation (most favorable forest growth scenario)	1990	1238.45	1317.92	79.46	1139.63	4244.94
	2060	2163.52	2303.80	140.28	2098.49	5325.31
Forest-based recreation (largest forest loss scenario)	1990	1238.45	1190.11	−48.34	−693.46	2411.85
	2060	2163.52	2078.16	−85.36	−1276.92	1949.90
Forest-based recreation (larger recreation response elasticity and largest forest loss scenario)	1990	1238.45	1141.33	−97.12	−1393.33	1711.98
	2060	2163.52	1992.22	−171.298	−2562.55	664.27
Forest-based recreation (proportionate response elasticity and moderate forest loss scenario)	1990	1238.45	1138.42	−100.03	−1435.13	1670.18
	2060	2163.52	2075.31	−88.208	−1319.71	1907.11
Beach recreation (Least favorable: 15% net beach loss)	1990	191.70	213.91	22.24	276.71	2687.12
	2060	256.10	285.81	29.71	369.73	2487.07
Golf recreation (Lower bound estimated temperature elasticity)	1990	487.95	508.979	21.03	446.86	1782.47
	2060	1119.68	1167.93	48.25	1025.39	352.41

Other climate change scenarios

Table 11.4 displays the effects of different climate change scenarios on aggregation of recreation activities. The coastal category represents beach recreation as well as coastal waterfowl and bird viewing. The freshwater category includes both stream and reservoir use. The forest recreation loss example assumes a 7 percent loss in forest cover. The results indicate that even using the middle forest loss scenario (7 percent forest cover loss) the gains in freshwater recreation and golf more than offset the loss in snowskiing and forest-based recreation in all climate scenarios. Generally speaking, the warmer and drier the scenario the more the gains to golf, beach, and reservoir recreation offset the losses in snowskiing and forest recreation. While the effect of higher temperatures on recreation is consistent with the pattern found by Mendelsohn and Markowski, they found much larger gains with the +5 °C scenarios than we did, as a result of much greater sensitivity of fishing (their linear and loglinear model) and boating (linear model only) to temperature (see Chapter 10).

11.4   Conclusion

For the outdoor recreation activities where we were able to quantitatively model the effect of temperature and precipitation, the general conclusion is one of gains in visitation and benefits. While our initial perception at the beginning of this study was that global climate change would have adverse effects on outdoor recreation, the direct temperature effects on many activities, particularly for golf and freshwater recreation, seem to be quite strongly positive. For activities like golf some of the gain undoubtedly comes from extending the season.

Using a +2.5 °C and +7 percent precipitation climate change scenario, our best estimate would result in a gross gain in recreation benefits of about $2.74 billion using 1990 use levels and $2.5 billion using use levels expected in the year 2060 when the impacts of effective CO_2 doubling are expected. The net gain in visitor days is 119 million under this scenario. This is quite reasonable as it implies about one more visitor day of recreation for each teenager and adult in the United States with the most likely activities being golf and reservoir recreation. This is quite plausible given the increased temperatures.

The scenario that has the assumptions and models least favorable to recreation shows a gross gain of only $578 million annually with 1990 base visitation levels, and losses of $262 million annually using baseline visitation expected in the year 2060. The reason for the difference across time periods is that the activities damaged by climate change (e.g. skiing) would otherwise be expected to grow in importance.

Table 11.4. *Recreation benefits across climate change scenarios modeled: (Middle forest loss assumption employed)*

Base year	Temperature (°C)	Precipitation (%)	Coastal (beach use waterfowl, hunting, and coastal bird viewing)	Freshwater (stream and reservoir use)	Forest (camp, hike, picnic)	Snowskiing (downhill and cross-country)	Golf	Total
				Change in economic value (millions of 1992 $)				
1990	2.5	7	326	2802	−357	−1440	1412	2744
	2.5	15	294	3863	−1205	−1302	1213	2863
	5.0	7	685	4616	−1590	−2210	3001	4501
	1.5	7	180	1630	−923	−823	778	843
2060	2.5	7	436	3822	−658	−4278	3241	2564
	2.5	15	393	5818	−2219	−3868	2784	2908
	5.0	7	915	6019	−2919	−6577	6888	4316
	1.5	7	241	2358	−1699	−2445	1786	241

The reader is cautioned to keep several caveats in mind with respect to these conclusions. First, the forest recreation results are sensitive to which of the underlying climate change scenario global climate circulation models are used as inputs for forest cover. Biogeographical (i.e. vegetation) models such as MAPSS when coupled with the UKMO climate change models show a decrease of 7.2 percent in forest cover and a reduction of about 2 percent in forest-based recreation activities. Other combinations of biogeographical and climate change models show larger losses to forest-based recreation, while others show gains. Alternatively, the gain in golf visitation was relatively large across all of the climate change scenarios. Our results are suggestive of the relative effect on different activities and the magnitude of effects for the 17 recreation activities we modeled but they are certainly not precise estimates.

Another important limitation of our analysis is that our estimates of the visitor gains and losses do not reflect transitional losses nor have they included the large adjustment costs necessary to move from the current equilibrium toward a new equilibrium in the year 2060. For many of the recreation increases to be realized additional investment in facilities or relocation of forest-based recreation facilities and access to areas gaining forests is necessary. This will no doubt involve hundreds of millions of dollars in costs. For example, with respect to beach recreation the reader should keep in mind the gains in beach visitation will only occur if the (cost-effective) beach restoration or public access is maintained. Future analyses should calculate these costs so that a net benefits analysis of recreation can be performed. In addition, our analysis ignores transition effects from the current climate to the future climate. This may be quite important in forest recreation as forest dieback may occur more quickly than new forests can establish themselves in more northern areas. During that transition, forest recreation losses could be substantially greater than those estimated here. Partially mitigating the overestimate of transitional losses is the possibility that these changes will be gradual enough that recreationists will be incrementally adjusting their mix of recreation activities to the new climate. For example, with forest dieback, people may substitute other nonforest-based activities or switch locations of picnics from forests to reservoirs resulting in a smaller reduction in recreation benefits than we have estimated. Empirically modeling this dynamic interaction which accounts for feedback of changing climate and recreational resources on activity mix will take much improved ecological models and may well require panel datasets as well.

Finally, this is clearly a partial analysis both in the number of climate-sensitive recreation activities that could be quantitatively modeled and in the ways in which climate affects the recreation activities that were modeled. For example, no increase in demand for forest-based recreation or stream recreation due to higher temperatures was included due to lack of data. This may have the effect of understating the gains from global warming. If and when the quantitative estimates of the effect of global climate change are refined, a more thorough analysis of its effect on outdoor recreation should be attempted.

References

Adamowicz, W., Fletcher, J. and Graham-Tomasi, T. 1989. Functional Form and the Statistical Properties of Welfare Measures. *American Journal of Agricultural Economics* **71**(2): 414–21.

Balogh, J.C. and Walker, W.J. (eds.). 1992. *Golf Course Management and Construction: Environmental Issues*. Far Hills, NJ: US Golf Association.

Bell, F.W. 1986. Economic Policy Issues Associated with Beach Nourishment. *Policy Studies Review* **6**(2): 374–81.

Bell, F.W. and Leeworthy, V.R. 1985. *An Economic Analysis of the Importance of Saltwater Beaches in Florida*. Tallahassee, FL: Department of Economics, Florida State University, Florida: Sea Grant College.

Bergstrom, J. and Cordell, K. 1991. An Analysis of the Demand and Value of Outdoor Recreation in the United States. *Journal of Leisure Research* **23**(1): 67–86.

Boyle, K., Reiling, S. and Phillips, M. 1990. Species Substitution and Question Sequencing in Contingent Valuation Surveys Evaluating the Hunting of Several Types of Wildlife. *Leisure Sciences* **12**:103–18.

Cato, J. and Gibbs, K. 1973. *An Economic Analysis Regarding the Effects of Weather Forecasts on Florida Coastal Recreationists*. Economics Report No. 50, Gainesville, FL: Food and Resource Economics Department, University of Florida.

Cline, W. 1992. *The Economics of Global Warming*. Washington DC: Institute for International Economics.

Cooper, J. and Loomis, J. 1991. Economic Value of Wildlife Resources in the San Joaquin Valley: Hunting and Viewing Values. In: *The Economics and Management of Water and Drainage in Agriculture*. Dinar, A. and Zilberman, D. (eds.). Kluwer Academic Publishers.

Cordell, K., Bergstrom, J., Hartman, L. and English, D. 1990. *An Analysis of the Outdoor Recreation and Wilderness Situation in the United States: 1989–2040*. General Technical Report RM-189. Fort Collins, CO: USDA Forest Service, Rocky Mountain Forest and Range Experiment Station.

Creel, M. and Loomis, J. 1990. Theoretical and Empirical Advantages of Truncated Count Data Estimators for Analysis of Deer Hunting in California. *American Journal of Agricultural Economics* **72**(2): 434–41.

Donnelly, D., Loomis J., Sorg, C. and Nelson, L. 1985. Net Economic Value of Recreational Fishing in Idaho. Resource Bulletin RM-9, Fort Collins, CO: Rocky Mountain Forest and Range Experiment Station, US Forest Service.

Ewert, A.W. 1991. Outdoor Recreation and Global Climate Change: Resource

Management Implications for Behaviors, Planning and Management. *Society and Natural Resources* **4**(4), 366.

Fankhauser, S. 1995. *Valuing Climate Change: The Economics of the Greenhouse Effect.* London: Earthscan.

Hay, M.J. 1988. *Net Economic Recreation Values for Deer Elk and Waterfowl Hunting and Bass Fishing.* Report 85-1. Washington, DC: US Department of the Interior, Fish and Wildlife Service.

Hay, M.J. and McConnell, K.E. 1979. An Analysis of Participation in Nonconsumptive Wildlife Recreation. *Land Economics* **55**(4): 460–70.

Hellerstein, D. and Mendelsohn, R. 1993. A Theoretical Foundation for Count Data Models. *American Journal of Agricultural Economics* **75**: 604–11.

Houghton, J., Callande, B. and Varney, S. 1992. *Climate Change 1992: The Supplementary Report to the IPCC Scientific Assessment.* WMO/UNEP, Intergovernmental Panel on Climate Change. Cambridge University Press, England.

Just, R., Hueth, D. and Schmidt, A. 1982. *Applied Welfare Economics.* Upper Saddle River, NJ: Prentice Hall.

Knetsch, J., Brown, R. and Hansen, W. 1974. *Estimating Expected Use and Value of Recreation Sites, in Planning for Tourism Development: Quantitative Approaches.* Gearing, C., Swart, W. and Var, T. (eds.) New York: Praeger.

Leatherman, S.P. 1989. National Assessment of Beach Nourishment Requirements Associated with Accelerated Sea Level Rise. In: *The Potential Effects of Global Climate Change on the United States, Appendix B: Sea Level Rise.* Smith, J. and Tirpak, D. (eds.). Washington, DC: US Environmental Protection Agency.

Leeworthy, V.R., Schruefer, D.S., Wiley, P.C., Meade, N.F. and Drazek, K. 1989, 1990. *A Socioeconomic Profile of Recreationists at Public Outdoor Recreation Sites in Coastal Areas.* Volumes 1–5. US Department of Commerce, National Oceanic and Atmospheric Administration.

Loomis, J., Roach, B., Ward, F. and Ready, R. 1995. Testing Transferability of Recreation Demand Models Across Regions: A Study of Corps of Engineers Reservoirs. *Water Resources Research* **31**(3): 721–30.

McConnell, K.E. 1977. Congestion and Willingness To Pay: A Study of Beach Use. *Land Economics* **53**(2): 185–95.

Miller, J.R. and Hay, M.J. 1981. Determinants of Hunter Participation: Duck Hunting in the Mississippi Flyway. *American Journal of Agricultural Economics* November, **63**(4): 677–84.

Morey, E. 1981. The Demand for Site-Specific Recreational Activities: A Characteristics Approach. *Journal of Environmental Economics and Management* **8**: 345–71.

Morey, E. 1994. What is Consumer's Surplus Per Day of Use, When Is it a Constant Independent of the Number of Days of Use and What Does it Tell us about Consumer Surplus. *Journal of Environmental Economics and Management* **26**(3): 257–70.

Mumy, G. and Hanke, S. 1975. Public Investment Criteria for Underpriced Public Products. *American Economic Review* **65**(4): 712–20.

National Academy of Sciences. 1992. *Policy Implications of Greenhouse Warming.* Washington DC.

NOAA 1993. *Compensation Tables for Natural Resource Damage Assessment under OPA: Oil Spills into Estuarine and Marine Environments, Volume III (Recreational Values.* 1993 Contract Report number ASA No. 91–97, Rockville, MD: NOAA.

Nordhaus, W. 1991. To Slow or Not to Slow: The Economics of The Greenhouse Effect. *Economic Journal* **101**: 920–37.

Rosenberg, N. 1995. Memo to Joel Smith, Hagler Bailly Inc. Regarding New Water Simulations from Climate Change Using Texas A&M HUMUS Model. Washington DC: Battelle Pacific Northwest Lab.

Rosenthal, D. 1987. The Necessity for Substitute Prices in Recreation Demand Analysis. *American Journal of Agricultural Economics* **64**(4): 828–37.

Silberman, J. and Klock, M. 1988. The Recreation Benefits of Beach Nourishment. *Ocean and Shoreline Management* **11**: 73–90.

Smith, J.B. and Tirpak, D.A. 1989. *The Potential Effects of Global Climate Change on the United States.* Report to Congress. Washington, DC: US Environmental Protection Agency.

Thayer, M. 1981. Contingent Valuation Techniques for Assessing Environmental Impacts. *Journal of Environmental Economics and Management* **8**: 27–44.

US Department of Agriculture. 1990. *Resource Pricing and Valuation Procedures for the Recommended 1990 RPA Program.* Washington, DC: US Forest Service.

US Department of Commerce, Bureau of the Census. 1990. Current Population Reports. Series P-26: No. 88-NE-SC; No. 88-ENC-SC; No. 88-WNC-SC; No. 88-S-SC; No. 88-W-SC, Washington, DC: US Government Printing Office.

US Department of Commerce, Bureau of the Census. 1992. *Statistical Abstract of the United States.* Washington, DC.

US Department of the Interior, Fish and Wildlife Service. 1988. *1985 National Survey of Fishing, Hunting, and Wildlife-Associated Recreation.* Washington DC: US Government Printing Office.

US Department of the Interior, Fish and Wildlife Service and US Department of Commerce, Bureau of the Census. 1993. *1991 National Survey of Fishing, Hunting, and Wildlife-Associated Recreation.* Washington, DC: US Government Printing Office.

US Department of the Interior. 1986. *1982–1983 Nationwide Recreation Survey.* Washington, DC: National Park Service.

US Water Resources Council. 1983. Economic and Environmental Principles and Guidelines for Water and Related Land Resources Implementation Studies. Washington, DC.

Vaughan, W.J. and Russell, C.S. 1982. *Freshwater Recreational Fishing: The National Benefits of Water Pollution Control.* Washington, DC: Resources for the Future.

VEMAP. 1995. Vegetation/Ecosystem Modeling and Analysis Project: Comparing Biogeography and Biogeochemistry Models in a Continental-Scale Study of Terrestrial Ecosystem Responses to Climate Change and CO_2 Doubling. *Global Biogeochemical Cycles* **9**(4): 407–37.

Walsh, R.G. and Olienyk. J.P. 1981. *Recreation Demand Effects of Mountain Pine Beetle Damage to the Quality of Forest Recreation Resources in the Colorado Front Range.* Ft. Collins, CO: Department of Economics, Colorado State University.

313

Walsh, R.G., Aukerman, R. and Milton, R. 1980a. *Measuring Benefits and The Economic Value of Water in Recreation on High Country Reservoirs*. Ft. Collins, CO: Colorado Water Resources Research Institute, Colorado State University.

Walsh, R.G., Ericson, R.K., Arosteguy, D.J. and Hansen, M.P. 1980b. *An Empirical Application of a Model for Estimating The Recreation Value of Instream Flow*. Ft. Collins, CO: Colorado Water Resources Research Institute, Colorado State University.

Walsh, R.G., Harpman, D.A., John, K.H., McKean, J.R. and LeCroy, D.L. 1987. *Wildlife and Fish Use Assessment: Long-Run Forecasts of Participating in Fishing, Hunting, and Nonconsumptive Wildlife Recreation*. Ft. Collins, CO: Colorado State University.

Walsh, R.G., Johnson, D. and McKean, J. 1992. Benefit Transfer of Outdoor Recreation Demand Studies, 1968–1988. *Water Resources Research* **28**(3): 707–14.

Ward, F., Roach, B., Loomis, J., Ready, R. and Henderson, J. 1995. *Regional Recreation Demand Models for Large Reservoirs*. Final Report to US Army Corps of Engineers, New Mexico State University, Las Cruces, NM: Waterways Experiment Station. Department of Agricultural Economics.

12 Synthesis and conclusions

ROBERT MENDELSOHN AND JAMES E. NEUMANN

This book has sought to improve the state of the art of economic impact assessment of climate change as well as the basis for understanding the potential impacts for the United States. The team of authors involved in this effort has developed several new approaches to measure the impact of climate change on markets and, to a limited extent, nonmarket resources. These new techniques more fully incorporate adaptation, involve dynamic analysis where needed, and provide a more comprehensive analysis of seven key climate-dependent sectors (agriculture, timber, water resources, energy, coastal property, outdoor recreation, and commercial fishing). The empirical studies, taken as a group, suggest that modest warming would have a small but beneficial impact on the US economy; these results are more optimistic about global warming than past studies. This analysis of US impacts, however, does not reflect several categories of nonmarket impacts, such as health effects, aesthetics, and some ecosystem impacts. Because these consequences are omitted, the analysis does not reveal how climate will affect the quality of life. However, the comprehensive analysis of sensitive market sectors and the consistency of the climate scenario and macro-economic assumptions provide an opportunity to synthesize and evaluate the overall impact of climate change on the US economy. In this chapter, we draw conclusions about the potential overall impact of climate change on the US economy based on the findings and uncertainties presented in the previous chapters.

As stressed in each of the preceding chapters, it is important to recognize the significant limitations involved in projecting climate, biophysical, and economic conditions over the next century. Although this book seeks to improve the arsenal of methodologies to measure the economic impact of climate change, none of the existing methods are perfect replicas of the experience that society will face if climate gradually warms over the next century. The methodologies provide analogies which shed light on climate sensitivity but may provide only a glimpse of the big picture. For example, future baseline conditions will never be known with certainty. Climate change can take many paths: it may unfold in a highly predictable gradual manner or in sharp sudden bursts and long lulls. Climate sensitivity may change over the decades with changing tastes and technologies. Society may react to climate change rationally or it may engage in politically expedient but unnecessarily costly responses. No single experiment or analysis will settle all these issues. However, these and other impact

315

studies shed new light on the potential magnitude and likelihood of different impacts occurring, and may serve to inform efforts to prepare for adaptation responses, where appropriate. Although one should always understand that these climate and economic forecasts are uncertain, there is important information in these studies which can and should inform and guide policy.

The remainder of this chapter is organized in three sections. In Section 12.1, we summarize the methodological improvements developed in this book. Section 12.2 presents the major results of the empirical analyses and gives an overview of the national and regional impacts for the United States. The final discussion places the new findings in perspective. This discussion highlights the implications of the results for abatement policy. The implication of the results for other countries is also discussed; extension of these methodologies and results to an international context is not a trivial task, especially in those settings (tropical and less developed nations) where the suite of potential impacts differs substantially from those that may be faced in the United States. The final section also discusses the future of impact analysis and identifies some of the research that remains to be done.

12.1   New methods

The studies presented in this book were carefully designed to address many of the shortcomings of the existing impacts literature. The studies were designed to be representative, comprehensive, dynamic (when necessary), empirical, and carefully science-based. Although the study as a whole took a partial equilibrium approach to studying each sector, care was taken to capture important interactions between sectors. For example, the studies were constructed to examine a common set of climate change and economic scenarios. This common set of assumptions makes it possible to add individual sector results together to estimate a total national impact. To ensure consistency, sectors that share a common resource were coordinated. For example, the irrigation assumptions in the water allocation and agricultural models were made consistently. Forestry and agricultural assumptions about land use were also made consistently.

The studies utilized a common set of economic assumptions to explore the implications of a range of climate scenarios. Because the effects of an altered climate will occur in the future, the study examined the impacts on both the current economy and an economy projected to 2060. The growth assumptions for the 2060 economy were based on the IPCC analysis (IPCC, 1994). A suite of climate scenarios were examined to understand how impacts would change depending upon the extent of climate change. Uniform increases of 1.5, 2.5, and 5.0 °C were examined with 0, 7, and 15 percent precipitation increases for a total of nine scenarios. This range of scenarios was selected to reflect the range of possibilities cited in recent IPCC reports (IPCC, 1990, 1996a). The only exception to this rule concerns the timber study which was linked directly to three GCM runs. Specific GCM runs were selected to try to approximate the scenarios used for the other sectors. In order to be comparable, results for each sector had to be expressed in similar units. In this synthesis, we shift from the present value estimates made in the dynamic sea level and timber studies, and examine the annual impacts in 2060.

The studies attempt to carefully address the issue of efficient adaptation. It was assumed throughout the book that private individuals would adjust their behavior as climate changed if, through adaptation, damages are mitigated or if adaptation improves overall welfare. Compared to past studies (IPCC, 1996b, c), these methodological improvements remove constraints on the actions of affected individuals in the economy. Farmers can adjust crops and farming methods, building managers can shift fuels and usage, recreationists can choose new sites and visitation rates, coastal owners can depreciate or protect buildings in the path of sea level rise, and timber owners can plant new species and harvest old ones before they die back. These adjustments reduce the cost of change compared to continuing old behaviors. In certain circumstances, they can turn change into an advantage as actors seize new opportunities presented by the new climate. The inclusion of efficient private adaptation is one of the most important methodological advances of the book.

Within each sector, the studies were designed to be comprehensive and thus represent the entire sector. For example, energy includes not only electricity but also oil, natural gas, and other fuels. The recreation impacts include not only skiing but also a host of summer activities. The agriculture study includes not only grains but vegetables, fruits, and livestock. In each sector, care was taken to analyze the impact across the entire sector, not just the part of the sector most vulnerable to climate damages.

The study attempts to address the impact of climate change on all sensitive sectors of the economy. The book also begins the difficult task of measuring nonmarket effects with studies of recreation and water quality. Methods to measure the nonmarket impacts of climate change, however, are not yet sufficiently developed to obtain a comprehensive estimate of some potential costs to society. Conspicuously absent are health, aesthetic, and nonmarket ecosystem impacts (such as species loss and the loss of coastal wetlands). These omitted impacts need to be included before a comprehensive total impact estimate can be obtained for the United States.

In Chapters 5 and 7, coastal property and timber, damages depend on the management of long-lived capital assets. Assessments of impacts in these two sectors employ dynamic simulation approaches. The sea level study projects inundation rates across a representative sample of developed coastal sites. Decisions are then made to either protect or abandon structures in the immediate path of inundation. With perfect anticipation, property can be depreciated just prior to abandonment. With imperfect information, only partial depreciation occurs prior to abandonment. The model then simulates a series of protection and abandonment choices through time on a site-specific basis. The timber model also simulates a series of rational decisions on the part of the key economic agents. Given a path of forest dieback, new growth, and expansion, forest owners choose which trees to harvest and which to replant. This process may reduce the damage which would otherwise be caused by dieback as trees which would die from climate change are harvested and replaced with species better suited to the new climate. The market interaction with the ecosystem hastens adaptation and allows forest managers to react to climate change and, in some cases, more quickly take advantage of new opportunities.

In several studies, the impact analysis is carefully linked with natural scientific models. For example, the Adams *et al.* simulation model of agriculture in Chapter 2 is based on careful agronomic results. Sohngen and Mendelsohn's analysis of timber in Chapter 5 is based on the ecosystem model comparison project (VEMAP, 1995). The Yohe *et al.* sea level rise analysis in Chapter 7 is based on earlier EPA models of inundation (Titus *et al.*, 1991). Ensuring a sound link between the economic and natural science models is a critical feature of environmental impact assessment.

12.2   New results

The effect of the methodological choices on the results presented in the preceding chapters is best understood in the context of previous estimates. Table 12.1 presents the range of previous US sector level estimates from the climate change impacts literature (IPCC, 1996c). The authors relied on a range of climate scenarios which varied across sectors so that it is difficult to discern exactly what is assumed about climate change in each estimate. Comparing these estimates it is clear that there is a wide range of opinions about the magnitude of specific sectoral damages. The aggregate size of predicted market damages vary from $14 billion (Nordhaus, 1991) to $68 billion (Titus, 1992). The fraction of total damages attributed to market impacts also varies across authors from 25 percent (Nordhaus, 1991; Tol, 1995) to 73 percent (Cline, 1992). The range of total impacts as a fraction of GDP range from 1 to 2.5 percent.

We wish to compare these older estimates with the results from this study. Table 12.2 summarizes the methodological innovations made in each sector and compares

Table 12.1. *Published estimates of US climate change impacts for doubling of CO_2* *(billions of 1990$)*

Sector	Nordhaus 1991 (3°C)	Cline 1992 (2.5°C)	Fankhauser 1995 (2.5°C)	Tol 1995 (2.5°C)	Titus 1992 (4°C)
Market impacts:					
Agriculture	−1.1	−17.5	−8.4	−10.0	−1.2
Energy	−1.1	−9.9	−7.9	−	−5.6
Sea level	−12.2	−7.0	−9.0	−8.5	−5.7
Timber	−	−3.3	−0.7	−	−43.6
Water	−	−7.0	−15.6	−	−11.4
Total market	−14.4	−44.7	−41.6	−18.5	−67.5
Nonmarket impacts:					
Human life	−	−5.8	−11.4	−37.4	−9.4
Migration	−	−0.5	−0.6	−1.0	−
Extreme events	−	−0.8	−0.2	−0.3	−
Human amenity	−	−	−	−12.0	−
Recreation	−	−1.7	−	−	−
Species loss	−	−4.0	−8.4	−5.0	−
Urban infrastructure	−	−0.1	−	−	−
Air pollution	−	−3.5	−7.3	−	−27.2
Water quality	−	−	−	−	−32.6
Mobile air conditioning	−	−	−	−	−2.5
Total nonmarket	−41.1	−16.4	−27.9	−55.7	−71.7
Total (market and nonmarket sectors)	−55.5	−61.1	−69.5	−74.2	−139.2
% of 1990 GDP	−1.0	−1.1	−1.3	−1.5	−2.5

Source: Derived from IPCC (1996c), Table 6.4 on page 203.

the new results to previous estimates. The new results reported in Table 12.2 reflect a uniform incremental climate change across the United States with an increase in temperature of 2.5 °C, a 7 percent increase in precipitation, and an increase to 530 ppm atmospheric carbon dioxide. This study treats this combination as the central climate scenario for the next century (although it may be somewhat more severe than the most recent scientific assessment in IPCC (1996a)).

There are two important conclusions to draw from the results in Table 12.2. First, both past and current studies suggest that the US economy is not likely to be devastated by modest climate change. The predictions of aggregate market effects for the

Table 12.2. *Estimated annual impact of effective doubling of CO_2 (billions of 1990$)*

Sector	New estimate: +2.5°C, +7% precipitation		Previous estimate 1990 economy	Methodological improvements
	2060 economy	1990 economy		
Market sector impact estimates:				
Agriculture	+$41.4	+$11.3	−$1 to −$18	Inclusion of additional crops and adaptation opportunities.
Timber	+$3.4	+$3.4	−$1 to −$44	Dynamic climate, ecological, and timber modeling.
Water resources – market only	−$3.7	−$3.7	−$7 to −$16	Integrated hydrologic and economic models.
Energy	−$4.1	−$2.5	−$1 to −$10	Includes all space conditioning fuels.
Coastal structures	−$0.1	−$0.1	−$6 to −$12	Dynamic analysis of representative sites.
Commercial fishing	−$0.4 to +$0.4	−$0.4 to +$0.4	NA	First estimates.
Total (market sectors)	+$36.9 (+0.2% of 2060 GDP)	+$8.4 (+0.2% of 1990 GDP)	−$14 to −$68 (−0.3% to −1.2% of 1990 GDP)	Totals are for above market sectors only.
Nonmarket sector impact estimates:				
Water quality	−$5.7	−$5.7	−$32.6	Basin-based regional estimates.
Recreation	+$3.5	+$4.2	−$1.7	Includes summer activities and empirical evidence.

Note:
All estimates apply to an effective doubling of CO_2 in the atmosphere. Previous estimates from Table 12.1. Nonmarket effects include all recreation sector impacts as well as most nonconsumptive components of the water resources estimates (excluding hydroelectric production, which is included in the market effects estimate).

next century in the previous literature ranges from 0.25 percent to 1.2 percent of GDP. This study suggests the market impacts are closer to 0.2 percent of GDP. Second, the new models and methods predict that mild warming will result in a net benefit rather than a net loss to the economy. The likely warming over the next century is expected to make the US economy better off on average.

There are several explanations for the more optimistic results of the current study in comparison to previous work, as indicated in Table 12.2. First, several of the new studies provide more comprehensive measures of sectoral impacts. The newly measured components of these sectors generally benefit from warming (citrus and vegetable crops in the agriculture assessment, heating reductions in the energy sector, summer activities in the recreation sector). These benefits were omitted in earlier analyses.

Second, the greater allowance for adaptive responses has the expected effect of increasing benefits and reducing damages. The increased effort in modeling adaptation accounts for a significant portion of the difference between the new and previous estimates for agriculture. The authors of Chapter 2 estimate that the improvements in modeling adaptation in the agricultural sector increase their national welfare benefit estimate by 20 percent, controlling for differences in scenarios. Better modeling of adaptation is also a significant factor in the differences in new and previous estimates for the timber and coastal property sectors, where adaptation takes a dynamic path over time. However, the estimates did not change in every sector. In the water resource and energy sectors, the inclusion of adaptation did not change existing estimates of damages a great deal, possibly reflecting a limited range of substitution possibilities in these sectors.

Third, some of the science surrounding impacts has changed. Sectors dependent upon the ecosystem appear not to be as threatened by climate change as first thought. Agronomic studies suggest that carbon fertilization is likely to offset some if not all of the damages from warming. Models of forest ecology suggest that increases in productivity and in the land base of productive forests in the Southeast could more than offset any reductions elsewhere in timber production regions (VEMAP, 1995). Further, estimates of the magnitude of climate change have moderated over the last decade (IPCC, 1990, 1996a). Whereas earlier impact studies examined the implications of climate changes of 4.5 °C or more with sea levels rising a full meter, doubling projections now center on 2 °C with sea level rising 0.4 m or less for the next century.

Estimates of climate sensitivity

A broad range of possible climate changes were examined in order to reveal the climate sensitivity of each sector. Estimates for nine temperature–precipitation combinations are displayed in Table 12.3. The estimates are presented in two cate-

Table 12.3. *Expected economic impacts of climate change in 2060ᵃ (billions of 1990$)*

| | Climate change | | | | | | | | |
| | +1.5°C | | | +2.5°C | | | +5.0°C | | |
Sector	0% precip	7% precip	15% precip	0% precip	7% precip	15% precip	0% precip	7% precip	15% precip
Market impacts									
Farming	37.2	45.1	53.6	32.6	41.4	49.1	9.5	22.3	31.7
Timberᵇ	2.0	2.8	3.1	2.3	3.4	5.4	2.8	7.4	6.5
Coastalᶜ	-0.1	-0.1	-0.1	-0.1	-0.1	-0.1	-0.2	-0.2	-0.2
Energy	-1.9	-1.9	-1.9	-4.1	-4.1	-4.1	-12.8	-12.8	-12.8
Water	-4.2	-1.7	0.8	-6.3	-3.7	-1.1	-11.7	-9.5	-6.5
Total market	33.0	44.2	55.5	24.4	36.9	49.2	-12.4	7.2	18.7
% of GNP	0.2%	0.2%	0.3%	0.1%	0.2%	0.2%	-0.1%	0.0%	0.1%
Nonmarket impacts									
Recreation	-0.6	1.1	3.7	1.7	3.5	6.2	20.0	22.4	26.2
Water	-2.8	0.1	9.0	-8.7	-5.7	-2.1	-31.4	-22.2	-11.4

Notes:

ᵃ Positive numbers represent benefits and negative numbers represent damages. Estimates based on 530 ppm of CO_2 and uniform expected climate change.

ᵇ Timber uses GCM not uniform climate scenarios and 710 ppm of CO_2.

ᶜ Sea level scenario assumes 33 cm rise by 2100 for +1.5°C and +2.5°C and 66 cm rise for +5.0°C.

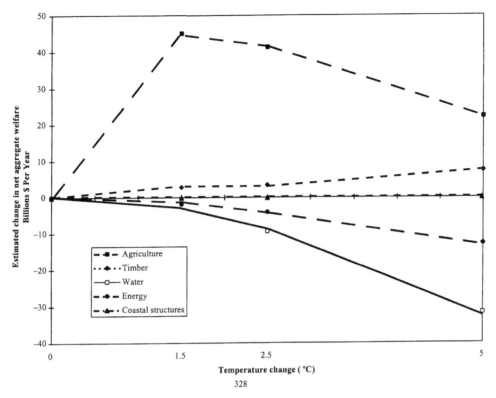

328

Figure 12.1 Sector-level market impacts of climate change (2060 economy, 7% precipitation increase).

gories: market sector effects, including agriculture, timber, coastal structures, energy, and most water resource sector impacts; and nonmarket sector effects, including recreation and water quality. As shown in Table 12.3, a small increase in temperature from current conditions (1.5°C, +7 percent precipitation) leads to aggregate net market benefits in 2060 of $44 billion per year. As temperatures rise (2.5°C, 7 percent precipitation increase), aggregate net market benefits shrink slightly to $37 billion. With a 5°C, 7 percent precipitation increase, market sector effects shrink to $7 billion in benefits. The results suggest that modest warming generates small aggregate market benefits but warming beyond 1.5°C is increasingly harmful. Table 12.3 also demonstrates that additional precipitation would be beneficial. Higher precipitation levels increase net benefits overall yielding large benefits in the agriculture, water, and recreation sectors. For example, increasing precipitation from +7 percent to +15 percent with 2.5°C warming increases net market benefits by $12 billion.

Figure 12.1 illustrates the temperature sensitivity of each market sector. Note that most of the market sectors exhibit a hill-shaped response function (although one may only see part of the hill in this temperature range). For example, farming benefits rise to a peak at $+1.5\,°C$ and then fall. Water and energy welfare decrease more rapidly as temperatures rise (indicating falling off the top of a hill). Coastal damages have a similar shape with respect to sea level rise, increasing rapidly as the seas rise (although, because these damages are relatively modest, the shape of the curve is not discernible in Figure 12.1). Timber is the only market sector to continue to increase as temperatures approach $+5\,°C$. This result originates from ecological modeling (VEMAP, 1995) which suggests that the migration of highly productive southern-type forests into more northern regions more than offsets any reduction in productivity per acre. However, the predicted forest benefits of warming beyond $2.5\,°C$ should be treated cautiously as this result may be an artifact of comparisons across alternative climate scenarios which have different seasonal and spatial patterns.

Warming has different effects on the two studied nonmarket sectors: water quality and outdoor recreation. Water quality is lessened by climate change because of predicted reductions in mean runoff. Recreation largely benefits from warming because of the relatively large increases in fishing and boating benefits associated with prolonged summer seasons. The impacts from these two sectors largely offset each other. Little can be inferred about the effect of warming on the overall quality of life from the water and recreation analyses because several important nonmarket impacts (health, species loss, and human amenities) have yet to be quantified by studies that explicitly model the effects of adaptation.

This set of studies also compares the impact of climate change on a 2060 economy with that of a 1990 economy. It is more difficult to predict what would happen to a 2060 economy because of the challenges in projecting economic progress that far in advance. However, climate change will not occur for decades, making the 2060 projections more relevant. The inclusion of the 1990 economic results reveals how sensitive the climate impacts are to the economic growth assumptions. With a $+2.5\,°C$, $+7$ percent precipitation scenario, the market benefits associated with a 2060 economy are four times higher than with a 1990 economy. The increase in the estimate is proportional to the projected increase in GDP. Both analyses suggest that a $2.5\,°C$ warming would result in small market benefits for the United States of approximately 0.2 percent of GDP (see Table 12.2).

Regional level results

Finally, the analysis sheds some light on regional effects within the United States. Regional results were estimated for the energy, agriculture, recreation, timber, and water resource sectors. For the most part, these results support the intuitively plausible hypothesis that colder more northern states will enjoy higher than average benefits from warming. For example, in the energy sector, the northern states will have only negligible increased cooling costs and yet they will enjoy large heating benefits. Symmetrically, warmer southern states are more vulnerable to increased temperatures; in these states, warming will provide only small heating benefits but will increase cooling costs considerably (see Figures 8.3 and 8.4 in Chapter 8).

In the agricultural sector, the simulation modeling approach employed in Chapter 2 suggests a general pattern of expansion in northern agricultural regions and a corresponding decline in southern regions, as shown in Figure 12.2. These results are consistent with those of previous analyses using the agronomic–economic model (Adams et al., 1990, 1995) and the Ricardian model (Mendelsohn et al., 1994, 1996). As indicated by the Ricardian models, potential losses in major grain crops in the Southeast region are more than offset by potential increases in production of high-value, heat-tolerant fruit and vegetable crops.

In the water resources sector, the most striking differences in the four regional case studies reflect an east–west difference in uses and value of water and, by extension, the nature and degree of impacts. In the western case study basins, both the absolute welfare changes and percentage differences from the baseline are larger than in the eastern study basins. Impacts in the western basins due to decreasing runoffs are dominated by negative effects on the hydropower sector and effects to the relatively large agricultural irrigation sector. By contrast, impacts in the eastern basins are driven by changes in water quality.

Regional effects are also an important aspect of the timber results. The surprisingly positive results are based on forest ecology models (VEMAP, 1995) which predict significant gains in the productivity and land base of southeastern forests. Because of their economic importance, these estimated gains in southeastern forests more than offset estimated reductions in the less productive northern and western timber regions.

Looking across all sectors, the dominant regional results for those market sectors that reflect a north–south difference in impacts are consistent with the quadratic response functions estimated in many of the chapters. If one starts on the cooler lower part of the response function, warming lifts one towards the optimum. However, if one starts at the optimum or beyond, warming moves one away from the optimum and so causes increasing damage.

Uncertainty and limitations

This analysis did not attempt to capture all the sources of uncertainty inherent in global warming predictions. The climate scenario and future economic conditions were taken as given. Predicting the 2060 economy, based here on the eco-

Scenario: +2.5°C, +7 percent precipitation. Both illustrations measure impacts from the 2060 baseline. See Table 2.4 in Chapter 2 for more detail.

Scenario: +5.0°C, no change in precipitation

Figure 12.2 Estimated changes in regional farm production under uniform climate change scenarios.

nomic scenarios of the International Panel on Climate Change (IPCC, 1994), is a highly challenging task itself. The reliance on a selected set of uniform temperature and precipitation changes also fails to capture the wide range of potential climate scenarios which might occur. Geographic climate variability is evident in the inconsistent predictions across General Circulation Models. Interannual climate variability and extreme events are also likely to affect economic impacts (see Chapter 3 and Mearns *et al.*, 1992). Recent discoveries such as the role of sulfur particles have yet to be incorporated into climate impact models (Mitchell *et al.*, 1995). The uncertainty inherent in the climate predictions could contribute to a broad range of outcomes.

Even given the climate and economic scenarios tested, the impact estimates remain uncertain. First, the climate sensitivity of each sector is uncertain because it is estimated using limited experimental evidence. Society does not engage in controlled experiments upon entire economies just to measure climate sensitivity. Sectors may prove to be more or less sensitive to climate change than current methods imply because future events may not resemble these simulated experiments. Efforts to simulate impacts in the future can be no more accurate than the analyst's model. If important phenomena have been omitted, the impacts could prove to be more harmful or beneficial. However, some effects are likely to remain small even given large uncertainties about climate sensitivity. Effects on commercial fishing, energy (space conditioning), and timber are all likely to remain relatively small because these sectors make up only a small part of the economy. On the other hand, even with large expected benefits, agriculture is likely to remain a concern for society if only because some individuals will continue to be heavily dependent on local crops for their subsistence.

Second, it is not clear how much adaptation is likely to occur. Economic theory would predict that firms and households would adapt efficiently to what they experience. It is less clear how well they will anticipate what will happen in the future, especially if climate change does not proceed along a gradual orderly path. How well governments will respond to the challenges of climate change is even more uncertain. For example, if government policies allocate scarce water to low-valued users or subsidize owners of coastal structures to stay near advancing seas, damages from warming would be higher.

Third, this analysis focuses on domestic climate change. The actual impacts in every country, however, will undoubtedly be affected by climate change in the rest of the world. If the United States is less (more) sensitive than the rest of the world to climate change, for example, world prices would change more (less) than projected. If the impact is damaging and the United States is less sensitive, the larger price increases would increase damages to domestic consumers but offset damages to domestic producers. It is consequently difficult to predict national impacts without first modeling global impacts. However, given that national impact estimates are available primarily for the United States, it will be some time before accurate world estimates will be available.

12.3   Discussion

The analyses presented here suggest three major types of conclusions. First, the aggregate and sectoral results of these analyses indicate climate change could have a less severe economic impact than has previously been estimated. Second, several of the methodological improvements in modeling economic impacts had significant effects on the results and directly address the needs of policy-makers for better information on potential impacts. For example, the analyses provide insights into the sensitivity of economic sectors to varying levels of climate change. Third, the "lessons learned" from this exercise provide important insights for the methodological design of studies addressing nonmarket and international impacts.

Perhaps the most striking conclusion of this work is the new set of results. Improvements in the methods used to model the economic impact of climate change on the US economy suggest that moderate warming over the next century would result in a much diminished economic impact compared to the results from previous efforts. The aggregate US results are not the entire story, however; at the regional and sectoral levels there are likely to be winners and losers. The biggest winners appear to be in the agricultural sector, and the biggest losers in water resources and energy. Nonetheless, our forecast of beneficial impacts to the US economy for the modest warming predicted for the next century (IPCC, 1996a) suggests that the US economy may be more resilient than earlier assessments suggest (IPCC, 1996c). The results suggest that aggregate market impacts in the United States are not a motivating factor for near-term action to reduce emissions of greenhouse gases.

One of the more powerful implications of the analyses presented in this book is that methodologies to measure climate impacts on market economies across the world are now available. The methods demonstrated in this study, particularly the cross-sectional studies, could be applied directly in other developed countries and with careful adjustment to developing countries. Although empirical studies may require cross-national comparisons in order to evaluate sufficiently different climate zones, there are many regional natural experiments which could be done across the world to measure climate sensitivities in different places. Each region could explore how sensitive their individual economic sectors are to climate change. Regional estimates based on careful empirical research could replace the current estimates based largely on professional judgment.

328

These new methods also provide insight into conducting nonmarket studies. The studies in this book strongly suggest that human adaptation can substantially affect impact estimates. Although providing "potential" impact estimates based on rigid behavioral assumptions is useful as a first step in impact analyses, these models systematically overestimate the magnitude of expected impacts. Studies of health, species loss, and aesthetics must begin to include adaptation or they will exaggerate the consequences of warming the planet. Developing realistic estimates of health, species loss, and aesthetic impacts should be a priority of domestic research on impacts in the United States.

This study does not measure the climate sensitivity of the economies of other countries, especially developing countries. This is a problem even if one were just interested in US welfare. The United States is likely to be affected by what happens to countries abroad either because these countries are current suppliers to US consumers or because these countries are consumers of US products. Although this study measures only impacts inside the United States, one could speculate about what is likely to happen in other countries given the US climate response functions. The other countries which make up the developed world (OECD) are likely to have quite similar response functions. Those closer to the poles are likely to enjoy slightly more benefits than the United States from the same scenarios. Those countries closer to the equator are likely to have more damages than the United States. However, the results for the United States provide a reasonable guide to how the OECD as a whole could respond to climate change.

What is much less certain is how the rest of the world will fare. The bulk of the developing world has higher current temperatures, larger fractions of their economy in vulnerable sectors, more primitive technologies, and lower incomes or resources for adaptation. All of these factors would suggest that the economies of developing countries will be more vulnerable to climate change than the US economy. In addition, these countries could experience a suite of nonmarket effects that would not be represented in analyses of developed countries, for example, disease epidemics, local famines, and desertification. One of the highest priorities for climate change research is to measure the climate sensitivity of developing countries. The results of these studies could have a major impact on the design of international greenhouse gas policies.

The results of this research effort demonstrate that it is possible to estimate solid empirical response functions to climate. With additional research on omitted nonmarket sectors, a respectable aggregate estimate of US sensitivity to climate change could be available. Extending this research to other countries, where appropriate, could improve the understanding of climate change impacts on a global basis. Armed

329

with these new estimates, policy analysts and politicians across the world would be better prepared to make sustainable long-term commitments to greenhouse gas policies. Although there remains a great deal of research to complete, our rapidly improving understanding of this potentially massive problem is bound to lead to more rational, better supported decision-making in the long run.

References

Adams, R.M., Rosenzweig, C., Peart, R.M., Ritchie, J.T., McCarl, B.A., Glyer, J.D., Curry, R.B., Jones, J.W., Boote, K.J. and Allen, L.H. Jr. 1990. Global Climate Change and US Agriculture. *Nature* **345(6272)**: 219–24.

Adams, R.M., Fleming, R., McCarl, B.A. and Rosenzweig, C. 1995. A Reassessment of the Economic Effects of Climate Change on US Agriculture. *Climatic Change* **30**: 147–67.

Cline, W. 1992. *The Economics of Global Warming*. Washington, DC: Institute of International Economics.

Fankhauser, S. 1995. *Valuing Climate Change–The Economics of The Greenhouse*. London: EarthScan.

IPCC. 1990. *Climate Change: The IPCC Scientific Assessment*, Houghton, J.T, Jenkins, G.J. and Ephraums, J.J. (eds.). Cambridge, UK: Cambridge University Press.

IPCC. 1994. *Climate Change 1994: Radiative Forcing of Climate Change and an Evaluation of the IPCC IS92 Emission Scenarios*. Houghton, J., Meira Filho, L., Bruce, J., Hoesung, Lee, Callander, B., Haites, E., Harris, N. and Maskell, K. (eds.). Cambridge, UK: Cambridge University Press.

IPCC. 1996a. *Climate Change 1995: The Science of Climate Change*. Houghton, J.T., Filho, L.G., Callander, B.A., Harris, N., Kattenberg, A. and Maskell, K. (eds.). Cambridge, UK: Cambridge University Press.

IPCC. 1996b. *Climate Change 1995: Impacts, Adaptations, and Mitigation of Climate Change: Science–Technical Analyses*. Watson, R., Zinyowera, M., Moss, R. and Dokken, D. (eds.). Cambridge, UK: Cambridge University Press.

IPCC. 1996c. *Climate Change 1995: Economic and Social Dimensions of Climate Change*. Bruce, J., Lee, H. and Haites, E. (eds.). Cambridge, UK: Cambridge University Press.

Mearns, L.O., Rosenzweig, C. and Goldberg, R. 1992. Effect of Changes in Interannual Climatic Variability of CERES – Wheat Yields: Sensitivity and $2 \times CO_2$ General Circulation Model Studies. *Agricultural and Forest Meteorology* **62**: 159–89.

Mendelsohn, R., Nordhaus, W. and Shaw, D. 1994. The Impact of Global Warming on Agriculture: A Ricardian Analysis. *American Economic Review* **84**: 753–71.

Mendelsohn, R., Nordhaus, W. and Shaw, D. 1996. Climate Impacts on Aggregate Farm Values: Accounting for Adaptation. *Agriculture and Forest Meteorology* **80**: 55–67.

Mitchell, J.F.B., Johns, T.C., Gregory, J.M., and Tett, S.F.B. 1995. Climate Response to Increasing Levels of Greenhouse Gases and Sulphate Aerosols. *Nature* **376**: 501–4.

Nordhaus, W. 1991. To Slow or Not to Slow: The Economics of The Greenhouse Effect. *Economic Journal* **101**: 920–37.

Smith, J. and Tirpak, D. 1989. *The Potential Effects of Global Climate Change on the United States: Report to Congress*. EPA-230-05-89-050. Washington DC: US Environmental Protection Agency.

Titus, J.G., 1992. The Cost of Climate Change to the United States. In: *Global Climate Change: Implications, Challenges, and Mitigation Measures*. Majumdar, S.K., Kalkstein, L.S., Yarnal, B., Miller, E.W., and Rosenfeld, L.S. (eds.). Easton, PA: Pennsylvania Academy of Science.

Titus, J., Park, R., Leatherman S., Weggel, J., Greene, M., Mausel, P., Brown, S., Gaunt, C., Trehan, M., and Yohe, G. 1991. Greenhouse Effect and Sea Level Rise: The Cost of Holding Back the Sea. *Coastal Management* **19**: 171–204.

Tol, R. 1995. The Damage Costs of Climate Change Toward More Comprehensive Calculations. *Environmental and Resource Economics* **5**: 353–74.

VEMAP. 1995. Vegetation/Ecosystem Modeling and Analysis Project: Comparing Biogeographic and Biogeochemistry Models in a Continental-Scale Study of Terrestrial Ecosystem Response to Climate Change and CO_2 Doubling. *Global Biogeochemical Cycles* **9**: 407–37.

331

Printed in the United States
By Bookmasters